Ensuring safety and quality in the production of beef

Volume 1: Safety

It is widely recognised that agriculture is a significant contributor to global warming and climate change. Agriculture needs to reduce its environmental impact and adapt to current climate change whilst still feeding a growing population, i.e. become more 'climate-smart'. Burleigh Dodds Science Publishing is playing its part in achieving this by bringing together key research on making the production of the world's most important crops and livestock products more sustainable.

Based on extensive research, our publications specifically target the challenge of climate-smart agriculture. In this way we are using 'smart publishing' to help achieve climate-smart agriculture.

Burleigh Dodds Science Publishing is an independent and innovative publisher delivering high quality customer-focused agricultural science content in both print and online formats for the academic and research communities. Our aim is to build a foundation of knowledge on which researchers can build to meet the challenge of climate-smart agriculture.

For more information about Burleigh Dodds Science Publishing simply call us on +44 (0) 1223 839365, email info@bdspublishing.com or alternatively please visit our website at www.bdspublishing.com.

Related titles:
Ensuring safety and quality in the production of beef Volume 2: Quality
Print (ISBN 978-1-78676-060-9); Online (ISBN 978-1-78676-062-3, 978-1-78676-063-0)

Achieving sustainable production of milk Volume 1: Milk composition, genetics and breeding
Print (ISBN 978-1-78676-044-9); Online (ISBN 978-1-78676-046-3, 978-1-78676-047-0)

Achieving sustainable production of milk Volume 2: Safety, quality and sustainability
Print (ISBN 978-1-78676-048-7); Online (ISBN 978-1-78676-050-0, 978-1-78676-051-7)

Achieving sustainable production of milk Volume 3: Dairy herd management and welfare
Print (ISBN 978-1-78676-052-4); Online (ISBN 978-1-78676-054-8, 978-1-78676-055-5)

Chapters are available individually from our online bookshop: https://shop.bdspublishing.com

climate•SMART•publishing
IN AGRICULTURAL SCIENCE

BURLEIGH DODDS SERIES IN AGRICULTURAL SCIENCE

NUMBER 11

Ensuring safety and quality in the production of beef

Volume 1: Safety

Edited by Professor Gary R. Acuff, Texas A&M University, USA and Professor James S. Dickson, Iowa State University, USA

burleigh dodds
SCIENCE PUBLISHING

Published by Burleigh Dodds Science Publishing Limited
82 High Street, Sawston, Cambridge CB22 3HJ, UK
www.bdspublishing.com

Burleigh Dodds Science Publishing, 1518 Walnut Street, Suite 900, Philadelphia, PA 19102-3406, USA

First published 2017 by Burleigh Dodds Science Publishing Limited

Library of Congress Control Number: 2016960975

British Library Cataloguing in Publication Data
A catalogue record for this book is available from the British Library

ISBN 978-1-78676-056-2 (print)
ISBN 978-1-78676-058-6 (online)
ISBN 978-1-78676-059-3 (online)
ISSN 2059-6936 (print)
ISSN 2059-6944 (online)

Typeset by Deanta Global Publishing Services, Chennai, India
Printed by Lightning Source

Contents

Part 2 Ensuring safety at slaughter

Series list

Title	Series number
Achieving sustainable cultivation of maize - Vol 1 From improved varieties to local applications Edited by: Dr Dave Watson, CGIAR Maize Research Program Manager, CIMMYT, Mexico	001
Achieving sustainable cultivation of maize - Vol 2 Cultivation techniques, pest and disease control Edited by: Dr Dave Watson, CGIAR Maize Research Program Manager, CIMMYT, Mexico	002
Achieving sustainable cultivation of rice - Vol 1 Breeding for higher yield and quality Edited by: Prof. Takuji Sasaki, Tokyo University of Agriculture, Japan	003
Achieving sustainable cultivation of rice - Vol 2 Cultivation, pest and disease management Edited by: Prof. Takuji Sasaki, Tokyo University of Agriculture, Japan	004
Achieving sustainable cultivation of wheat - Vol 1 Breeding, quality traits, pests and diseases Edited by: Prof. Peter Langridge, The University of Adelaide, Australia	005
Achieving sustainable cultivation of wheat - Vol 2 Cultivation techniques Edited by: Prof. Peter Langridge, The University of Adelaide, Australia	006
Achieving sustainable cultivation of tomatoes Edited by: Dr Autar Mattoo, USDA-ARS, USA & Prof. Avtar Handa, Purdue University, USA	007
Achieving sustainable production of milk - Vol 1 Milk composition, genetics and breeding Edited by: Dr Nico van Belzen, International Dairy Federation (IDF), Belgium	008
Achieving sustainable production of milk - Vol 2 Safety, quality and sustainability Edited by: Dr Nico van Belzen, International Dairy Federation (IDF), Belgium	009
Achieving sustainable production of milk - Vol 3 Dairy herd management and welfare Edited by: Prof. John Webster, University of Bristol, UK	010
Ensuring safety and quality in the production of beef - Vol 1 Safety Edited by: Prof. Gary Acuff, Texas A&M University, USA & Prof.James Dickson, Iowa State University, USA	011
Ensuring safety and quality in the production of beef - Vol 2 Quality Edited by: Prof. Michael Dikeman, Kansas State University, USA	012
Achieving sustainable production of poultry meat - Vol 1 Safety, quality and sustainability Edited by: Prof. Steven C. Ricke, University of Arkansas, USA	013
Achieving sustainable production of poultry meat - Vol 2 Breeding and nutrition Edited by: Prof. Todd Applegate, University of Georgia, USA	014
Achieving sustainable production of poultry meat - Vol 3 Health and welfare Edited by: Prof. Todd Applegate, University of Georgia, USA	015

Acknowledgements

We wish to acknowledge the following for their help in reviewing particular chapters:

- Chapter 5: *Dr Mike Dugan, Agriculture and Agri-Food Canada*
- Chapter 12: *Professor Fidel Toldrá, IATA-CSIC, Spain; and Dr Steven Lehotay, ARS-USDA, USA*
- Chapter 10: *Dr Mick Bosilevac ARS-USDA, USA; Dr Xianqin Yang, Agriculture and Agri-Food Canada; Professor Robert Delmore, Colorado State University, USA*

Introduction

Beef production faces a range of challenges. There is an ongoing need to ensure safety in the face of threats from zoonoses and other contaminants, particularly in more intensive beef production systems and with more complex supply chains (allowing potentially broader transmission). At the same time, consumers have ever higher expectations of sensory quality. There has been a wealth of research to address these challenges, for example in developing more rapid and sensitive methods for detecting pathogens. These challenges are addressed in the two volumes of *Ensuring safety and quality in the production of beef production*. Volume 2 reviews developments related to quality. This first volume reviews current research on identifying and managing key hazards in the supply chain for fresh beef, from the farm through to slaughter, packaging, retail display and consumer handling.

Part 1 Ensuring safety on the farm

The first group of chapters review current research on zoonoses related to cattle. Chapter 1 summarises what we currently know about zoonotic parasites, viruses and pathogens. Pathogenic organisms are highly variable and adapt well to a changing environment. Surveillance and eradication programmes have worked well to control some zoonotic diseases, particularly where the illnesses are acute and symptoms in animals are easily recognisable. However, some zoonotic remain less easy to recognise. More research is still needed to reduce pathogens in the cattle production environment and minimise transmission to humans.

In the case of each pathogen, the chapter provides a concise review of current knowledge on growth conditions, mechanisms of transmission and causes of recent outbreaks. The chapter begins with zoonotic diseases related to cattle: anthrax, bovine spongiform encephalopathy (BSE), brucellosis and tuberculosis. It then covers zoonotic parasites and viruses: *Cryptosporidium*, *Giardia* and haemorrhagic fevers. Finally, it reviews the major zoonotic pathogens related to cattle: *Escherichia coli*, *Salmonella*, *Campylobacter*, *Leptospira*, *Listeria* and other pathogens.

Good animal husbandry has significantly improved the safety of beef supply. The chapter discusses pathogen control using animal management and biosecurity practices. These aim to prevent the introduction of infection, the survival and spread of infection within the herd or flock and, where necessary, reduce or eliminate an established infection. After a detailed review of the effectiveness of the use of vaccines, the chapter discusses non-traditional interventions such as probiotics, direct-fed microbials (DFM), competitive exclusion (CE) cultures and prebiotics to reduce pathogens in livestock animals. As the chapter points out, the bacterial microbiota in the gastrointestinal tract is highly complex and the scientific community is just only starting to understand the beneficial and antagonistic interactions between microorganisms in the gut and opportunistic pathogens.

As discussed in Chapter 1, the main food-borne pathogens of concern in the beef chain are Shiga toxin-producing *Escherichia coli* (STEC) and *Salmonella*. Other pathogens, including *Listeria monocytogenes* and *Campylobacter* spp. are also important. The occurrence and development of antimicrobial resistant (AMR) pathogens in the beef chain is also a concern. Though both accurate and reliable, traditional culture-based methods are laborious and time consuming. Rapid methods, including nucleic acid-, immunologic-, and biosensor-based techniques can be very sensitive and specific and provide more timely information regarding the presence of pathogens in the beef chain. Chapters 2 and 3 review these methods as well as recent advances in next generation sequencing technologies.

As the chapter describes, many types of rapid methods, including nucleic acid- and immunologic-based methods, have been developed and are commercially available for detection of pathogens, including STEC and *Salmonella*. Nucleic-acid-based methods rely on detection of specific DNA or RNA sequences of the target organism. The most commonly used DNA-based methods utilise the polymerase chain reaction (PCR) that involves amplification of specific sequences of the pathogen and detection of the ethidium bromide (or other stain)-stained PCR products by agarose gel electrophoresis. Nucleic acid-based methods, particularly PCR methods, are commonly used since they have high sensitivity and specificity, can be automated, allow detection of multiple pathogens, and provide reliable results. However, sample processing is necessary to remove inhibitors of the PCR assay or other nucleic acid-based method.

Immunological methods involve the use of monoclonal or polyclonal antibodies that bind to an antigen of the bacterium or to a toxin. Types of immunoassays include enzyme-linked immunosorbent assays (ELISA) or enzyme immunoassays, fluoroimmunoassays, radioimmunoassays, chemiluminescence immunoassays and lateral flow immunoassays. Immunologic-based assays can also be designed to be specific, can be automated, are easy to perform, and allow detection of bacterial toxins. However, sensitivity may be lower than that of nucleic acid-based assays, and cross-reactivity of the antibodies with closely related antigens can occur. Improvements in these rapid methods will continue to be made, with the goal of being able to obtain real-time or near real-time results.

Biosensor-based methods consist of devices that have a bioreceptor that can be an antibody, enzyme, nucleic acid, or cells and a transducer that converts the interactions into a measureable signal. Examples are optical, electrochemical, or mass-based biosensors, surface plasmon resonance, multianalyte array/evanescent wave, quartz crystal microbalance or surface-enhanced Raman scattering biosensors. As noted, rapid methods are very useful as screening methods targeting specific pathogens in large numbers of samples. However, positive results are still often regarded as presumptive and need to be confirmed by traditional culture methods which remain the gold standard. More still needs to be done to refine sensitivity, specificity and reliability.

Chapter 2 also reviews methods for detecting AMR in pathogens. As the chapter points out, the scientific study of the factors influencing AMR occurrence in cattle production and processing environments is in its infancy. No single method, culture or culture-independent, can fully illuminate these factors. There is a need for well-designed studies of cattle production and processing environments, using a combination of culture-dependent and culture-independent methods, if we are to understand and identify the factors responsible for the occurrence of AMR.

Building on Chapter 2, Chapter 3 provides a detailed review of the best methods for detecting particular pathogens, including *E. coli* O157:H7 and non-O157 STEC, *Salmonella, Listeria, Campylobacter, Clostridium, Bacillus cereus* and *Mycobacterium avium subsp. paratuberculosis* (MAP). It discusses sampling regimes, best practice and ways of improving sensitivity and specificity. The chapter looks in particular detail at methods for detecting *E. coli* O157:H7. Methods for detection of *E. coli* O157H7 are used throughout the beef chain, from monitoring the pre-harvest colonisation of cattle and their environment to harvest and post-harvest processing of meat as well as final beef cuts and products. The chapter describes sampling methods, sample enrichment and screening using rapid methods based on immunoassays or molecular detection (DNA or RNA amplification), often followed by culture isolation.

As Chapter 3 points out, to monitor *E. coli* O157:H7 in the farm environment, only the most robust rapid methods are useful. In addition to having a high interfering background, samples collected on farms (pastures or feedlots) contain a number of complex molecules that inhibit many PCR reactions. However, properly performed, *E. coli* O157H7 detection assays can both determine if the organism is present and provide quantitative results. This aids, for example, in identifying 'super-shedder' cattle that harbour and shed high levels of *E. coli* O157:H7 and can spread contamination across a whole herd. The chapter concludes by looking at developments in standardised methods for pathogen detection, looking at the example of the US Food Safety and Inspection Service (FSIS) Microbiology Laboratory Guidebook for testing for non-O157 STEC.

As well as effective methods of detection, the production of safe, high-quality beef requires both good farming practices and additional measures targeted against specific hazards to human or animal health. Building on Chapter 1, Chapter 4 provides an overview of current best practice for managing safety on beef cattle farms. Good farming practices (GFPs) need to control infectious and parasitic agents (biological hazards), chemical contaminants and residues (chemical hazards) as well as physical hazards such as foreign bodies. GFP guidelines typically cover such areas as: general farm management, animal health, using veterinary medicines, and management of animal feed as well as animal and product handling.

The chapter puts particular emphasis on biosecurity, animal welfare and the importance of clean cattle policies. In the latter case, the chapter reviews programmes targeted on preventing contamination of hides and grading of cattle sent to slaughter. Based on this assessment, additional measures are taken such as washing, clipping and separation of cattle for slaughter. There is also a detailed review of the effectiveness of using vaccines. The chapter then discusses hazard-specific control measures for one important hazard: *E. Coli* O157. Finally, it reviews key components of quality assurance programmes for beef production.

Because of their ruminant digestive system, beef cattle are able to eat and digest a wide range of feed products, including by-product or alternative feeds not utilised in other livestock production systems. It is important to ensure that the various feed products provided to cattle do not pose a food safety risk to the people who consume beef products. Chapter 5 provides an overview of the three major areas where food safety risks can be introduced via feed material: purchase of contaminated feed products, contamination during storage and contamination of feed due to mixing or delivery errors. In each case, measures to prevent unacceptable risk are reviewed. Echoing Chapter 4, it highlights the importance of hazard analysis and critical control point (HACCP) systems, prerequisite programmes, standard operating procedures (SOPs) and good production practices in managing feed safety.

Echoing chapters 2 and 3, Chapter 6 highlights the importance of effective detection of contaminants, this time in cattle feed. The chapter reviews methods for sampling and detection of antibiotic residues in animal feed, using the example of distillers' grains. The chapter reviews the use of distillers' grains as animal feed and provides an overview of multiresidue methods for analysing residues in feed, with particular attention to ion-trap tandem mass spectrometry and orbitrap high-resolution mass spectrometry. It also discusses the extent to which antibiotic residues have been found to survive in feed, and considers how the risk of antimicrobial resistance can be assessed.

Part 2 Ensuring safety at slaughter

The second part of the book focuses on maintaining meat safety during slaughter and the preparation of fresh meat. Written by a former meat inspector, Chapter 7 begins by providing an overview of the purposes and importance of meat inspection programmes in the United States in addressing hazards to both human and animal health. Subsequent sections cover the design, implementation and results of ante-mortem and post-mortem inspection, and provide a review of some important procedures used by veterinary services to inspect cow heads, viscera and carcasses.

Chapter 8 reviews the slaughter process and the mechanisms of bacterial attachment to meat tissue. Contamination of animal carcasses during slaughtering procedures is undesirable, but unavoidable in the conversion of live animals to meat for consumption. Internal muscle tissues are essentially sterile, and most initial contamination of red meat carcasses is contributed by the hide during removal. The chapter reviews decontamination methods such as knife trimming, water washing and steam vacuuming, and evaluate their effectiveness. The chapter also considers hot water, organic acid and other decontamination treatments, reviews the potential for contamination during fabrication, and finally discusses the role of packaging, storage and shelf-life estimates in ensuring the safety of meat delivered to consumers.

Beef spoilage is primarily caused by bacteria, and maximising shelf-life is dependent on a multiple hurdle approach to generate conditions that inhibit microbial growth. This chapter begins by reviewing the factors which affect beef spoilage and microbial shelf-life, and the bacteria and processes involved in spoilage. Subsequent sections then consider methods for maximising the microbial shelf-life of beef products, including minimising the initial microbial burden in beef, chilling of beef carcasses, and the wide variety of packaging technologies. The latter include modified atmosphere packaging and low oxygen packaging. Packaging technologies also include active and intelligent packaging using oxygen scavengers, carbon dioxide scavengers/emitters, chlorine dioxide generators, moisture control agents and antimicrobial compounds.

While the industry has made great strides in reducing bacteria on raw beef products, improving safe consumer handling of beef remains a significant challenge. Factors such as the decline of some traditional cooking skills and the pressure to prepare food quickly will continue to challenge food safety educators in creating good and consistent food safety habits among consumers. Chapter 10 review the history of consumer beef preparation practices and food safety advice, research on compiling the best advice to give to consumers, and how awareness of food safety issues and food preparation behaviour has changed over time.

Traceability, or rapid access to knowledge of the history, treatment and location of cattle and beef products through supply chains, is of paramount importance to food safety investigations, corrective actions and product recalls. Chapter 11 defines traceability as applied to beef before describing in detail methods of unique identification for cattle, such as radio frequency identification (RFID) and DNA profiling. The chapter also explores the traceability of beef offered by one- and two-dimensional barcoding. Cloud-based, distributed information systems for data entry, storage and retrieval are emerging as methodologies of choice for participants in the beef supply chain. The chapter provides a case study of cattle and beef traceability through a supply chain from farm to retail using a distributed network, the EPCglobal Network for cattle and beef traceability.

Part 1

Ensuring safety on the farm

Chapter 1

Pathogens affecting beef

James E. Wells and Elaine D. Berry, US Meat Animal Research Center, USDA-ARS, USA

1 Introduction

Prior to the early 1900s in the United States, the handling of foods, specifically meat, was not consistent and oversight was according to laws in each state in which a facility was located. The development and maturation of transportation allowed the rapid movement of live animals and distribution of meat products, and, as a consequence, the meat industry concentrated in major midwestern U.S. cities where labour, particularly immigrant labour, was abundant. Federal oversight was negligible for domestic meat production until *The Jungle* written by Upton Sinclair was published in 1906. The intent of the book was to highlight the plight of immigrants in the Chicago slaughterhouses, but the vivid description of the unhygienic slaughterhouse operations and the improper preparation and handling of product led to a public outcry for government regulation. Following a government investigation, the Federal Meat Inspection Act (FMIA) and the Pure Food and Drug Act were quickly passed in the same year. The FMIA mandated the inspection of livestock before slaughter (antemortem), the carcasses after slaughter (postmortem)

http://dx.doi.org/10.19103/AS.2016.0008.01

and sanitary standards as well as authorized the U.S. Department of Agriculture (USDA) to monitor and inspect facilities for slaughter and processing. The inspections required by FMIA provided consumers with assurances that animals were healthy and meat from these animals was wholesome. Following a 1992–3 outbreak of illnesses linked to *Escherichia coli* O157:H7, a Shiga-toxin-producing strain of *Escherichia coli* (STEC), in ground beef, the Food Safety Inspection Service (FSIS) of the USDA established that *E. coli* O157:H7 was to be considered an adulterant in raw ground beef and sale of this contaminated raw product was prohibited (FSIS, 1994). This ruling by the FSIS was revolutionary in that a specific pathogen was regulated in raw meat, and subsequently FSIS established that other STEC in addition to *E. coli* O157:H7 were to be considered as adulterants of raw ground beef and non-intact raw beef products (FSIS, 1999, 2012).

Microscopic organisms, or microorganisms, are ubiquitous in nature. Microorganisms are diverse in form and function and include protozoa, bacteria, archaea (previously known as archaebacteria) and viruses. Microorganisms can affect animals in multiple ways. Close associations (Fig. 1) between animal hosts and microorganisms are symbiotic relationships and the type of interaction could be commensal (little benefit or cost), mutualistic (benefits one or both) and parasitic/pathogenic (cost/death to the host). Pathogenic microorganisms are typically either bacterial or viral, and not only affect the health and well-being of an animal, but may lead to death. In addition to pathogenic microorganisms, viroids and prions are misfolded proteins that can be infectious to a host, with prions affecting animals and viroids typically affecting plants. Cattle may harbour a number of pathogenic agents that can infect humans, and most of the agents are bacteria.

Cattle are reservoirs for a number of diseases and, depending on the pathogen, transmission can occur via animal contact or consumption of meat or milk (Table 1). Animal

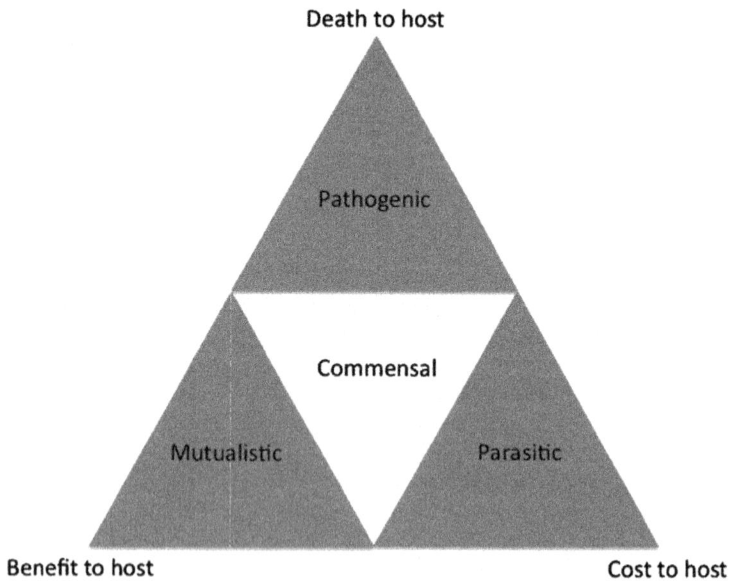

Figure 1 Representation of interactions between an animal host and microorganisms associated with that host. Interactions can be beneficial, costly or deadly to the host.

Table 1 Pathogenic agents affecting cattle and cattle production that may impact human health. Traditional pathogenic agents are those zoonotic agents that through history of animal domestication have affected cattle and humans, but today in most developed countries, transmission to humans is rare.

Pathogenic agent from cattle	Disease(s) in humans	Type of agent	Traditional agent of concern	Contemporary agent of concern
Bacillus anthracis	Anthrax	Bacterial	X	
Brucella abortus	Brucellosis	Bacterial	X	
Campylobacter spp. (*C. coli* and *C. jejuni*)	Campylobacteriosis	Bacterial		X
Cryptosporidium spp. (*Cryp. parvum*)	Cryptosporidiosis	Protozoan		X
Giardia spp. (*G. duodenalis* and *G. enterica*)	Giardiasis	Protozoan		X
Leptospira spp. (*Lept. borgpetersenii Hardjo* and *Lept. interrogans Pomona*)	Leptospirosis	Bacterial	X	
Listeria spp. (*List. monocytogenes* and *List. ivanovii*)	Listeriosis	Bacterial		X
Nairovirus	Crimean-Congo haemorrhagic fever	Viral		X[a]
Phlebovirus	Rift Valley haemorrhagic fever	Viral		X[a]
PrP^{sc}	Bovine spongiform encephalopathy	Prion protein		X
Salmonella enterica subsp. *enterica* serotypes[b]	Salmonellosis	Bacterial		X
Shiga-toxigenic, *Escherichia coli* (*E. coli* O157, *E. coli* O26, *E. coli* O45, *E. coli* O103, *E. coli* O111, *E. coli* O121 and *E. coli* O145)	Haemorrhagic colitis and haemolytic uraemic syndrome	Bacterial		X
Mycobacterium tuberculosis complex spp. (*M. bovis* and *M. caprae*)	Tuberculosis	Bacterial	X	

Contemporary pathogenic agents are those zoonotic agents that are recently recognized as zoonotic organism from cattle production, and many of these agents are of primary concern in many developed countries.

[a]These agents are not currently found in cattle production in developed countries and are limited to certain regions in the world, but are a growing concern as human illness has spread.

[b]There are nearly 1500 *Salmonella enterica* serotypes and Table 2 lists the specific serotypes associated with most human illnesses in the United States and the specific serotypes often observed in cattle production.

contact includes occupational contact during meat animal production (farming and harvest) or recreational direct contact such as in petting zoos. Other indirect vehicles for pathogen transmission from cattle production include contact with contaminated water, manures, soils or foods that are cross-contaminated during production, processing or preparation. These zoonotic pathogens can be viewed as either traditional or contemporary. Traditional bovine pathogenic agents are those zoonotic agents that through history of animal domestication have affected cattle and humans, but today in most developed countries, transmissions to humans are rare. Contemporary bovine pathogenic agents are those zoonotic agents that are more recently recognized as associated with cattle production and can result in human illnesses. Many contemporary agents are primary concerns for foodborne transmissions in developed countries.

In the United States, the Centers for Disease Control and Prevention (CDC) monitors the incidences of disease based on clinical observations and categorizes the various illnesses. Foodborne illnesses are the most common form of illness in the United States, and the CDC will attribute known illnesses to bacterial, parasitic or viral causes (if known) and estimate the incidence in the United States on a per annual basis (Scallan et al., 2011). Among the foodborne bacterial illnesses monitored by CDC, *Campylobacter* spp., Shiga-toxin-producing *E. coli* O157:H7 (STEC O157), non-O157 STEC, diarrhoeagenic *E. coli* (other than STEC and enterotoxigenic *E. coli*), *Listeria monocytogenes*, *Mycobacterium bovis* and *Salmonella* spp. (non-typhoidal) are pathogens that are observed in beef production. However, beef production is not the sole source for these pathogens and the pathogens *List. monocytogenes* and *M. bovis* are rarely directly linked to illnesses from beef consumption.

The CDC monitors illnesses caused by the parasitic organisms *Cryptosporidium* spp., *Giardia intestinalis* (syn. *G. lamblia* or *G. duodenalis*) and *Toxoplasma* spp. that may originate from beef cattle production, though these organisms are rarely foodborne. In the United States, in addition to the foodborne illnesses, a number of important zoonotic illnesses require notification to federal agencies. Furthermore, there are a number of human illnesses caused by pathogens of cattle origin that may be emerging or less prevalent in developed nations and not monitored in the United States. In many developed countries, the incidence of human disease has been minimized because of interventions used in animal production. The following sections discuss many of the zoonotic human illnesses, both historical and emerging, that may be linked to a bovine reservoir for the causative pathogen.

2 Zoonotic diseases related to cattle: anthrax, bovine spongiform encephalopathy (BSE), brucellosis and tuberculosis

2.1 Anthrax

One of the first diseases documented to affect animals, particularly herbivores, and humans is anthrax. The disease has been known by many names, but the term is derived from a Greek word 'anthrakites' (meaning coal-like) to describe the dark sunken skin lesions observed with some forms of the disease (Sternbach, 2003). Studies with anthrax led Robert Koch to identify the causative agent in which he solidified support for the germ theory of disease and developed what is commonly called Koch's postulates for confirmation of disease-causing microbes. Louis Pasteur's studies with anthrax led to the development of one of the first effective vaccines against bacterial disease, and as a consequence, the wide acceptance of vaccination to prevent disease. In the last century, concern for anthrax has diminished as a zoonotic pathogen from cattle, but the threat of the agent being used as a biological weapon has increased.

The causative agent for anthrax is a Gram-positive, rod-shaped, spore-forming aerobic bacterium known as *Bacillus anthracis*. Grazing animals ingest or inhale spores from contaminated soils; vegetative cells are not pathogenic to animals (Beyer and Turnbull, 2009; Hugh-Jones and Blackburn, 2009; Fasanella et al., 2010). Contamination may originate from distribution by other sick animals, improperly buried diseased animals or distribution of offal and/or bonemeal as a fertilizer made from diseased animals. Spores of *B. anthracis* can remain dormant indefinitely, particularly in rich, slightly alkaline soils, and infectious dose can be less than 100 spores. Spores can enter the body through micro-wounds in the pulmonary or gastrointestinal tract, or through open wounds in the skin, and are engulfed by macrophages. The macrophages carry the bacterium through the lymph system and into the bloodstream. During this process, the spores will sense available nutrients and germinate, which results in a vegetative cell that can reproduce and produce toxins. Animal death can occur rapidly, depending on the level of pathogen exposure. The virulence factors are encoded on two plasmids, pXO1 and pXO2. The plasmids are typically stable in the bacterial cell, but can be eliminated when cultures are grown above 42°C, causing the bacteria to lose pathogenicity. Vegetative cells are thought to sporulate when exposed to oxygen, and contaminated meat or hides can be a source for human transmission. In developed countries, animal management has reduced the incidence of anthrax in animals to negligible incidents and animal to human transmission is extremely rare.

2.2 Bovine spongiform encephalopathy (BSE)

Transmissible spongiform encephalopathies (TSE) can cause neurodegenerative diseases in mammals that are fatal (Imran and Mahmood, 2011; Lee et al., 2013). The diseases are caused by a transmissible misfolding disorder of the prion protein (PrP) and, based on current evidence, are not caused by a bacteria or virus. Scrapie is now recognized as a TSE disease affecting sheep and goats and was known for nearly three centuries. In cattle, no TSE disease was recognized until 1986 when the outbreak of the TSE 'Mad Cow' disease, now known as BSE, was observed. Prions are typically host-adapted, with the exception of prions of BSE origin, which have been linked with diseases in susceptible humans (variant Cruetzfeldt–Jacob disease) and cats (Feline Spongiform Encephalopathy).

Prion proteins in cattle are encoded by the PRNP gene on chromosome 13 and the normally folded prion (PrP^{c}) is found in brain and nervous system, digestive tract and lymphoid system, with currently no known function but is degradable and turned over in the host tissues (Balkema-Buschmann et al., 2011; Murdoch and Murdoch, 2015). Typical BSE disease arises after cattle have ingested feed contaminated with misfolded prion protein (PrP^{sc}), and misfolded proteins are extremely resistant to degradation. Ingested PrP^{sc} accumulates in the Peyer's Patch of the terminal ileum and, through a process of catalysing the misfolding of endogenous PrP^{c} to form PrP^{sc}, spreads through the nerve cells to the brain. TSEs are degenerative neurological diseases, and the infected brains of BSE cattle develop sponge-like lesions, and the animals eventually succumb to neurological symptoms, followed by paralysis and death. With time, the PrP^{sc} spreads into peripheral tissues. Epidemiological studies of the Great Britain BSE outbreak in 1986 suggest that specific risk material from either sheep (or possibly a bovine animal that developed BSE from a spontaneous misfolding of the prion protein) was likely used in meat and bonemeal products fed as a dietary protein source to dairy cattle.

2.3 Brucellosis

Infertility and abortions in production animals can be caused by bacterial infections due to different types of *Brucella* (Seleem et al., 2010; Godfroid et al., 2011). The disease brucellosis has been documented in animals since ancient times and is known to affect cattle, sheep, goats and swine. There are ten species of *Brucella* and three of the species associated with cattle are also highly pathogenic to humans. *B. abortus* is the host-adapted *Brucella* species causing disease in cattle and infection is usually through the gastrointestinal tract (Carvalho Neta et al., 2010). The *Brucella* infection spreads to the lymph nodes and the bacteria invade phagocytes, leading to a systemic infection and localized colonization of the uterus, genital organs and mammary glands. Cattle can also be infected with *B. melitensis* (sheep and goat adapted species) and *B. suis* (swine adapted species), but these two species do not cause clinical signs in the infected animals. All three species of these species are highly pathogenic to humans.

The *Brucella* species are Gram-negative, facultative coccobacilli bacteria (Carvalho Neta et al., 2010; Seleem et al., 2010). The bacterium is non-motile and is an intracellular pathogen that replicates in the host cell. The bacteria do not appear to carry plasmids and no virulence factors have been reported. Historically, most infection of humans resulted from consumption of contaminated dairy and meat products, but in the developed countries most infections result from contact with *Brucella*-contaminated tissues and manures, or from processing an infected beef carcass (Galińska and Zagórski, 2013). Human acute illness is characterized by weakness, undulant fever, muscle and joint pain and gastrointestinal distress, which can lead to death, if not treated. Human-to-human transmission is rare.

2.4 Tuberculosis

Bovine tuberculosis is a widespread pulmonary disease in cattle caused by the bacterium *Mycobacterium bovis* (Michel et al., 2010; Pesciaroli et al., 2014). *M. bovis* has also been associated with disease in humans and the bacterial species is closely related to *M. tuberculosis* (Rodriguez-Campos et al., 2014). There are nine species of *Mycobacterium* that are known to cause tuberculosis in animals and humans, and these species are collectively referred to as *M. tuberculosis* complex (MTBC). Numerous MTBC species have been associated with human disease, but only *M. bovis*, *M. tuberculosis* and *M. caprae* have been observed in cattle disease. *M. caprae* is often associated with tuberculosis in goats, but increasingly is observed in diseased cattle. *M. tuberculosis* is the common pathogen associated with human tuberculosis, but this bacterium has been observed in diseased animals, including cattle. *M. bovis* is typically shed as an aerosol from the lungs of coughing or sneezing infected cattle and spread by inhalation of the aerosols or contaminated dust (Menzies and Neill, 2000; Wedlock et al., 2002). MTBC bacteria are engulfed by macrophages and the infected macrophages are walled-off to form a granuloma. The granulomas are predominantly in lymph nodes of the respiratory tract, particularly in the thoracic cavity followed by head and lungs. Infection can progress over years and in subsequently significantly impair lung function.

Mycobacterium are Gram-intermediate, non-motile, slow-growing aerobic bacteria with a waxy outer cell envelope containing mycolic acids atypical to other bacteria (Barka et al., 2015). Bovine tuberculosis is worldwide, but surveillance and eradication in many developed countries have reduced the disease prevalence (Thoen et al., 2006). Human illnesses typically arise from consumption of unpasteurized contaminated dairy products, but contact from infected cattle during production and processing is also a significant hazard. Less than 10% of all tuberculosis cases in humans have been associated with *M. bovis* (Michel et al., 2010), but it is possible that nearly one-third of these illnesses may be due to infections from *M. caprae* (Thoen et al., 2006).

3 Zoonotic parasites and viruses related to cattle: *Cryptosporidium*, *Giardia* and haemorrhagic fevers

3.1 *Cryptosporidium*

Parasitic protozoa are often overlooked eukaryotic microorganisms that can cause disease, and *Cryptosporidium* is one of the most common causes of intestinal illnesses worldwide. *Cryptosporidium* was first reported in 1907 in the stomach of mice, but the first human intestinal infections were not reported until 1976 (Sponseller et al., 2014). In recent years, *Cryptosporidium* has been implicated in respiratory disease of birds and humans. *Cryptosporidium* has a life cycle of asexual and sexual development stages in the host and is excreted as oocytes in the faeces that can be ingested by the new host and initiate the next infection.

Currently, there are as many as 26 species of *Cryptosporidium* and most exhibit a high degree of host adaptation (Ryan et al., 2014). Cattle-associated *Cryptosporidium* species include *Cryp. parvum*, *Cryp. andersonii*, *Cryp. bovis*, *Cryp. ryanae* and *Cryp. ubiquitum* and infection in cattle as young as one week of age has been observed (Budu-Amoaka et al., 2011). Most infections of cattle are observed in calves, with *Cryp. parvum* infections typically observed in one to three weeks of age and *Cryp. bovis* and *Cryp. ryanae* infections in three to eight months of age. Older cattle are typically infected with the species *Cryp. andersonii*. Human cryptosporidiosis has been linked to six species, but two *Cryptosporidium* species are most predominant, *Cryp. hominis* (human specific) and *Cryp. parvum* (cattle associated). In North America, *Cryp. hominis* is the most prevalent species associated with human illnesses, but in Europe *Cryp. parvum* is the most prevalent. There is little evidence of significant transmission of *Cryptosporidium* through beef consumption to humans. However, *Cryptosporidium* oocysts can be excreted in the faeces of cattle, which can survive well in the environment and are resistant to disinfection, and contaminate water and produce.

3.2 *Giardia*

Giardia are parasitic anaerobic flagellated protozoa that colonize the intestinal epithelial surface, and these cyst-forming eukaryotic parasites are common enteric pathogens of humans and domestic livestock (Adam, 2001). *Giardia* were first observed 350 years ago by Anton van Leeuwenhoek, but it was not until 200 years later that the stable cyst forms were observed and a life cycle related back to the vegetative form of the protozoa was postulated. The naming of species within *Giardia* historically has been complex, particularly in regard to human protozoal pathogen, which has been commonly named either *G. duodenalis* or *G. lamblia*, with host-adapted strains clustered into 'assemblages'. Recent proposals have considered renaming the *G. duodenalis* (syn *G. lamblia*) assemblages and expanding *Giardia* from six species into 11 species (Monis et al., 2009). Cattle-associated *Giardia* species, particularly calves from 1 to 2 months of age, include *G. duodenalis* (former *G. duodenalis* assemblage A group), *G. enterica* (former *G. duodenalis* assemblage B group) and *G. bovis* (former *G. duodenalis* assemblage E group), but of these three species only *G. duodenalis* and *G. enterica* are of significant risk to infect and cause gastroenteritis in humans (Budu-Amoako et al., 2011). Most cattle infections are associated with the non-zoonotic *G. bovis*, and there is little evidence of significant transmission of the other bovine-associated strains through beef consumption to humans. However, *Giardia* cysts can be excreted in the faeces of cattle, which can survive well in the environment and contaminate water and produce.

3.3 Haemorrhagic fever

Humans are susceptible to a number of viruses that can cause haemorrhagic fevers, also known as viral haemorrhagic fever, and animals are often asymptomatic reservoirs for some of the viral pathogens. In humans, the viral infections lead to fever with flu-like symptoms and bleeding disorders, and in severe cases patients exhibit high fever that can progress into shock and death. There are five families of viruses that currently are recognized to cause haemorrhagic fever, but cattle appear to only be associated with the family *Bunyaviridae* (Ikegami, 2012; Leblebicioglu et al., 2015).

Rift Valley fever is a haemorrhagic fever and is caused by the Rift Valley fever virus (RVFV). This virus belongs to the viral genus *Phlebovirus* and was first identified in 1930 associated with an epidemic of abortions of pregnant ewes and newborn lamb deaths followed by human illnesses that occurred in the Rift Valley region of Kenya (Ikegami, 2012). Since 1930, a number of outbreaks have been reported as the virus has spread throughout Africa and into the Arabian Peninsula. Cattle are also infected and the only symptom is abortion in cows (Bird and Nichol, 2012). The virus is typically spread from infected cattle or sheep to other animals and humans by the floodwater mosquito *Aedes* sp., but infected animals may have high viral loads and contact with contaminated tissue is a significant risk factor.

The Crimean-Congo haemorrhagic fever is different from Rift Valley fever and is caused by the Crimean-Congo haemorrhagic fever virus (VVHFV). This arbovirus belongs to the viral genus *Nairovirus* and was first recognized in Tajikistan in 1944 (Leblebicioglu et al., 2015; Tishkova et al., 2012). Currently, this haemorrhagic fever is found in southwest and central Asia, southeast Europe and throughout Africa. Cattle are a significant reservoir for the virus but are asymptomatic. The virus is spread from bites by the *Hyalomma anatolicum* tick, but other species of *Hyalomma* tick have also tested positive for the virus. Direct contact with blood or tissues from infected animals has also been identified as a significant source for human infection.

4 Major zoonotic pathogens related to cattle: *Escherichia coli*, *Salmonella* and *Campylobacter*

4.1 *Escherichia coli*

In 1885, Theodor Escherich isolated from the human colon and described a bacterium he called, *Bacterium coli commune*, and the bacterium was later named *Escherichia coli* in his honour (Doyle and Padhye, 1989). The bacterium *E. coli* is near ubiquitous in nature, and is commonly isolated from faeces of mammals. Most strains of *E. coli* are considered commensal in the gastrointestinal tract and do not cause disease. However, some strains have assimilated colonization and virulence genes and are capable of causing disease. Pathogenic strains of *E. coli* have been associated with a variety of diseases in animals, including septicemia, mastitis, meningitis, urinary infections and pulmonary infections, but varying forms of gastroenteritis are the most common illnesses. The highly numerous strains of *E. coli* are phylogenetically related but diverse with regard to pathogenicity, which in combination with widespread distribution makes identification and association with disease difficult. The common classification method for *E. coli* strains is to serotype using the O (cell surface lipopolysaccharide), H (flagella) and K (cell capsule) antigens, but typically only the O and H results are denoted for identification.

E. coli are Gram-negative, motile (most isolates), facultative anaerobic, rod-shaped bacteria. Strains of pathogenic *E. coli* associated with gastrointestinal infection can be categorized on the basis of colonization and virulence mechanisms into specific pathotypes (Doyle and Padhye, 1989). Enteroinvasive *E. coli* (EIEC) and Enteroaggregative *E. coli* (EAEC) can cause diarrhoea, but both types of *E. coli* are host-restricted to humans and have not been observed in other animals. Enterotoxigenic E. coli (ETEC) is one of the most common causes of diarrhoea in humans, and this toxin-producing type of bacteria can also cause disease in other animals. Enteropathogenic *E. coli* (EPEC) can tightly colonize the intestinal epithelium and deform the colonized epithelial cell, and this 'attaching and effacing' can result in diarrhoea in the host. Strains of both ETEC and EPEC are observed in cattle, particularly in calves, but the specific role these pathotypes may play in cattle disease is controversial (Kolenda et al., 2015). Enterohaemorrhagic *E. coli* (EHEC) is a pathotype with colonization mechanisms similar to EPEC, but EHEC strains also harbour genes that produce shiga toxins, also known as vero toxins (Smith et al., 2014). Production of shiga toxins is associated with severe illnesses in infected humans and the toxin genes are found in a variety of *E. coli* strains. However, the most common disease-causing shiga-toxigenic *E. coli* (STEC) are the strains characterized as EHEC.

E. coli O157:H7 is commonly found in the gastrointestinal tract of cattle (Berry and Wells, 2010). Faecal shedding of the pathogen is seasonal and during the summer months when prevalence is the highest, most feedlot cattle will shed this pathogen. *E. coli* O157:H7 can be enumerated by direct plating onto selective agar, and the number of this pathogen in faeces can be highly variable. Animals shedding this pathogen at levels equal to or greater than 10^4 cells per gram are typically defined as super-shedders, but the percentage of super-shedders in a pen is typically low and the period of shedding is limited. A site of colonization in the bovine gastrointestinal tract has been shown to be at the rectal-anal junction. Once shed in the environment, *E. coli* O157:H7 can persist for weeks in the feedlot surface material. Hide contamination is commonplace, and carriage on the hide at harvest has been a source of carcass contamination.

In cattle, STEC/EHEC is not associated with disease in the host animal, but in humans the pathogens can cause haemorrhagic colitis, haemolytic uraemic syndrome (HUS) and possibly death. The first STEC/EHEC associated with human disease was *E. coli* O157:H7 in two outbreaks in 1982 (Doyle and Padhye, 1989). In 1993, an *E. coli* O157:H7 outbreak of over 500 confirmed human infections in the western United States was linked to consumption of undercooked hamburger (Center for Disease Control and Prevention, 1993). The associations between human disease and *E. coli* O157:H7 for ground beef products has resulted in significant changes in beef processing to reduce transmission to carcasses and meat. Since the reported outbreak in 1993, additional pathogenic STEC/EHEC have been identified and characterized (Smith et al., 2014). *E. coli* O157:H7 accounts for the majority of STEC illnesses, but the STEC/EHEC serogroups O26, O45, O103, O111, O121 and O145 account for the majority of non-STEC O157 illnesses in the United States.

4.2 *Salmonella*

Most animals are susceptible to one or more diseases caused by the bacterium genus *Salmonella*. Diseases can be classified as typhoidal and non-typhoidal. Typhoid, or typhoid fever, is a systemic disease caused by *S. enterica* Typhi or *S. enterica* Paratyphi that are host-adapted specifically to humans, and is most prevalent in developing countries (D'Aoust, 1989). In contrast, enteritis is a non-typhoidal disease caused by a variety of *S. enterica* subspecies that affects a variety of animals, including cattle and humans. Typhoidal *Salmonella* was first isolated from human typhoid patients in 1884 by Georg Gaffky, a Prussian army surgeon (Yoshikawa et al., 1980), and non-typhoidal *Salmonella* was first isolated in 1885 by Theobald Smith, a U.S. Department of Agriculture (USDA) Bureau of Animal Industry microbiologist (Schultz, 2008). In 1900, *Salmonella* was proposed as a name in honour of Daniel Salmon, the director of the USDA Bureau of Animal Industry that isolated the first non-typhoidal strain, and the name was officially adopted in 1933.

Salmonella are Gram-negative, motile, facultative anaerobic, rod-shaped bacteria and infections from this pathogen are reported worldwide. *Salmonella* has been recognized as a pathogen for more than 130 years, but due to phenotypic diversity, nomenclature of *Salmonella* was unclear until the last decade. Genetically, *Salmonella* are highly related and there are only two recognized species, *S. enterica* and *S. bongori* (Brenner et al., 2000). Within the *S. enterica* species, there are six subspecies (*enterica*, *salamae*, *arizonae*, *diarizonae*, *houtenae* and *indica*). Furthermore, *Salmonella* are differentiated on the basis of serotyping of the O (cell surface lipopolysaccharide) and H (flagella) antigens, and currently there are more than 2500 serotypes of *Salmonella*. The subspecies *S. enterica* subsp. *enterica* accounts for nearly 1500 of the *Salmonella* serotypes and include all of the serotypes that affect warm-blooded animals, such as serotypes Choleraesuis, Enteritidis, Newport, Paratyphi, Typhi and Typhimurium. Most of the remaining serotypes are assigned to the subspecies groupings *S. enterica* subsp. *salamae* and *S. enterica* subsp. *diarizonae*, and these serotypes, many of which can still cause human illness, are associated with cold-blooded animals and the environment.

Illnesses caused by *Salmonella* can be variable with regard to host susceptibility and illnesses observed. Collectively, *Salmonella* have assimilated and evolved a vast repertoire of genes to bind, invade, colonize and survive in the host, but the presence of specific genes can be highly variable (Desai et al., 2013; LaRock et al., 2015; Wiedemann et al., 2015). *Salmonella* can be host-restricted, such as *S. enterica* Typhi and *S. enterica* Paratyphi that cause typhoid in humans (Silva et al., 2014). In chickens, *S. enterica* Gallinarum can cause typhoid-like illnesses with high mortality and *S. enterica* Pullorum can cause 'bacillary white diarrhoea' also with high mortality; however, both of these serotypes can infect other fowl, but neither appear to cause disease in humans or other mammals. Host-adapted serotypes of *Salmonella* are commonly observed in one animal reservoir but have the capability of causing disease in other animals, whereas broad-host-adapted serotypes lack a specific reservoir. Examples of host-adapted include strains include *S. enterica* Choleraesuis in swine and *S. enterica* Dublin in cattle (Costa et al., 2012). In cattle, *S. enterica* Dublin can cause systemic infections and may lead to abortions in pregnant cows. Broadly adapted strains of *Salmonella* can infect and cause illness to a variety of animals (mammals, birds and reptiles). Specific broad-host-adapted serotypes for the beef industry include *S. enterica* Typhimurium, which is associated with enteric disease in young calves (Mohler et al., 2009), and *S. enterica* Newport, which can harbour multiple drug resistance.

The USDA National Veterinary Services Laboratory (NVSL) determines serotypes for *Salmonella* isolates from clinically ill animals (clinical isolates) submitted by veterinary diagnostic laboratories across the United States and for isolates from production animal (non-clinical isolates) and food surveillance programmes (CDC, 2013). Many of the clinical and non-clinical *Salmonella* isolates characterized by USDA-NVSL originate from poultry production samples. The most recent CDC update summarized the NVSL predominant clinical strains in the United States (Table 2), and within the *S. enterica* serotypes Cerro, Dublin, Kentucky, Newport and the serotype I 4,[5],12:i:- isolates from cattle represented a significant percentage of each serotype's total isolates, which can reflect some host-adaptation within cattle (CDC, 2014). Cattle clinical isolates were also found at high percentages within the *S. enterica* serotypes Anatum, Mbandaka, Montevideo, Muenchen and Typhimurium. In non-clinical isolates, the *S. enterica* serotypes Kentucky, Montevideo, Muenster and Sandiego are represented with significant percentages of isolates from cattle within each serotype, possibly indicating a potential tolerance by bovine animals with little exhibited clinical signs of illness.

For humans, most *Salmonella* illnesses are non-typhoidal and can arise from consumption of contaminated food and water. Not all *Salmonella* found in food and water will cause disease in humans, and for those serotypes that do promote human illness, there is variation in the severity of the illness. In humans, some degree of gastroenteritis is the most common symptom of *Salmonella* infection; however, some of these same *Salmonella* serotypes can invade and lead to extraintestinal disease (Gal-Mor et al., 2014). Based on recent data from the CDC (2014), the top 20 predominant serotypes accounted for nearly 70% of all human illnesses and three serotypes were each greater than 10%. *S. enterica* Enteritidis is the most common human isolate and this serotype is common to poultry and eggs. An interesting observation regarding this serotype was that *S. enterica* Enteritidis was rarely observed in chicken until the host-restricted *S. enterica* serotypes Gallinarum and Pullorum were eradicated from commercial chicken flocks (Silva et al., 2014). *S. enterica* Typhimurium is the second most common human isolate, and this serotype is not limited to any species and can be found in most food animals. *S. enterica* Newport is the third most common human isolate and this serotype, while considered broad-host-adapted, is commonly observed in cattle. Thirteen serotypes accounted for 1–6% of the human isolates associated with disease, but only one of these serotypes (*S. enterica* Montevideo) is specifically observed at a high incidence in cattle over other

Table 2 Listing of major *Salmonella enterica* serotypes based on CDC (human illness isolates) and NVSL (farm animal illness isolates and non-clinical isolates) Top 20 rankings as part of the National Enteric Disease Surveillance, 2012 (CDC, 2014)

	CDC		NVSL clinical		NVSL non-clinical	
Serotype	2012, rank	% of total[a]	2012, rank	% from cattle[b]	2012, rank	% from cattle[b]
Agona	19	0.7	3	13.8	18	7.8
Anatum	17	0.8	12	19.5	13	37.3
Bareilly	10	1.8		-		–
Berta	20	0.6		3.5		0.3
Blockley		–		1.2		0.0
Braenderup	11	1.7		7.2	10	1.8
Cerro		–	6	–	17	–
Derby		–	4	3.3		2.6
Dublin		–	2	–	19	–
Enteritidis	1	14.5	7	2.4	2	0.5
Hadar		-		1.8	14	0.3
Heidelberg	9	2.0	10	2.1	3	0.2
I 4,[5],12:i:-	5	3.6	9	35.9		3.6
I 4,12:i:-		-	14	–		-
Infantis	7	2.3	8	8.7	15	4.8
Javiana	4	5.9		4.1		1.0
Johannesburg		–	20	–		–
Kentucky		–	16	–	1	–
Litchfield		–		12.1		1.2
Liverpool		–		–	16	–
Mbandaka		–	17	21.9	6	16.5
Mississippi	15	1.3		6.8		0.0
Montevideo	6	2.5	11	25.6	7	43.8
Muenchen	8	2.1	15	23.4	20	7.9
Muenster		–		–	9	–
Newport	3	10.4	5	49.3	12	18.6
Ohio		–	19	–		–
Oranienburg	14	1.5		11.4		16.6
Panama		–		11.9		18.8
Paratyphi B var. L(+) tartrate+	16	1.0		–		–
Poona		–		8.2		13.6
Reading		–		6.1		14.3
Saintpaul	13	1.6		2.8		0.7
Sandiego		–		1.9		80.9
Schwarzengrund		–		5.3	11	1.2
Senftenberg		–	13	3.3	4	1.7
Stanley		–		6.8		0.0
Thompson	12	1.7		9.6	8	2.5
Typhi	18	0.7		0.0		0.0
Typhimurium	2	11.6	1	33.7	5	15.7
Worthington		–	18	–		–

[a]CDC percentages are from 49000 human isolates reported in 2012 (CDC, 2014).
[b]NVSL percentages within serotype attributed to beef and dairy cattle, and are cumulative data 1968–2011 for top 30 CDC *Salmonella enterica* serotypes (CDC, 2013).

‌\$ 130 –

food animals, and three serotypes are common to most food animals including cattle (*S. enterica* serotype I 4,[5],12:i:-, Muenchen, and Oranienburg). *Salmonella* on cattle hides may contaminate beef carcasses during processing, but recent evidence suggests that lymph nodes may harbour *Salmonella* and serve as a source of *Salmonella* in ground beef products (Arthur et al., 2008).

4.3 *Campylobacter*

Theodor Escherich in 1886 observed bacteria similar to *Campylobacter* spp. in the colon contents of infants, but was unable to isolate these fastidious microorganisms (Park, 2002). In 1906, *Campylobacter*-like microorganisms were reported in the uterine mucus of pregnant sheep and in 1913 similar bacteria were isolated from aborted bovine foetuses (Silva et al., 2011). Bacteria associated with diarrhoea in cattle were isolated in 1927 and named *Vibrio jejuni*. In 1944, a different 'vibrio' was isolated from swine with diarrhoea and named *V. coli*. The animal isolates differed from other *Vibrio* strains and were renamed *Campylobacter* in 1963, and subsequently recognized to be associated with human diarrhoea. Although first reported as a cause of disease in animals in 1913, *Campylobacter* spp. have only been recognized as a specific cause of human disease since 1980.

Since 1963, a variety of bacterial species have been placed into the genus *Campylobacter*, but in recent years some of the species have been phylogenetically reclassified into the related genera *Arcobacter* and *Helicobacter* (Silva et al., 2011). Within the *Campylobacter* genus there currently are 18 recognized species with 6 subspecies (Humphrey et al., 2007). *Campylobacter* spp. are observed in cattle and can impact cattle health (Michi et al., 2016). *C. fetus* subspecies *fetus* is associated with spontaneous abortions and *C. fetus* subspecies *venerealis* is associated with infertility in bulls and cows. *C. hyoinstestinalis* subspecies *hyoinstestinalis* has been associated specifically with bovine enteritis, but *C. coli* and *C. jejuni* subspecies *jejuni* may also cause enteritis in a variety of animals including cattle (Humphrey et al., 2007). *C. sputorum* and *C. lanienae* have also been found to be associated with cattle, but are not known to cause illnesses in cattle. *C. jejuni* and *C. coli* are the predominant species recovered from cattle (Horrocks et al., 2009).

The *Campylobacter* species are Gram-negative, motile, microaerophilic, small spiral-shaped bacteria. The fastidious microorganism grows best in an atmosphere of 10% CO_2 and 5% O_2. Most pathogenic *Campylobacter* grow between 30 and 46°C, and based within the *Campylobacter* genus these strains are classed as 'thermophilic campylobacters'. *Campylobacter* pathogenesis differs from most other pathogens and virulence mechanisms for *Campylobacter* are not fully understood (Silva et al., 2011). The flagella are required for intestinal colonization, but species differ in flagella synthesis, location and components. *C. jejuni*, *C. coli* and *C. fetus* all harbour three genes required for the synthesis and delivery of cytolethal distending toxin (CDT) and the toxin is considered a key virulence factor. In humans, *Campylobacter* species are recognized as a leading cause of diarrhoeal disease worldwide. Campylobacteriosis in humans is typically limited to gastroenteritis, and, although humans are susceptible to most *Campylobacter* species, these illnesses are associated with *C. jejuni*, which is most often associated with poultry products but sometimes with beef, and *C. coli*. Systemic infections resulting in bacteraemia in humans are rare, but fatal, and these infections are commonly associated with *C. fetus* (Wagenaar et al., 2014).

5 Additional pathogens related to cattle: *Leptospira*, *Listeria* and other pathogens

5.1 *Leptospira*

Currently, one of the most widespread zoonotic diseases is leptospirosis, and a variety of production animals are susceptible to infection by certain species *Leptospira* (Spickler and Leedom Larson, 2013). Cattle are a reservoir and symptoms of the disease can be age dependent. Calves can exhibit acute illness with symptoms of fever, anorexia, diarrhoea, jaundice, anaemia and meningitis, which could result in death. In older cattle, infertility and reproductive problems, including abortions, are observed. Not all of the 21 species of *Leptospira* are pathogenic, but currently 13 pathogenic species are recognized, comprising more than 200 known serovars (Adler and de la Pena Moctezuma, 2010). Not all 13 species are pathogenic to cattle, but most serovars of the pathogenic species can cause disease in humans to some degree. In cattle, *Lept. borgpetersenii* Hardjo is the host-adapted strain, but cattle may be asymptomatic carriers for *Lept. interrogans* Pomona (swine-adapted). *Leptospira* spp. invade the host animal and humans through cuts and abrasions in skin and mucosal membrane surfaces and eventually colonize tissue and cell surfaces (Adler and de la Pena Moctezuma, 2010; Adler, 2014). The pathogen can resist phagocytosis, but if engulfed by a macrophage, the bacterium can survive and induce apoptosis to escape. The renal proximal tubular epithelial cells of the kidneys are major colonization sites and the pathogen is excreted in the urine.

Leptospira are Gram-negative, motile, obligate aerobic, tightly coiled (spirochete) bacteria that appear elongated with bent ends (Guerra, 2009; Adler and de la Pena Moctezuma, 2010). The bacterium can only replicate in the mammalian host and does not appear to replicate in the environment. Most infectious serovars of the pathogen can survive in soil and water environments, particularly when temperatures are warm and the environment is slightly alkaline. In tropical climates, human illnesses are consistently observed throughout the year whereas temperate climates the illnesses are observed in warmer months. Humans can acquire the disease from contact with diseased animals or contamination from production environments, or from processing an infected beef carcass. In humans, the disease can range from asymptomatic to severe and humans tend to not be chronic carriers (Spickler and Leedom Larson, 2013). Human leptospirosis is typically biphasic, initiated with acute symptoms such as fever, chills, headache and myalgia lasting for up to one week, and then it abates or disappears. Many patients recover after the initial phase, but in some patients a second phase may develop similar in many ways to meningitis with severe headaches and stiffness in the neck. With the second phase, up to 10% of the patients may experience progressive symptoms that may be associated with multi-organ failure that could result in death. Human-to-human transmission is rare.

5.2 *Listeria*

A bacilli associated with listeriosis-like illness was first described in 1891, but the causative agent was not isolated until 1926 after an epidemic in laboratory animals occurred (Lovett, 1989). A symptom of the disease was an increase in blood monocytes, thus the bacterium was named *Bacterium monocytogenes*. The bacterium was renamed *Listeria monocytogenes* in 1940 in honour of Sir Joseph Lister, a British surgeon who introduced the widespread use of antiseptic surgery. Human cases of listeriosis were documented in 1929, but it was not until 1983 that an outbreak associated with contaminated food was reported (Cossart, 2007). *Listeria* spp. are nearly ubiquitous in nature and can be found in the intestinal tract of a variety of animals, including humans. The genus *Listeria* contains ten species of bacteria, but only two are considered pathogenic (Roberts and Wiedmann, 2003). *List. ivanovii* is typically limited to ruminants, particularly sheep and cattle, whereas *List. monocytogenes* is observed in a variety of animals. Both of these pathogenic strains are commonly found in cattle, and if ingested in high numbers can cause septicemia in cattle that can result in abortions. The pathogens can be shed in faeces of infected cattle but also can invade the mammary gland and be secreted into milk.

Listeria are Gram-positive, motile, facultative anaerobic, non-spore-forming bacilli. The pathogenic strains *List. monocytogenes* and *List. ivanovii* are typically invasive to the host, but infections due to *List. monocytogenes* are more widely recognized in humans and death can be common (Roberts and Wiedmann, 2003; Lecuit, 2007). *List. monocytogenes* is unique among pathogens in that the bacterium can circumvent the intestinal, blood–brain and placental barriers. *List. monocytogenes* harbours a variety of virulence genes to invade and survive in the host cells, but also to aid in escape from the host cells (Cossart and Toldeo-Arana, 2008). In humans, *List. monocytogenes* is most often associated with gastroenteritis, but it can be associated with more severe disease such as septicemia, meningitis, meningo-encephalitis and abortions. *List. monocytogenes* are differentiated based on serotyping of the O (cell surface lipopolysaccharide) and H (flagella) antigens, and currently there are more than five serotypes associated with human illness (Liu, 2013). When *Listeria* infections advance beyond gastroenteritis, *List. monocytogenes* 4b is more common associated with infections of central nervous system, whereas *List. monocytogenes* 1/2a and *List. monocytogenes* 1/2c are more often associated with bacteraemia. The serotype *List. monocytogenes* 1/2b is more often associated with neonatal illnesses and abortions. Foodborne transmission is the most common route of transmission to humans, but *List. monocytogenes* is rarely observed in raw meat products (Swaminathan and Gerner-Smidt, 2007; Todd and Notermans, 2011).

5.3 Other pathogens

A number of pathogens that cause gastroenteritis in humans may be contaminants via a link to beef products when the food is not handled or prepared properly. *Clostridium perfringens* is a common inhabitant in the gastrointestinal tract of most animals and this Gram-positive spore-forming, toxin-producing bacterium is often observed on raw and frozen meat and human illness occurs from consumption of temperature-abused cooked meat products (Labbe, 1989). *Citrobacter* spp. are commonly enriched from the faeces of animals and these Gram-negative bacteria have been associated with illnesses from meat (Stiles, 1989). *Coxiella burnetti* is a Gram-negative, obligate intracellular bacterium found in ruminants that causes Q ('Query') fever in humans, but most human illnesses are from occupational contact during animal production. *Toxoplasma gondii* is a parasite that causes toxoplasmosis and this pathogen has been recently recognized as a foodborne pathogen associated with meat from range-fed animals, but cattle, unlike other ruminants, are resistant to the colonization by this pathogen and a direct link to beef has not been demonstrated (Jones and Dubey, 2012).

6 Pathogen control: good animal management and biosecurity practices

Zoonotic pathogens from cattle represent a significant risk to humans, and there are various transmission routes by which humans can get infected. The risk of foodborne transmission to beef consumers can be minimized during harvest (Wheeler et al., 2014). However, pathogens are still present in the animal production environment and there are still significant risks for pathogen transmissions to not only other cattle but also to humans. Animal producers, handlers, animal health professionals, abattoir personnel and butchers are at risk for direct transmission from infected cattle. Run-off from cattle fields and feedlot pens can contaminate surface water and this water is subsequently used for irrigation of crops or recreation. Manure and waste from cattle production is rich in nitrogen and minerals and this co-product of cattle production can be used for as a fertilizer, particularly for natural and organic crop producers. Co-products from cattle at harvest, such as hides, are also potential sources of pathogens in downstream product processing. Good husbandry and management practices by cattle producers can reduce pathogens in beef production and reduce many of these zoonotic risks.

Good animal management practices are fundamental for protecting cattle health and limiting their exposures to zoonotic pathogens. Core objectives of a programme aimed at these goals include (1) preventing the introduction of infection, (2) preventing the survival and spread of infection within the herd or flock and (3) reducing or eliminating an established infection (WHO, 2001). Specific approaches for accomplishing these objectives depend upon the particular disease agent (such as the infectiousness and severity of the disease threat, or any economic or regulatory considerations for its presence in the cattle herd) and the tools and resources available to the producer.

Providing a clean, well-drained environment and clean, high-quality feed and water is fundamental to maintain animal health and requires attention to prevent pathogen introduction and transmission by these routes (BIFSCO, 2013). Contaminated feed has been demonstrated as a vehicle for the introduction of *Salmonella* and *E. coli* O157:H7 into cattle herds (Crump et al., 2002; Davis et al., 2003). Ingestion of improperly fermented silage contaminated with *List. monocytogenes* is a primary cause of listeriosis for cattle and other ruminants (Fenlon et al., 1996; García et al., 2015; Wiedmann et al., 1996). Wild animals, birds and insects can carry and transmit zoonotic pathogens to feed, water and livestock, and although difficult, their presence in cattle operations should be discouraged to the extent possible (Ahmad et al., 2007; Callaway et al., 2014b; Cizek et al., 1999; FSIS, 2014; Gaukler et al., 2009; Palmer et al., 2012; Sanad et al., 2013; Swirski et al., 2014). As discussed further below, vaccines may prevent infection or reduce the severity of the disease and limit the survival and spread of disease agents in the herd.

Biosecurity practices to prevent the introduction of pathogens to the herd or the spread of pathogens within the herd include the isolation of sick animals from healthy animals, and the isolation or quarantine of new animals that are brought onto the farm (Buhman et al., 2000; Smith, 2011). Controlling traffic onto the cattle operation, including people, equipment and vehicles, can limit pathogen introduction and control farm-to-farm spread of disease-causing organisms (Buhman et al., 2000). Likewise, controlling traffic within a cattle operation can reduce the spread of pathogens from sick pens to where healthy animals are housed, or reduce the risk for contaminating feed. Hygienic practices and regular sanitation contribute to these efforts. As reviewed by Buhman et al. (2000), there should be regular cleaning and disinfection of any equipment that is introduced into the mouths of cattle or that is used to handle feedstuffs. For example, instruments used to treat animals should be cleaned and disinfected between animals, especially when used to treat sick cattle. As another example, loaders used to move manure or dead cattle should be cleaned before using to move feed. Regular cleaning of feed bunks and water tanks, and regular removal and/or treatment of manure and bedding can minimize the spread of pathogens between animals (Smith, 2011). For disease eradication programmes like those targeting brucellosis and bovine tuberculosis in the U.S. cattle, more rigorous biosecurity practices are employed, including active surveillance and testing, movement restrictions and quarantine, removal and destruction of infected animals, and/or maintenance of closed herds (APHIS, 2014; Palmer and Waters, 2011; Ragan, 2002). These practices have been necessary for cattle producers to achieve disease-free status for their herds and for reducing the transmission of *B. abortus* and *M. bovis* to humans.

Surveillance and classification of dairy and beef cattle herds according to their risk of *Salmonella* Dublin infection, coupled with animal movement records and trade restrictions, are important biosecurity elements of Denmark's *Salmonella* eradication campaign, which has achieved a reduction in the prevalence of *Salmonella* test-positive herds (Nielsen, 2009; Warnick et al., 2006). Subsequent research and analyses of data collected in the Danish programme associated successful *Salmonella* control with avoidance of purchasing of animals from test-positive herds and good calving management, including good separation between calf pens (solid walls versus bars), not using the calving area for sick animals, and frequent cleaning and bedding of the calving area (Nielsen et al., 2012; Nielsen and Dohoo, 2012, 2013). Risk factors or predictors for a cattle herd that test positive for *Salmonella* included increasing herd sizes, having a neighbouring *Salmonella*-positive herd within 5 km and organic farming (Nielsen and Dohoo, 2012, 2013). *Salmonella* control in infected herds was improved with implementation of management procedures to reduce transmission in calving and young calf areas (Nielsen and Nielsen, 2012).

E. coli O157:H7 became a focus of pre-harvest food safety efforts in the years following its declaration as an adulterant in ground beef (FSIS, 1994). Research to date suggests that good animal production and management practices alone will not eliminate this pathogen in cattle. Several studies provide evidence that contaminated drinking water is a means to transmit *E. coli* O157:H7 to cattle, which suggested drinking water as a potential intervention target to reduce this pathogen (Faith et al., 1996; Shere et al., 2002; Van Donkersgoed et al., 2001). However, measures to reduce *E. coli* O157:H7 in drinking water have not been demonstrated to reduce its prevalence in cattle (Ellis-Iversen et al., 2008, 2009). There were no differences in the prevalence of *E. coli* O157:H7 shed by cattle between feedlot pens supplied with chlorinated (1 ppm residual free chlorine) or nonchlorinated drinking water (LeJeune et al., 2004). The accumulation of organic matter in cattle water tanks (e.g. feed, faeces) can reduce the bactericidal activity of many water treatments (LeJeune et al., 2004; Stevenson et al., 2004; Zhao et al., 2006). As reviewed by LeJeune and Wetzel (2007), there likewise is no clear evidence that reducing exposure of cattle to *E. coli* O157:H7 by feed hygiene, reducing animal density or separately housing calves from older animals significantly reduces the prevalence of this organism in cattle herds. However, high-level contamination and persistence of *E. coli* O157:H7 in the cattle production environment may be involved its transmission to other animals. The results of several studies examining the impact of cattle that shed high numbers of *E. coli* O157:H7 (10^4 CFU per gram of faeces or greater; 'super-shedders') suggest that this typically small proportion of cattle are responsible for a large proportion of *E. coli* O157:H7 contamination in a production environment (e.g. farm, feedlot pen) and this may in turn drive the *E. coli* O157:H7 prevalence of cattle in that environment (Bach et al., 2005a; Chase-Topping et al., 2007, 2008; Cobbold et al., 2007; Matthews et al., 2006; Stephens et al., 2009). As another example, *E. coli* O157:H7 persists longer in manure from cattle fed 20 and 40% corn wet distillers grains with solubles (WDGS) compared to corn and this persistence may contribute to the higher prevalence and levels of this pathogen in faeces and on hides of cattle-fed WDGS (Varel et al., 2008; Wells et al., 2009). On the other hand, monthly feedlot pen cleaning during a 140-day feeding period did not affect the prevalence of *E. coli* O157:H7 in steer calves (Folmer et al., 2003).

Although the practice of good animal management and biosecurity may not eliminate disease agents, their application will provide a good foundation for additional control efforts to successfully reduce zoonotic pathogens in cattle (Wheeler et al., 2014).

7 Pathogen control: use of vaccines

Vaccines have a long history of controlling many of the diseases that can result in morbidity or mortality in cattle. Where successfully used, cattle vaccination also has subsequently reduced many of the associated zoonotic human diseases. Louis Pasteur's public demonstrations in the 1880s of the ability of the anthrax vaccine to dramatically reduce disease and death in sheep, goats and cattle represent one of the earliest successes of immunization, and spurred continued development of vaccinations to control disease (Swartz, 1990). Anthrax is caused by *B. anthracis*, results in acute mortality in ruminants and other livestock and has caused devastating losses of food animals, as well as human illness, throughout recorded human history (Bengis and Frean, 2014). The development and adoption of the very effective Sterne spore vaccine in the 1930s sharply reduced the incidence of anthrax in livestock; in turn, the naturally acquired incidence of human anthrax by ingestion, inhalation or cutaneous exposure is now rare in those countries that routinely practice vaccination and surveillance (WHO, 2008). Indeed, control of natural anthrax (as opposed to anthrax from biowarfare or bioterrorism purposes) in both humans and food animals is achieved largely upon its prevention in livestock, with vaccines being key components of control programmes (Bengis and Frean, 2014; WHO, 2008). Anthrax is also a serious disease for many wildlife species, and outbreaks in wildlife provide a source of transmission and environmental contamination for infection of both livestock and humans (Bengis and Frean, 2014; WHO, 2008). Additionally, *B. anthracis* spores can survive for years in soils (Dragon and Rennie, 1995; WHO, 2008). Thus, vigilance in livestock vaccination is necessary to control anthrax. Potential vaccine-related improvements to anthrax control include the development of vaccines with longer duration of protection, vaccines that require only single or fewer doses, and/or effective oral vaccines, which would in particular simplify vaccine administration to wildlife species (Bengis and Frean, 2014; WHO, 2008).

There are vaccines for many other zoonotic diseases of cattle, although global usage may vary because of regulatory approval, availability, economic incentive and/or vaccine efficacy to protect animals (APHIS, 2013; Videnova and Mackay, 2012). Typically, vaccination combined with specific management practices is most effective at controlling animal disease. For example, vaccination against *B. abortus*, along with surveillance, quarantine and depopulation of cattle herds has been a major contributor to the success of the U.S. State-Federal Brucellosis Eradication Program that was initiated in the 1950s (APHIS, 2014; Ragan, 2002). A decline in the U.S. incidence of human brucellosis reflects the reductions in cattle brucellosis; 6400 cases of human brucellosis were reported in 1947, while only about 100 cases of human brucellosis have been reported annually since the 1990s (APHIS, 2014; Ragan, 2002). In contrast, vaccines against *M. bovis* currently are not used in bovine tuberculosis control and eradication programmes, including the highly successful U.S. Cooperative State-Federal Tuberculosis Eradication Program (Palmer and Waters, 2011). Cattle slaughter surveillance and live animal testing using the tuberculin skin test have been fundamental to the progress in *M. bovis* eradication, but the current best-candidate vaccine *M. bovis* Bacille Calmette-Guérin (BCG) interferes with the specificity of this test (Whelan et al., 2011). In combination with active disease surveillance and testing, intensive biosecurity measures including removal and destruction of infected animals, movement restrictions, quarantine and adoption of the accredited tuberculosis-free herd concept and animal identification practices, have been successfully used to control *M. bovis* in cattle; likewise, these measures along with milk and colostrum pasteurization have substantially reduced the transmission of *M. bovis* to humans (Conlan et al., 2015; Palmer and Waters, 2011; Pérez-Lago et al., 2014). Despite the successes, *M. bovis* elimination from cattle has remained elusive, due in part to the maintenance of this pathogen in wildlife and its transmission to cattle (Palmer et al., 2012). Thus, vaccination of cattle is under consideration for continued progress to eradicate bovine tuberculosis, as is field vaccination of the wildlife species that serve as reservoir maintenance hosts of *M. bovis*, including white-tailed deer in the United States, badgers in the United Kingdom and the Republic of Ireland, brushtail possum in New Zealand and wild boar in the Iberian Peninsula (Conlan et al., 2015; Palmer et al., 2012; Waters and Palmer, 2015). Additional zoonotic pathogens for which cattle vaccines are available include *Leptospira* (Guerra, 2009), *Salmonella* (Mohler et al., 2009; Smith et al., 2015), *C. fetus* (CFSPH, 2013; Truyers et al., 2014) and Rift Valley fever virus (Bird et al., 2011; Bird and Nichol, 2012).

While the historic use of livestock vaccines against zoonotic pathogens has been to protect both animal and human hosts from disease and death, there has been increased attention in recent years on the development and use of vaccines as pre-harvest food safety interventions. *E. coli* O157:H7 is a good example of this new focus; this pathogen causes moderate-to-severe illness in humans, while its colonization of adult cattle results in no apparent detriment to the animal. Vaccination to reduce *E. coli* O157:H7 in cattle has received particular interest because of its regulatory status as an adulterant in raw ground beef and non-intact raw beef products (FSIS, 1999, 2015). There are two *E. coli* O157:H7 vaccines that have undergone extensive field-testing in North America beef cattle feedlots. One vaccine targets components of the *E. coli* O157:H7 Type III secreted protein system (TTSP) that are involved in intestinal colonization (Naylor et al., 2005; Smith, 2014). The TTSP vaccine currently is not licensed for use in U.S. cattle, although it is approved for use in Canada and other countries (FSIS, 2014). The second vaccine targets siderophore receptor and porin proteins (SRP) of *E. coli* O157:H7, thereby limiting the ability of the target cells to uptake iron needed for growth and survival (Thomson et al., 2009; Thornton et al., 2009). The SRP vaccine has a conditional license for use in the United States to reduce *E. coli* O157:H7 in cattle (FSIS, 2014). Both the TTSP and SRP vaccines have been demonstrated to reduce, though not eliminate, natural *E. coli* O157:H7 colonization and shedding by beef cattle in feedlot production settings (Fox et al., 2009; Moxley et al., 2009; Smith et al., 2009; Thomson et al., 2009; Vogstad et al., 2013; as reviewed by Smith [2014]). For both vaccines, three-dose regimens more effectively reduced the pathogen compared to two-dose regimens (Thomson et al., 2009; Vogstad et al., 2013). Predictive models indicate that reductions in beef contamination and human disease cases can be achieved by vaccinating cattle against *E. coli* O157:H7 (Hurd and Malladi, 2012; Matthews et al., 2013; Signorini and Tarabla, 2010). As an example, Matthews et al. (2013) demonstrate that a cattle vaccine reducing *E. coli* O157:H7 shedding frequency by 50% could reduce human cases by 83%. Despite this potential public health benefit, *E. coli* O157:H7 vaccines for cattle have seen limited use (APHIS, 2013; Matthews et al., 2013). *E. coli* O157:H7 does not cause clinical illness in cattle, so the cost and labour associated with vaccinations, combined with the lack of economic incentives for their use, has limited their adoption by cattle producers to date (Matthews et al., 2013; Smith, 2014; Tonsor and Schroeder, 2015). Research objectives to improve *E. coli* O157:H7 vaccines for cattle in ongoing work include the identification of additional antigens and/or optimization of antigen combinations for improved vaccine formulations, and testing and development for vaccines that provide cross-protection against other disease-causing non-O157 STEC (García-Angulo et al., 2014; McNeilly et al., 2015).

Likewise, vaccination of livestock reservoirs of *Salmonella* has been explored to reduce animal infection and contamination of meat or milk products, and subsequent human foodborne illness (Barrow, 2007; De Ridder et al., 2014; Desin et al., 2013; Dodd et al., 2011; Hermesch et al., 2008). *Salmonella* is an important pathogen that can cause disease in both humans and livestock animals, including cattle. As discussed above, various host range adaptations of different *Salmonella* serotypes have resulted in large differences in virulence, host adaptation and host specificities. Hence, while salmonellosis in food animals may be caused by host-restricted, host-adapted or broad-host-range *Salmonella* serotypes, these same serotypes may or may not cause illness in humans.

Vaccines that are currently available for cattle in the United States include products targeted against cattle-adapted *S. enterica* Dublin and broad-host-range *S. enterica* Newport and *S. enterica* Typhimurium (Hermesch et al., 2008; Mohler et al., 2009; Smith et al., 2015). These serotypes can cause disease in both cattle and humans (Costa et al., 2012; Maguire et al., 1992; Mohler et al., 2009; Taylor et al., 1982; Centers for Disease Control and Prevention, 2013). A *S. enterica* Newport bacterial extract vaccine with SRP® technology (*Salmonella* Newport SRP® vaccine; Epitopix LLC, Willmar, MN), directed against siderophore receptor and porin proteins as described above for *E. coli* O157:H7, has been examined for impacts on health, performance and shedding of *Salmonella* by dairy and beef cattle (Dodd et al., 2011; Hermesch et al., 2008). There was no significant difference in *Salmonella* faecal shedding by dairy cows vaccinated with the *Salmonella* Newport SRP® vaccine compared to unvaccinated control cows (Hermesch et al., 2008). *Salmonella* faecal shedding decreased during the study in both vaccinated and control cows, which suggested an effect of herd immunity, although the experiment was not designed to determine this aspect. Interestingly, milk production was greater and somatic cell counts were lower in the vaccinated cows, even though neither treatment group showed clinical signs of salmonellosis. Similarly, Heider et al. (2008) saw no differences in *S. enterica* faecal shedding by vaccinated and non-vaccinated dairy cows and saw no reduction in the prevalence of shedding when using the *Salmonella* Newport SRP® vaccine. In a separate study examining the epidemiology of *Salmonella* in cull dairy cows, prevalence of the pathogen was lower in cattle from dairies that practised whole-herd vaccination using the *Salmonella* Newport SRP® vaccine, compared to dairies that did not use this vaccine (Loneragan et al., 2012). In feedlot cattle, there were no effects of vaccination on *Salmonella* faecal prevalence, or cattle health and performance were detected when the *Salmonella* Newport SRP® vaccine was administered (Dodd et al., 2011). The effects of this vaccine to reduce this pathogen were inconsistent in experiments examining *Salmonella* infection of bovine peripheral lymph nodes (Edrington et al., 2013).

The successful reduction of *Salmonella* in livestock by using vaccines has been challenging because of the large number of serotypes of this pathogen, with their associated differences in virulence and host adaptation. In addition, the limited use of *Salmonella* vaccines in cattle is likely a reflection of the lack of demonstrated efficacy in field use. A recent report (APHIS, 2013) estimated that 6.5% of large feedlots (those feedlots with a capacity of 1000 or more head) used vaccines against *Salmonella*. These challenges suggest considerations for the continued development of food animal vaccines for this pathogen. Vaccines that protect against diverse *Salmonella* serotypes would be useful for prophylactic immunization strategies that are aimed at reducing the risk for both animal and human infection (Barrow, 2007; Desin et al., 2013; Heithoff et al., 2008; Mohler et al., 2008). With the continuing discussion of the regulation of this pathogen in meat products, the prevention of human infection by *Salmonella* and other zoonotic pathogens through the use of vaccines for food animals will remain a significant growth area for research and the animal health industry (Desin et al., 2013; Wheeler et al., 2014).

8 Pathogen control: non-traditional interventions

Recent research has explored additional, less traditional approaches to reduce zoonotic pathogens in cattle, with the goal to improve the microbial safety of beef products. Some of these pre-harvest interventions are currently available for use in cattle production, while others require more evaluation and/or scientific validation as effective at reducing pathogens in cattle.

The potential for probiotics, direct-fed microbials (DFM), competitive exclusion (CE) cultures and prebiotics to reduce pathogens in livestock animals has been reviewed (Berry and Wells, 2010; Callaway et al., 2008a, 2013; LeJeune and Wetzel, 2007; Oliver et al., 2009; Wheeler et al., 2014). These products are fed to animals to elicit positive effects on production and health by promoting beneficial bacteria in the gastrointestinal tract at the expense of harmful or pathogenic bacteria, by competition for attachment sites or for nutrients, by production of antimicrobial compounds or by promotion of immune function (Callaway et al., 2013). Probiotic microorganisms and DFM have long been fed to cattle to improve production efficiency (e.g. improve health, increase growth rate, increase milk production). The evaluation of select probiotic cultures suggests that feeding them to cattle may reduce faecal shedding of *E. coli* O157:H7 and *Salmonella* (Stephens et al., 2007; Tabe et al., 2008).

Bacteriophages are host specific, and their ability to selectively eliminate their host pathogens from a mixed bacterial population makes bacteriophages appealing as a potential pre-harvest treatment to reduce pathogens in the gastrointestinal tract of food animals (Greer, 2005; Johnson et al., 2008). *E. coli* O157:H7-specific bacteriophages have been examined as a means to reduce this pathogen in the gastrointestinal tract of ruminants. Oral administration of a cocktail of bacteriophages reduced faecal shedding of *E. coli* O157:H7 by sheep (Callaway et al., 2008b). In feedlot cattle, oral application of bacteriophages was more effective than rectal administration for reducing faecal shedding of *E. coli* O157:H7 (Rozema et al., 2009). The USDA-FSIS has approved the application of bacteriophage preparations as a spray or wash to reduce *Salmonella*, *E. coli* O157:H7 and non-O157 STEC on the hides of live cattle in holding pens before slaughter (FSIS, 2014).

Regulatory approval and associated concerns can be additional barriers to the implementation of pre-harvest food safety interventions. The addition of sodium chlorate to water or feed of cattle immediately before shipping and harvest has been proposed for reduction of *E. coli* O157:H7, and has been demonstrated to reduce *E. coli* O157:H7 and/or total *E. coli* in bovine faeces and on the hides (Anderson et al., 2005; Callaway et al., 2002). *E. coli* O157:H7 has the intracellular enzyme nitrate reductase, which reduces chlorate to chlorite, which then accumulates to lethal levels in the bacterial cell (Callaway et al., 2014a). The use of this compound to reduce pathogens in livestock currently is not approved, but an application for sodium chlorate use in cattle currently is under review by the U.S. Food and Drug Administration.

In addition to the above-mentioned interventions, a variety of other potential approaches have been explored for pathogen reduction in cattle and other livestock, such as dietary supplementation with organic acids/fatty acids, essential oils and other phenolic compounds, use of antimicrobials and the implementation of dietary shifts and other feeding strategies. These pre-harvest food safety studies have been topics of many recent reviews (Berry and Wells, 2010; Callaway et al., 2008a, 2009, 2013, 2014a; Jacob et al., 2009; LeJeune and Wetzel, 2007; Oliver et al., 2009; Wheeler et al., 2014). For cattle, much of this research has focused on reducing the pathogen *E. coli* O157:H7. However, there are both other zoonotic pathogens of concern and other food animal species, and with more of these pathogens likely to become subject to regulation in meat and meat products, pre-harvest food safety will remain a high-priority research area. Both the demonstrated reduction of zoonotic pathogens from cattle and their production environment, along with the demonstrated reduction of pathogens in the final consumer beef products, are needed to significantly reduce the risks to public health.

9 Summary and future trends

Pathogenic organisms are highly variable and adapt well to a changing environment. Surveillance and eradication programmes have worked well to control some zoonotic diseases, particularly where the illnesses are acute and symptoms in animals are easily recognizable. Vaccination has also been effective as a preventative action for many of these same diseases. However, some zoonotic pathogens may occur in cattle and infection may not be easily recognized. More research is still needed to reduce pathogens in the cattle production environment and minimize transmission to humans.

Many of the predominant zoonotic foodborne pathogens recognized in the United States may colonize the animal with few if any recognizable symptoms, such as with STEC, salmonellae and *Campylobacter* spp. In the case of STEC, a number of pre-harvest interventions have been approved, but the results are not clear. Many of these zoonotic pathogens must survive host and microbial interactions in animals to attach, colonize and multiply (Berry and Wells, 2016). The bacterial microbiota in the gastrointestinal tract is highly complex and the scientific community is only starting to understand the beneficial and antagonistic interactions between community members and the opportunistic pathogen. Furthermore, the sites of pathogen attachment and colonization in the host may require specific gene expression on the part of the host, or the host may indirectly affect pathogen survival by modulating the indigenous gastrointestinal microbiota. Many of the scientific tools to study and define these complex host–microbial interactions have only recently been developed.

10 Where to look for further information

The reader should note that several excellent reviews specific to each bacterial pathogen are referenced within each section. The book 'Foodborne Bacterial Pathogens' edited by Dr. Michael P. Doyle (1989, Marcel Dekker, Inc.) is an exhaustive overview of most known bacterial foodborne pathogens and each chapter was expertly prepared for specific pathogenic bacterial species. In the U. S. beef industry, *E. coli* and *Salmonella* are two pathogenic species of greatest concern and, with regard to the pre-harvest environment, the publications by Callaway et al. (2009, 2013, and 2014a) and Berry and Wells (2010 and 2016) provide reviews of current knowledge of research from around the world. The USDA APHIS has an excellent website for animal diseases (https://www.aphis.usda.gov/aphis/ourfocus/animalhealth/animal-disease-information), and the CDC has an excellent website for information regarding foodborne pathogens (http://www.cdc.gov/foodsafety/cdc-and-food-safety.html), and both websites are updated regularly.

11 References

Adam, R. D. (2001). Biology of *Giardia lamblia*. *Clin. Microbiol. Rev.* 14(3): 447–75.

Adler, B. (2014). Pathogenesis of leptospirosis: cellular and molecular aspects. *Vet. Microbiol.* 172(3–4): 353–8.

Adler, B. and de la Peña Moctezuma, A. (2010). *Leptospira* and leptospirosis. *Vet. Microbiol.* 140(3–4): 287–96.

Ahmad, A., Nagaraja, T. G. and Zurek, L. (2007). Transmission of *Escherichia coli* O157:H7 to cattle by house flies. *Prev. Vet. Med.* 80(1): 74–81.

Anderson, R. C., Carr, M. A., Miller, R. K., King, D. A., Carstens, G. E., Genovese, K. J., Callaway, T. R., Edrington, T. S., Jung, Y. S., McReynolds, J. L., Hume, M. E., Beier, R. C., Elder, R. O. and Nisbet, D. J. (2005). Effects of experimental chlorate preparations as feed and water supplements on *Escherichia coli* colonization and contamination of beef cattle and carcasses. *Food Microbiol.* 22(5): 439–47.

APHIS (Animal and Plant Health Inspection Service, U.S. Department of Agriculture) (2013). Vaccine usage in U.S. feedlots. Available at: https://www.aphis.usda.gov/animal_health/nahms/feedlot/downloads/feedlot2011/Feed11_is_VaccineUsage.pdf. Accessed 21 December 2015.

APHIS (Animal and Plant Health Inspection Service, U.S. Department of Agriculture) (2014). Facts about brucellosis. Available at: https://www.aphis.usda.gov/animal_health/animal_diseases/brucellosis/downloads/bruc-facts.pdf. Accessed 15 December 2015.

Arthur, T. M., Brichta-Harhay, D. M., Bosilevac, J. M., Guerini, M. N., Kalchayanand, N., Wells, J. E., Shackelford, S. D., Wheeler, T. L. and Koohmaraie, M. (2008). Prevalence and characterization of *Salmonella* in bovine lymph nodes potentially destined for use in ground beef. *J. Food Prot.* 71(8): 1685–8.

Bach, S. J., Selinger, L. J., Stanford, K. and McAllister, T. A. (2005). Effect of supplementing corn- or barley-based feedlot diets with canola oil on faecal shedding of *Escherichia coli* O157:H7 by steers. *J. Appl. Microbiol.* 98(2): 464–75.

Balkema-Buschmann, A., Fast, C., Kaatz, M., Eiden, M., Ziegler, U., McIntyre, L., Keller, M., Hills, B. and Groschup, M. H. (2011). Pathogenesis of classical and atypical BSE in cattle. *Prev. Vet. Med.* 102(2): 112–7.

Barka, E. A., Vatsa, P., Sanchez, L., Gaveau-Vaillant, N., Jacquard, C., Klenk, H. P., Clément, C., Ouhdouch, Y. and van Wezel, G. P. (2015). Taxonomy, physiology, and natural products of *Actinobacteria*. *Microbiol. Mol. Biol. Rev.* 80(1): 1–43.

Barrow, P. A. (2007). *Salmonella* infection: immune and non-immune protection with vaccines. *Avian Pathol.* 36(1): 1–13.

BIFSCO (Beef Industry Food Safety Council) (2013). Production best practices (PBP) to aid in the control of foodborne pathogens in groups of cattle. Available at: http://www.bifsco.org/CMDocs/BIFSCO/Best%20Practices/Production%20Best%20Practices.pdf. Accessed 25 January 2016.

Bengis, R. G. and Frean, J. (2014). Anthrax as an example of the One Health Concept. *Rev. Sci. Tech.* 33(2): 593–604.

Berry, E. D. and Wells, J. E. (2010). *Escherichia coli* O157:H7: recent advances in research on occurrence, transmission and control in cattle and the production environment. *Adv. Food Nutr. Res.* 60: 67–117.

Berry, E. D. and Wells, J. E. (2016). Reducing foodborne pathogen persistence and transmission in animal production environments: challenges and opportunities, in Pre-Harvest Food Safety, eds. S. Thakur and K. E. Kniel-Tolbert. ASM Press, Washington DC. *In press.*

Beyer, W. and Turnbull, P. C. (2009). Anthrax in animals. *Mol. Aspects Med.* 30(6): 481–9.

Bird, B. H., Maartens, L. H., Campbell, S., Erasmus, B. J., Erickson, B. R., Dodd, K. A., Spiropoulou, C. F., Cannon, D., Drew, C. P., Knust, B., McElroy, A. K., Khristova, M. L., Albariño, C. G. and Nichol, S. T. (2011). Rift Valley Fever Virus vaccine lacking the NSs and NSm genes is safe, nonteratogenic, and confers protection from viremia, pyrexia, and abortion following challenge in adult and pregnant sheep. *J. Virol.* 84(24): 12901–9.

Bird, B. H. and Nichol, S. T. (2012). Breaking the chain: Rift Valley fever control via livestock vaccination. *Curr. Opin. Virol.* 2(3): 315–32.

Brenner, F. W., Villar, R. G., Angulo, F. J., Tauxe, R. and Swaminathan, B. (2000). *Salmonella* nomenclature. *J. Clin. Microbiol.* 38(7): 2465–7.

Budu-Amoako, E., Greenwood, S. J., Dixon, B. R., Barkema, H. W. and McClure, J. T. (2011). Foodborne illness associated with *Cryptosporidium* and *Giardia* from livestock. *J. Food Prot.* 74(11): 1944–55.

Buhman, M., Dewell, G. and Griffin, D. (2000). Biosecurity basics for cattle operations and good management practices (GMP) for controlling infectious diseases. University of Nebraska-Lincoln Extension, Institute of Agriculture and Natural Resources, NebGuide G1411. Available at: http://extensionpublications.unl.edu/assets/pdf/g1411.pdf. Accessed 25 January 2016.

Callaway, T. R., Anderson, R. C., Genovese, K. J., Poole, T. L., Anderson, T. J., Byrd, J. A., Kubena, L. F. and Nisbet, D. J. (2002). Sodium chlorate supplementation reduces *E. coli* O157:H7 populations in cattle. *J. Anim. Sci.* 80(6): 1683–9.

Callaway, T. R., Edrington, T. S., Anderson, R. C., Harvey, R. B., Genovese, K. J., Kennedy, C. N., Venn, D. W. and Nisbet, D. J. (2008a). Probiotics, prebiotics and competitive exclusion for prophylaxis against bacterial disease. *Anim. Health Res. Rev.* 9(2): 217–25.

Callaway T. R., Edrington, T. S., Brabban, A. D., Anderson, R. C., Rossman, M. L., Engler, M. J., Carr, M. A., Genovese, K. J., Keen, J. E., Looper, M. L., Kutter, E. M. and Nisbet, D. J. (2008b). Bacteriophage isolated from feedlot cattle can reduce *Escherichia coli* O157:H7 populations in ruminant gastrointestinal tracts. *Foodborne Pathog. Dis.* 5(2): 183–91.

Callaway, T. R., Carr, M. A., Edrington, T. S., Anderson, R. C. and Nisbet, D. J. (2009). Diet, *Escherichia coli* O157:H7, and cattle: A review after 10 years. *Curr. Issues Mol. Biol.* 11(2): 67–80.

Callaway, T. R., Anderson, R. C., Edrington, T. S., Genovese, K. J., Harvey, R. B., Poole, T. L. and Nisbet, D. J. (2013). Novel methods for pathogen control in livestock preharvest: an update. In J. Sofos (ed.), *Advances in Microbial Food Safety*, 1st edn., pp. 275–304. Woodhead Publishing Limited, Sawston, Cambridge, UK.

Callaway, T. R., Edrington, T. S. and Nisbet, D. J. (2014a). Meat science and muscle biology symposium: ecological and dietary impactors of foodborne pathogens and methods to reduce fecal shedding in cattle. *J. Anim. Sci.* 92(4): 1356–65.

Callaway, T. R., Edrington, T. S. and Nisbet, D. J. (2014b). Isolation of *Escherichia coli* O157:H7 and *Salmonella* from migratory brown-headed cowbirds (*Molothrus ater*), common grackles (*Quiscalus quiscula*), and cattle egrets (*Bubulcus ibis*). *Foodborne Pathog. Dis.* 11(10): 791–4.

Carvalho Neta, A. V., Mol, J. P., Xavier, M. N., Paixão, T. A., Lage, A. P. and Santos, R. L. (2010). Pathogenesis of bovine brucellosis. *Vet. J.* 184(2): 146–55.

CDC (Centers for Disease Control and Prevention) (1993). Update: multistate outbreak of *Escherichia coli* O157:H7 infections from hamburgers – western United States, 1992–3. *MMWR Morb. Mortal. Wkly. Rep.* 42(14): 258–63.

CDC (Centers for Disease Control and Prevention) (2013). An atlas of *Salmonella* in the United States, 1968–2011: laboratory-based enteric disease surveillance. US Department of Health and Human Services, CDC, Atlanta, GA.

CDC (Centers for Disease Control and Prevention) (2014). National *Salmonella* Surveillance Annual Report, 2012. US Department of Health and Human Services, CDC, Atlanta, GA.

CFSPH (Center for Food Security and Public Health, Iowa State University, Ames, IA) (2013). Zoonotic Campylobacteriosis. Available at: http://www.cfsph.iastate.edu/Factsheets/pdfs/campylobacteriosis.pdf. Accessed January 2016.

Chase-Topping, M. E., McKendrick, I. J., Pearce, M. C., MacDonald, P., Matthews, L., Halliday, J., Allison, L., Fenlon, D., Low, J. C., Gunn, G. and Woolhouse, M. E. J. (2007). Risk factors for the presence of high-level shedders of *Escherichia coli* O157 on Scottish farms. *J. Clin. Microbiol.* 45(5): 1594–603.

Chase-Topping, M. E., Gally, D., Low, C., Matthews, L. and Woolhouse, M. (2008). Supershedding and the link between human infection and livestock carriage of *Escherichia coli* O157. *Nat. Rev. Microbiol.* 6(12): 904–12.

Cizek, A., Alexa, P., Literak, I., Hamrik, J., Novak, P. and Smola, J. (1999). Shiga toxin-producing *Escherichia coli* O157 in feedlot cattle and Norwegian rats from a large-scale farm. *Lett. Appl. Microbiol.* 28(6): 435–9.

Cobbold, R. N., Hancock, D. D., Rice, D. H., Berg, J., Stilborn, R., Hovde, C. J. and Besser, T. E. (2007). Rectoanal junction colonization of feedlot cattle by *Escherichia coli* O157:H7 and its association with supershedders and excretion dynamics. *Appl. Environ. Microbiol.* 73(5): 1563–8.

Conlan, A. J. K., Brooks Pollock, E., McKinley, T. J., Mitchell, A. P., Jones, G. J., Vordermeier, M. and Wood, J. L. N. (2015). Potential benefits of cattle vaccination as supplementary control for bovine tuberculosis. *PLoS Comput. Biol.* 11: e1004038. doi: 10.1371/journal.pcbi.1004038.

Cossart, P. (2007). Listeriology (1926–2007): the rise of a model pathogen. *Microbes Infect.* 9(10): 1143–6.

Cossart, P. and Toledo-Arana, A. (2008). *Listeria monocytogenes*, a unique model in infection biology: an overview. *Microbes Infect.* 10(9): 1041–50.

Costa, L. F., Paixão, T. A., Tsolis, R. M., Bäumler, A. J. and Santos, R. L. (2012). Salmonellosis in cattle: advantages of being an experimental model. *Res. Vet. Sci.* 93(1): 1–6.

Crump, J. A., Griffin, P. M. and Angulo, F. J. (2002). Bacterial contamination of animal feed and its relationship to human foodborne illness. *Clin. Infect. Dis.* 35(7): 859–65.

D'Aoust, J.-V. (1989). *Salmonella*. In M. P. Doyle (ed.), *Foodborne Bacterial Pathogens*, pp. 327–445. Merkel and Dekker, Inc., New York, NY.

Davis, M. A., Hancock, D. D., Rice, D. H., Call, D. R., DiGiacomo, R., Samadpour, M. and Besser, T. E. (2003). Feedstuffs as a vehicle of cattle exposure to *Escherichia coli* O157:H7 and *Salmonella enterica*. *Vet. Microbiol.* 95(3): 199–210.

De Ridder, L., Maes, D., Dewulf, J., Butaye, P., Pasmans, F., Boyen, F., Haesebrouck, F. and Van der Stede, Y. (2014). Use of a live attenuated *Salmonella enterica* serovar Typhimurium vaccine on farrow-to-finish pig farms. *Vet. J.* 202(2): 303–8.

Desai, P. T., Porwollik, S., Long, F., Cheng, P., Wollam, A., Bhonagiri-Palsikar, V., Hallsworth-Pepin, K., Clifton, S. W. Weinstock, G. M. and McClelland, M. (2013). Evolutionary Genomics of *Salmonella enterica* Subspecies. *MBio.* 4: e00579–12.

Desin, T. S., Köster, W. and Potter, A. A. (2013). *Salmonella* vaccines in poultry: past, present and future. *Expert Rev. Vaccines.* 12(1): 87–96.

Dodd, C. C., Renter, D. G., Thomson, D. U. and Nagaraja, T. G. (2011). Evaluation of the effects of a commercially available *Salmonella* Newport siderophore receptor and porin protein vaccine on fecal shedding of *Salmonella* bacteria and health and performance of feedlot cattle. *Am. J. Vet. Res.* 72(2): 239–47.

Doyle, M. P. and Padhye, V. V. (1989). *Salmonella.* In M. P. Doyle (ed.), *Foodborne Bacterial Pathogens,* pp. 235–81. Merkel and Dekker, Inc., New York, NY.

Dragon, D. C. and Rennie, R. P. (1995). The ecology of anthrax spores: tough but not invincible. *Can. Vet. J.* 36(5): 295–301.

Edrington, T. S., Loneragan, G. H., Hill, J., Genovese, K. J., Brichta-Harhay, D. M., Farrow, R. L., Krueger, N. A., Callaway, T. R., Anderson, R. C. and Nisbet, D. J. (2013). Development of challenge models to evaluate the efficacy of a vaccine to reduce carriage of *Salmonella* in peripheral lymph nodes of cattle. *J. Food Prot.* 76(7): 1259–63.

Ellis-Iversen, J., Smith, R. P., Van Winden, S., Paiba, G. A., Watson, E., Snow, L. C. and Cook, A. J. C. (2008). Farm practices to control *E. coli* O157 in young cattle – A randomised controlled trial. *Vet. Res.* 39(1): 3. doi: 10.1051/vetres:2007041.

Ellis-Iversen, J., Cook, A. J. C., Smith, R. P., Pritchard, G. C. and Nielen, M. (2009). Temporal patterns and risk factors for *Escherichia coli* O157 and *Campylobacter* spp. in young cattle. *J. Food Prot.* 72(3): 490–6.

Faith, N. G., Shere, J. A., Brosch, R., Arnold, K. W., Ansay, S. E., Lee, M.-S., Luchansky, J. B. and Kaspar, C. W. (1996). Prevalence and clonal nature of *Escherichia coli* O157:H7 on dairy farms in Wisconsin. *Appl. Environ. Microbiol.* 62(5): 1519–25.

Fasanella, A., Galante, D., Garofolo, G. and Jones, M. H. (2010). Anthrax undervalued zoonosis. *Vet. Microbiol.* 140(3–4): 318–31.

Fenlon, D. R., Wilson, J. and Donachie, W. (1996). The incidence and level of *Listeria monocytogenes* contamination of food sources at primary production and initial processing. *J. Appl. Bacteriol.* 81(6): 641–50.

Folmer, J., Macken, C., Moxley, R., Smith, D., Brashears, M., Hinkley, S., Erickson, G. and Klopfenstein, T. (2003). Intervention strategies for reduction of *E. coli* O157:H7 in feedlot steers. 2003 Nebraska Beef Cattle Report, paper 227. Available at: http://digitalcommons.unl.edu/animalscinbcr/227/?utm_source=digitalcommons.unl.edu%2Fanimalscinbcr%2F227&utm_medium=PDF&utm_campaign=PDFCoverPages. Accessed 28 January 2016.

Fox, J. T., Thomson, D. U., Drouillard, J. S., Thornton, A. B., Burkhardt, D. T., Emery, D. A. and Nagaraja, T. G. (2009). Efficacy of *Escherichia coli* O157:H7 siderophore receptor/porin proteins-based vaccine in feedlot cattle naturally shedding *E. coli* O157. *Foodborne Pathog. Dis.* 6(7): 893–9.

FSIS (Food Safety and Inspection Service, U. S. Department of Agriculture) (1994). FSIS Notice 50–94, Microbiological testing program for *Escherichia coli* O157:H7 in raw ground beef.

FSIS (Food Safety and Inspection Service, U. S. Department of Agriculture) (1999). Beef products contaminated with *Escherichia coli* O157:H7. *Fed. Reg.* 64: 2803–5.

FSIS (Food Safety and Inspection Service, U. S. Department of Agriculture) (2012). Shiga toxin-producing *Escherichia coli* in certain raw beef products. *Fed. Reg.* 77: 31975–81.

FSIS (Food Safety and Inspection Service, U. S. Department of Agriculture) (2014). Pre-harvest management controls and intervention options for reducing Shiga toxin-producing *Escherichia coli* shedding in cattle: an overview of current research. Available at: http://www.fsis.usda.gov/wps/wcm/connect/d5314cc7-1ef7-4586-bca2-f2ed86d9532f/Reducing-Ecoli-Shedding-in-Cattle.pdf?MOD=AJPERES. Accessed 27 January 2016.

FSIS (Food Safety and Inspection Service, U. S. Department of Agriculture) (2015). FSIS Directive 10,010.1, Revision 4, Sampling verification activities for Shiga toxin-producing *Escherichia coli* (STEC) in raw beef products.

Galińska, E. M. and Zagórski, J. (2013). Brucellosis in humans – etiology, diagnostics, clinical forms. *Ann. Agric. Environ. Med.* 20(2): 233–8.

Gal-Mor, O., Boyle, E. C. and Grassl, G. A. (2014). Same species, different diseases: how and why typhoidal and non-typhoidal *Salmonella enterica* serovars differ. *Front. Microbiol.* 5: 391.

García, J. A., Micheloud, J. F., Campero, C. M., Morrell, E. L., Odriozola, E. R. and Moreira, A. R. (2015). Enteric listeriosis in grazing steers supplemented with spoiled silage. *J. Vet. Diagn. Invest.* 28(1): 65–9.

García-Angulo, V. A., Kalita, A., Kalita, M., Lozano, L. and Torres, A. G. (2014). Comparative genomics and immunoinformatics approach for the identification of vaccine candidates for enterohemorrhagic *Escherichia coli* O157:H7. *Infect. Immun.* 82(5): 2016–26.

Gaukler, S. M., Linz, G. M., Sherwood, J. S., Dyer, N. W., Bleier, W. J., Wannemuehler, Y. M., Nolan, L. K. and Logue, C. M. (2009). *Escherichia coli*, *Salmonella*, and *Mycobacterium avium* subsp. *paratuberculosis* in wild European starlings at a Kansas cattle feedlot. *Avian Dis.* 53(4): 544–51.

Godfroid, J., Scholz, H. C., Barbier, T., Nicolas, C., Wattiau, P., Fretin, D., Whatmore, A. M., Cloeckaert, A., Blascom, J. M., Moriyon, I., Saegerman, C., Muma, J. B., Al Dahouk, S., Neubauer, H. and Letesson, J. J. (2011). Brucellosis at the animal/ecosystem/human interface at the beginning of the 21st century. *Prev. Vet. Med.* 102(2): 118–31.

Greer, G. G. (2005). Bacteriophage control of foodborne bacteria. *J. Food Prot.* 68(5): 1102–11.

Guerra, M. A. (2009). Leptospirosis. *J. Am. Vet. Med. Assoc.* 234(4): 472–8.

Heider, L. C., Meiring, R. W., Hoet, A. E., Gebreyes, W. A., Funk, J. A. and Wittum, T. E. (2008). Evaluation of vaccination with a commercial subunit vaccine on shedding of *Salmonella enterica* in subclinically infected dairy cows. *J. Am. Vet. Med. Assoc.* 233(3): 466–9.

Heithoff, D. M., Shimp, W. R., Lau, P. W., Badie, G., Enioutina, E. Y., Daynes, R. A., Byrne, B. A., House, J. K. and Mahan, M. J. (2008). Human *Salmonella* clinical isolates distinct from those of animal origin. *Appl. Environ. Microbiol.* 74(6): 1757–66.

Hermesch, D. R., Thomson, D. U., Loneragan, G. H., Renter, D. R. and White, B. J. (2008). Effects of a commercially available vaccine against *Salmonella enterica* serotype Newport on milk production, somatic cell count, and shedding of *Salmonella* organisms in female dairy cattle with no clinical signs of salmonellosis. *Am. J. Vet. Res.* 69(9): 1229–34.

Horrocks, S. M., Anderson, R. C., Nisbet, D. J. and Ricke, S. C. (2009). Incidence and ecology of *Campylobacter jejuni* and *coli* in animals. *Anaerobe* 15(1–2): 18–25.

Hugh-Jones, M. and Blackburn, J. (2009). The ecology of *Bacillus anthracis*. *Mol. Aspects Med.* 30(6): 356–67.

Humphrey, T., O'Brien, S. and Madsen, M. (2007). Campylobacters as zoonotic pathogens: a food production perspective. *Int. J. Food Microbiol.* 117(3): 237–57.

Hurd, H. S. and Malladi, S. (2012). An outcomes model to evaluate risks and benefits of *Escherichia coli* vaccination in beef cattle. *Foodborne Pathog. Dis.* 9(10): 952–61.

Ikegami, T. (2012). Molecular biology and genetic diversity of Rift Valley fever virus. *Antiviral Res.* 95(3): 293–310.

Imran, M. and Mahmood, S. (2011). An overview of animal prion diseases. *Virol. J.* 8: 493.

Jacob, M. E., Callaway, T. R. and Nagaraja, T. G. (2009). Dietary interactions and interventions affecting *Escherichia coli* O157 colonization and shedding in cattle. *Foodborne Pathog. Dis.* 6(7): 785–92.

Johnson, R. P., Gyles, C. L., Huff, W. E., Ojha, S., Huff, G. R., Rath, N. C. and Donoghue, A. M. (2008). Bacteriophages for prophylaxis and therapy in cattle, poultry and pigs. *Anim. Health Res. Rev.* 9(2): 201–15.

Jones, J. L. and Dubey, J. P. (2012). Foodborne toxoplasmosis. *Clin. Infect. Dis.* 55(6): 845–51.

Kolenda, R., Burdukiewicz, M. and Schierack, P. (2015). A systematic review and meta-analysis of the epidemiology of pathogenic *Escherichia coli* of calves and the role of calves as reservoirs for human pathogenic *E. coli*. *Front. Cell. Infect. Microbiol.* 5: 23.

Labbe, R. (1989). *Clostridium perfringens*. In M. P. Doyle (ed.), *Foodborne Bacterial Pathogens*, pp. 191–234. Merkel and Dekker, Inc., New York, NY.

LaRock, D. L., Chaudhary, A. and Miller, S. I. (2015). Salmonellae interactions with host processes. *Nat. Rev. Microbiol.* 13(4): 191–205.

Leblebicioglu, H., Sunbul, M., Memish, Z. A., Al-Tawfiq, J. A., Bodur, H., Ozkul, A., Gucukoglu, A., Chinikar, S. and Hasan Z. (2015). Consensus report: preventive measures for Crimean-Congo Hemorrhagic Fever during Eid-al-Adha festival. *Int. J. Infect. Dis.* 38: 9–15.

Lecuit, M. (2007). Human listeriosis and animal models. *Microbes Infect.* 9(10): 1216–25.

Lee, J., Kim, S. Y., Hwang, K. J., Ju, Y. R. and Woo, H. J. (2013). Prion diseases as transmissible zoonotic diseases. *Osong Public Health Res. Perspect.* 4(1): 57–66.

LeJeune, J. T. and Wetzel, A. N. (2007). Preharvest control of *Escherichia coli* O157 in cattle. *J. Anim. Sci.* 85(13 Suppl.): E73–E80.

LeJeune, J. T., Besser, T. E., Rice, D. H., Berg, J. L., Stilborn, R. P. and Hancock, D. D. (2004). Longitudinal study of fecal shedding of *Escherichia coli* O157:H7 in feedlot cattle: predominance and persistence of specific clonal types despite massive cattle population turnover. *Appl. Environ. Microbiol.* 70(1): 377–84.

Liu, D. (2013). Molecular approaches to the identification of pathogenic and nonpathogenic listeriae. *Microbiol. Insights.* 6: 59–69.

Loneragan, G. H., Thomson, D. U., McCarthy, R. M., Webb, H. E., Daniels, A. E., Edrington, T. S., Nisbet, D. J., Trojan, S. J., Rankin, S. C. and Brashears, M. M. (2012). *Salmonella* diversity and burden in cows on and culled from dairy farms in the Texas High Plains. *Foodborne Pathog. Dis.* 9(6): 549–55.

Lovett, J. (1989). *Listeria monocytogenes*. In M. P. Doyle (ed.), *Foodborne Bacterial Pathogens*, pp. 283–310. Merkel and Dekker, Inc., New York, NY.

Maguire, H., Cowden, J., Jacob, M., Rowe, B., Roberts, D., Bruce, J. and Mitchell, E. (1992). An outbreak of *Salmonella dublin* infection in England and Wales associated with a soft unpasteurized cows' milk cheese. *Epidemiol. Infect.* 109(3): 389–96.

Matthews, L., Low, J. C., Gally, D. L., Pearce, M. C., Mellor, D. J., Heesterbeek, J. A. P., Chase-Topping, M., Naylor, S. W., Shaw, D. J., Reid, S. W. J., Gunn, G. J. and Woolhouse, M. E. J. (2006). Heterogeneous shedding of *Escherichia coli* O157 in cattle and its implications for control. *Proc. Natl. Acad. Sci. U.S.A.* 103(3): 547–52.

Matthews, L., Reeve, R., Gally, D. L., Low, J. C., Woolhouse, M. E. J., McAteer, S. P., Locking, M. E., Chase-Topping, M. E., Haydon, D. T., Allison, L. J., Hanson, M. F., Gunn, G. J. and Reid, S W. J. (2013). Predicting the public health benefit of vaccinating cattle against *Escherichia coli* O157. *Proc. Natl. Acad. Sci. U.S.A.* 110(40): 16265–70.

McNeilly, T. N., Mitchell, M. C., Corbishley, A., Nath, M., Simmonds, H., McAteer, S. P., Mahajan, A., Low, J. C., Smith, D. G. E., Huntley, J. F. and Gally, D. L. (2015). Optimizing the protection of cattle against *Escherichia coli* O157:H7 colonization through immunization with different Combinations of H7 flagellin, Tir, intimin-531 or EspA. *PLoS ONE* 10(5): e0128391. doi: 10.1371/journal.pone.0128391.

Menzies, F.D. and Neill, S. D. (2000). Cattle-to-cattle transmission of bovine tuberculosis. *Vet. J.* 160(2): 92–106.

Michel, A. L., Müller, B. and van Helden, P. D. (2010). *Mycobacterium bovis* at the animal-human interface: a problem, or not? *Vet. Microbiol.* 140(3–4): 371–81.

Michi, A. N., Favetto, P. H., Kastelic, J. and Cobo, E. R. (2016). A review of sexually transmitted bovine trichomoniasis and campylobacteriosis affecting cattle reproductive health. *Theriogenology* 85(5): 781–91.

Mohler, V. L., Heithoff, D. M., Mahan, M. J., Walker, K. H., Hornitzky, M. A., Shum, L. W. C., Makin, K. J. and House, J. K. (2008). Cross-protective immunity conferred by a DNA adenine methylase deficient *Salmonella enterica* serovar Typhimurium vaccine in calves challenged with *Salmonella* serovar Newport. *Vaccine* 26(14): 1751–8.

Mohler, V. L., Izzo, M. M. and House, J. K. (2009). *Salmonella* in calves. *Vet. Clin. Food Anim.* 25(1): 37–54.

Monis, P. T., Caccio, S. M. and Thompson, R. C. (2009). Variation in *Giardia*: towards a taxonomic revision of the genus. *Trends Parasitol.* 25(2): 93–100.

Moxley, R. A., Smith, D. R., Luebbe, M., Erickson, G. E., Klopfenstein, T. J. and Rogan, D. (2009). *Escherichia coli* O157:H7 vaccine dose-effect in feedlot cattle. *Foodborne Pathog. Dis.* 6(7): 879–84.

Murdoch, B. M. and Murdoch, G. K. (2015). Genetics of prion disease in cattle. *Bioinform. Biol. Insights.* 9(Suppl. 4): 1–10.

Naylor, S. W., Roe, A. J., Nart, P., Spears, K., Smith, D. G. E., Low, J. C. and Gally, D. L. (2005). *Escherichia coli* O157:H7 forms attaching and effacing lesions at the terminal rectum of cattle and colonization requires the LEE4 operon. *Microbiology* 151(Part 8): 2773–81.

Nielsen, L. R. (2009). Current *Salmonella*-control in cattle in Denmark. Available at: http://www.fodevarestyrelsen.dk/NR/rdonlyres/24AEB101-ABAC-405C–A0EF-693B8E4B36F2/15348/MicrosoftWordBilag7FINALCurrentSalmonellacontrolin.pdf. Accessed 26 January 2016.

Nielsen, L. R. and Dohoo, I. (2012). Survival analysis of factors affecting incidence risk of *Salmonella* Dublin in Danish dairy herds during a 7-year surveillance period. *Prev. Vet. Med.* 107(3–4): 160–9.

Nielsen, L. R. and Nielsen, S. S. (2012). A structured approach to control of *Salmonella* Dublin in 10 Danish dairy herds based on risk scoring and test-and-manage procedures. *Food Res. Int.* 45(2): 1158–65.

Nielsen, L. R. and Dohoo, I. (2013). Time-to-event analysis of predictors for recovery from *Salmonella* Dublin infection in Danish dairy herds between 2002 and 2012. *Prev. Vet. Med.* 110(3–4): 370–8.

Nielsen, T. D., Vesterbæk, I. L., Kudahl, A. B., Borup, K. J. and Nielsen, L. R. (2012). Effect of management on prevention of *Salmonella* Dublin exposure of calves during a one-year control programme in 84 Danish dairy herds. *Prev. Vet. Med.* 105(1–2): 101–9.

Oliver, S. P., Patel, D. A., Callaway, T. R. and Torrence, M. E. (2009). ASAS Centennial Paper: development and future outlook for preharvest food safety. *J. Anim. Sci.* 87(1): 419–37.

Palmer, M. V. and Waters, W. R. (2011). Bovine tuberculosis and the establishment of an eradication program in the United States: role of veterinarians. *Vet. Med. Int.* 2011: 186345. doi:10.4061/2011/816345.

Palmer, M. V., Thacker, T. C., Waters, W. R., Gortázar, C. and Corner, L. A. (2012). *Mycobacterium bovis*: a model pathogen at the interface of livestock, wildlife, and humans. *Vet. Med. Int.* 2012: 236205. doi:10.1155/2012/236205.

Park, S. F. (2002). The physiology of *Campylobacter* species and its relevance to their role as foodborne pathogens. *Int. J. Food Microbiol.* 74(3): 177–88.

Pérez-Lago, L., Navarro, Y. and García-de-Viedma, D. (2014). Current knowledge and pending challenges in zoonosis caused by *Mycobacterium bovis*: a review. *Res. Vet. Sci.* 97(Suppl): S94–S100.

Pesciaroli, M., Alvarez, J., Boniotti, M. B., Cagiola, M., Di Marco, V., Marianelli, C., Pacciarini, M. and Pasquali, P. (2014). Tuberculosis in domestic animal species. *Res. Vet. Sci.* 97(Suppl): S78–S85.

Ragan, V. E. (2002). The Animal and Plant Health Inspection Service (APHIS) brucellosis eradication program in the United States. *Vet. Microbiol.* 90(1–4): 11–18.

Roberts, A. J. and Wiedmann, M. (2003). Pathogen, host and environmental factors contributing to the pathogenesis of listeriosis. *Cell. Mol. Life Sci.* 60(5): 904–18.

Rodriguez-Campos, S., Smith, N. H., Boniotti, M. B. and Aranaz, A. (2014). Overview and phylogeny of *Mycobacterium tuberculosis* complex organisms: implications for diagnostics and legislation of bovine tuberculosis. *Res. Vet. Sci.* 97(Suppl): S5–S19.

Rozema, E. A., Stephens, T. P., Bach, S. J., Okine, E. K., Johnson, R. P., Stanford, K. and McAllister, T. A. (2009). Oral and rectal administration of bacteriophages for control of *Escherichia coli* O157:H7 in feedlot cattle. *J. Food Prot.* 72(2): 241–50.

Ryan, U., Fayer, R. and Xiao, L. (2014). *Cryptosporidium* species in humans and animals: current understanding and research needs. *Parasitology* 141(13): 1667–85.

Sanad, Y. M., Closs Jr., G., Kumar, A., LeJeune, J. T. and Rajashekara, G. (2013). Molecular epidemiology and public health relevance of *Campylobacter* isolated from dairy cattle and European starlings in Ohio, USA. *Foodborne Pathog. Dis.* 10(3): 229–36.

Scallan, E., Hoekstra, R. M., Angulo, F. J., Tauxe, R. V., Widdowson, M. A., Roy, S. L., Jones, J. L. and Griffin, P. M. (2011). Foodborne illness acquired in the United States–major pathogens. *Emerg. Infect. Dis.* 17(1): 7–15.

Schultz, M. (2008). Theobald Smith. *Emerg. Infect. Dis.* 14(12): 1940–2.

Seleem, M. N., Boyle, S. M. and Sriranganathan, N. (2010). Brucellosis: a re-emerging zoonosis. *Vet. Microbiol.* 140(3–4): 392–8.

Shere, J. A., Kaspar, C. W., Bartlett, K. J., Linden, S. E., Norell, B., Francey, S. and Schaefer, D. M. (2002). Shedding of *Escherichia coli* O157:H7 in dairy cattle housed in a confined environment following waterborne inoculation. *Appl. Environ. Microbiol.* 68(4): 1947–54.

Signorini, M. L. and Tarabla, H. D. (2010). Interventions to reduce verocytotoxigenic *Escherichia coli* in ground beef in Argentina: a simulation study. *Prev. Vet. Med.* 94(1–2): 36–42.

Silva, C., Calva, E. and Maloy, S. (2014). One health and food-borne disease: *Salmonella* transmission between humans, animals, and plants. *Microbiol Spectr.* 2(1): OH-0020–2013.

Silva, J., Leite, D., Fernandes, M., Mena, C., Gibbs, P. A. and Teixeira, P. (2011). *Campylobacter* spp. as a Foodborne Pathogen: a Review. *Front. Microbiol.* 2: 200.

Smith, D. R. (2014). Vaccination of cattle against *Escherichia coli* O157:H7. *Microbiol. Spectrum.* 2(6): EHEC-0006-2013. doi: 10.1128/microbiolspec.EHEC-0006-2013.

Smith, D. R., Moxley, R. A., Klopfenstein, T. J. and Erickson, G. E. (2009). A randomized longitudinal trial to test the effect of regional vaccination within a cattle feedyard on *Escherichia coli* O157:H7 rectal colonization, fecal shedding, and hide contamination. *Foodborne Pathog. Dis.* 6(7): 885–92.

Smith, G. W., Smith, F., Zuidhof, S. and Foster, D. M. (2015). Short communication: characterization of the serologic response induced by vaccination of late-gestation cows with a *Salmonella* Dublin vaccine. *J. Dairy Sci.* 98(4): 2529–32.

Smith, J. L. (2011). Healthy Farms, Healthy Agriculture. University of Vermont Extension. Available at: http://www.uvm.edu/~ascibios/. Accessed 26 January 2016.

Smith, J. L., Fratamico, P. M. and Gunther 4th, N. W. (2014). Shiga toxin-producing *Escherichia coli*. *Adv. Appl. Microbiol.* 86: 145–97.

Spickler, A. R. and Leedom Larson, K. R. (2013). Leptospirosis: Factsheet. Available at: http://www.cfsph.iastate.edu/DiseaseInfo/factsheets.php. Accessed29 January 2016.

Sponseller, J. K., Griffiths, J. K. and Tzipori, S. (2014). The evolution of respiratory Cryptosporidiosis: evidence for transmission by inhalation. *Clin. Microbiol. Rev.* 27(3): 575–86.

Stephens, T. P., Loneragan, G. H., Karunasena, E. and Brashears, M. M. (2007). Reduction of *Escherichia coli* O157 and *Salmonella* in feces and on hides of feedlot cattle using various doses of a direct-fed microbial. *J. Food Prot.* 70(10): 2386–91.

Stephens, T. P., McAllister, T. A. and Stanford, K. (2009). Perineal swabs reveal effect of super shedders on the transmission of *Escherichia coli* O157:H7 in commercial feedlots. *J. Anim. Sci.* 87(12): 4151–60.

Sternbach G. (2003). The history of anthrax. *J. Emerg. Med.* 24(4): 463–7.

Stevenson, S. M. L., Cook, S. R., Bach, S. J. and McAllister, T. A. (2004). Effects of water source, dilution, storage, and bacterial and fecal loads on the efficacy of electrolysed oxidizing water for the control of *Escherichia coli* O157:H7. *J. Food Prot.* 67(7): 1377–83.

Stiles, M. E. (1989). Less recognized or presumptive foodborne pathogenic bacteria. In M. P. Doyle (ed.), *Foodborne Bacterial Pathogens*, pp. 673–733. Merkel and Dekker, Inc., New York, NY.

Swaminathan, B. and Gerner-Smidt, P. (2007). The epidemiology of human listeriosis. *Microbes Infect.* 9(10): 1236–43.

Swartz, M. N. (1990). Aerobic spore-forming bacilli. In B. D. Davis, R. Dulbecco, H. N. Eisen and H. S. Ginsberg (eds), *Microbiology*, 4th edn., pp. 625–31. J. B. Lippincott Company, Philadelphia, PA.

Swirski, A. L., Pearl, D. L., Williams, M. L., Homan, H. J., Linz, G. M., Cernicchiaro, N. and LeJeune, J. T. (2014). Spatial epidemiology of *Escherichia coli* O157:H7 in dairy cattle in relation to night roosts of *Sturnus vulgaris* (European starling) in Ohio, USA (2007–9). *Zoonoses Public Health* 61(6): 427–35.

Tabe, E. S., Oloya, J., Doetkott, D. K., Bauer, M. L., Gibbs, P. S. and Khaitsa, M. L. (2008). Comparative effect of direct-fed microbials on fecal shedding of *Escherichia coli* O157:H7 and *Salmonella* in naturally infected feedlot cattle. *J. Food Prot.* 71(3): 539–44.

Taylor, D. N., Bied, J. M., Munro, J. S. and Feldman, R. A. (1982). *Salmonella* Dublin infections in the United States, 1979–80. *J. Infect. Dis.* 146(3): 322–7.

Thoen, C., Lobue, P. and de Kantor, I. (2006). The importance of *Mycobacterium bovis* as a zoonosis. *Vet. Microbiol.* 112(2–4): 339–45.

Thomson, D. U., Loneragan, G. H., Thornton, A. B., Lechtenberg, K. F., Emery, D. A., Burkhardt, D. T. and Nagaraja, T. G. (2009). Use of a siderophore receptor and porin proteins-based vaccine to control the burden of *Escherichia coli* O157:H7 in feedlot cattle. *Foodborne Pathog. Dis.* 6(7): 871–7.

Thornton, A. B., Thomson, D. U., Loneragan, G. H., Fox, J. T., Burkhardt, D. T., Emery, D. A. and Nagaraja, T. G. (2009). Effects of a siderophore receptor and porin proteins-based vaccination of fecal shedding of *Escherichia coli* O157:H7 in experimentally inoculated cattle. *J. Food Prot.* 72(4): 866–9.

Tishkova, F. H., Belobrova, E. A., Valikhodzhaeva, M., Atkinson, B., Hewson, R. and Mullojonova, M. (2012). Crimean-Congo hemorrhagic fever in Tajikistan. *Vector Borne Zoonotic Dis.* 12(9): 722–6.

Todd, E. C. D. and Notermans, S. (2011). Surveillance of listeriosis and its causative pathogen, *Listeria monocytogenes*. *Food Control* 22(9): 1484–90.

Tonsor, G. T. and Schroeder, T. C. (2015). Market impacts of *E. coli* vaccination in U.S. feedlot cattle. *Agric. Food Econ.* 3: 7. doi: 10.1186/s40100-014-0021-2.

Truyers, I., Luke, T., Wilson, D. and Sargison, N. (2014). Diagnosis and management of venereal campylobacteriosis in beef cattle. *BMC Vet. Res.* 10: 280. doi:10.1186/s12917-014-0280-x.

Van Donkersgoed, J., Berg, J., Potter, A., Hancock, D., Besser, T., Rice, D., LeJeune, J. and Klashinsky, S. (2001). Environmental sources and transmission of *Escherichia coli* O157 in feedlot cattle. *Can. Vet. J.* 42(9): 714–20.

Varel, V. H., Wells, J. E., Berry, E. D., Spiehs, M. J., Miller, D. N., Ferrell, C. L., Shackelford, S. D. and Koohmaraie, M. (2008). Odorant production and persistence of *Escherichia coli* in manure slurries from cattle fed zero, twenty, forty, or sixty percent wet distillers grains with solubles. *J. Anim. Sci.* 86(12): 3617–27.

Videnova, K. and Mackay, D. K. J. (2012). Availability of vaccines against major animal diseases in the European Union. *Rev. Sci. Tech.* 31(3): 971–8.

Vogstad, A. R., Moxley, R. A., Erickson, G. E., Klopfenstein, T. J. and Smith, D. R. (2013). Assessment of heterogeneity of efficacy of a three-dose regimen of a Type III secreted protein vaccine for reducing STEC O157 in feces of feedlot cattle. *Foodborne Pathog. Dis.* 10(8): 678–82.

Wagenaar, J. A., van Bergen, M. A., Blaser, M. J., Tauxe, R. V., Newell, D. G. and van Putten, J. P. (2014). *Campylobacter fetus* infections in humans: exposure and disease. *Clin. Infect. Dis.* 58(11): 1579–86.

Warnick, L. D., Nielsen, L. R., Nielsen, J. and Greiner, M. (2006). Simulation model estimates of test accuracy and predictive values for the Danish *Salmonella* surveillance program of dairy herds. *Prev. Vet. Med.* 77(3–4): 284–303.

Waters, W. R. and Palmer, M. V. (2015). *Mycobacterium bovis* infection of cattle and white-tailed deer: translational research of relevance to human tuberculosis. *ILAR J.* 56(1): 26–43.

Wedlock, D. N., Skinner, M. A., de Lisle, G. W. and Buddle, B. M. (2002). Control of *Mycobacterium bovis* infections and the risk to human populations. *Microbes Infect.* 4(4): 471–80.

Wells, J. E., Shackelford, S. D., Berry, E. D., Kalchayanand, N., Guerini, M. N., Varel, V. H., Arthur, T. M., Bosilevac, J. M., Freetly, H. C., Wheeler, T. L., Ferrell, C. L. and Koohmaraie, M. (2009). Prevalence and level of *Escherichia coli* O157:H7 in feces and on hides of feedlot steers fed diets with or without wet distillers grains with solubles. *J. Food Prot.* 72(8): 1624–33.

Wheeler, T. L., Kalchayanand, N. and Bosilevac, J. M. (2014). Pre- and post-harvest interventions to reduce pathogen contamination in the U.S. beef industry. *Meat Sci.* 98(3): 372–82.

Whelan, A. O., Coad, M., Upadhyay, B. L., Clifford, D. J., Hewinson, R. G. and Vordermeier, H. M. (2011). Lack of correlation between BCG-induced tuberculin skin test sensitisation and protective immunity in cattle. *Vaccine* 29(33): 5453–8.

Wiedemann, A., Virlogeux-Payant, I., Chaussé, A. M., Schikora, A. and Velge, P. (2015). Interactions of *Salmonella* with animals and plants. *Front. Microbiol.* 5: 791.

Wiedmann, M., Bruce, J. L., Knorr, R., Bodis, M., Cole, E. M., McDowell, C. I., McDonough P. L. and Batt, C. A. (1996). Ribotype diversity of Listeria monocytogenes strains associated with outbreaks of listeriosis in ruminants. *J. Clin. Microbiol.* 34(5): 1086–90.

WHO (World Health Organization, Department of Communicable Disease) (2001). Pre-harvest safety: Report of a WHO consultation with the participation of the Food and Agriculture Organization of the United Nations and the Office of International des Epizooties. Available at: http://apps.who.int/iris/handle/10665/68889. Accessed 24 January 2016.

WHO (World Organization for Animal Health, Food and Agriculture Organization of the United Nations, and World Health Organization) (2008). In P. C. Turnbull (ed.), *Anthrax in Humans and Animals*, 4th edition. World Health Organization, Geneva, Switzerland.

Yoshikawa, T. T., Herbert, P. and Oill, P. A. (1980). Salmonellosis. *West. J. Med.* 133(5): 408–17.

Zhao, T., Zhao, P., West, J. W., Bernard, J. K., Cross, H. G. and Doyle, M. P. (2006). Inactivation of enterohemorrhagic *Escherichia coli* in rumen content- or feces-contaminated drinking water for cattle. *Appl. Environ. Microbiol.* 72(5): 3268–73.

Chapter 2

Methods for detecting pathogens in the beef food chain: an overview

Pina M. Fratamico, Joseph M. Bosilevac and John W. Schmidt, United States Department of Agriculture, USA

1 Introduction

Pathogens of importance in beef include Shiga toxin-producing *Escherichia coli* (STEC), especially STEC that cause haemorrhagic colitis and haemolytic uraemic syndrome known as enterohaemorrhagic *E. coli* (EHEC). Beef and dairy cattle are an important reservoir of STEC and EHEC belonging to a variety of serotypes (Presad and Lejeune, 2014). *Salmonella* are also carried by beef cattle and can cause illness in cattle when they are stressed due to transportation, crowding, food deprivation and other stresses (Stevens et al., 2009). *Salmonella* of multiple serotypes can carry antibiotic resistance genes, and when these infect humans, treatment of illnesses is more difficult. Antimicrobial-resistant (AMR) bacteria in general have become a high priority for all those involved in animal agriculture. Other pathogens that can be associated with beef but are of less impact are *Campylobacter*, *Listeria monocytogenes*, *Mycobacterium avium* subsp. *paratuberculosis* and *Clostridium difficile*. Of these, the most rigorous testing of cattle, their environments and beef products is for *E. coli* O157:H7, *Salmonella* and non-O157 STEC. Furthermore, cattle and environments are often tested for AMR bacteria and *Campylobacter*.

Several challenges in testing for pathogens in meat and environmental samples arise for reasons including the following:

1 specificity for the target organism in samples with high bacterial load such as faeces or feedlot surface material;
2 sensitivity for detecting organisms with a zero tolerance policy such as *E. coli* O157:H7 in ground beef;

http://dx.doi.org/10.19103/AS.2016.0008.04

3 non-homogeneous distribution of pathogens in food necessitating testing of larger samples or multiple samples to ensure the probability of obtaining a positive sample from contaminated lots of beef;
4 different sample types requiring different processing protocols.

For example, faecal samples can inhibit molecular assays for pathogen detection (Luedtke et al., 2014), high-fat meats can interfere with immunocapture and molecular assays (Bosilevac et al., 2010), and some sample types require special processing to access pathogens, such as *Salmonella* harboured by lymph nodes of beef cattle (Brichta-Harhay et al., 2012). Furthermore, rapid tests must be simple for the user to perform, and results should be accurate and easy to interpret. Since the levels of target pathogens in beef samples may be very low, enrichment of the samples in selective or non-selective growth medium is needed. However, enrichment increases the time needed for detection; therefore, there is a need to develop sampling and processing procedures that recover and concentrate the target pathogens in order to eliminate or decrease enrichment times (Dwivedi and Jaykus, 2011). Ideally, there is a need for sensitive and accurate biosensor-based or culture-independent technologies that can be used for testing beef samples or other types of food for pathogens on the processing line in real-time to screen out negative samples.

For polymerase chain reaction (PCR)-based methods, selection of appropriate target sequences for specific detection of pathogens is important. For example, for detection of non-O157 STEC, some methods target the *stx*1/*stx*2 (Shiga toxin) and the *eae* (intimin) virulence genes, as well as serogroup-specific genes. These assays are usually performed as multiplex PCR assays (Fratamico et al., 2014). However, a problem with performing PCR targeting multiple STEC genes on beef enrichments is that false-positive results are often obtained, as the positive signals obtained for target genes does not necessarily indicate that the genes are all carried by one bacterium. While some pathogens associated with beef, including *Salmonella* and *Listeria* have quite unique and universal genetic targets that allow for rapid screening, it is more difficult to identify genetic targets for STEC, particularly non-O157 STEC. Being a single serotype, *E. coli* O157:H7 has a number of unique regions that allow for the design of simple detection assays; however, the non-O157 STEC represent numerous serogroups and serotypes, as well as distinct genetic lineages. Selection of targets to just detect the top six non-O157 STEC (serogroups O267, O45, O103, O111, O121 and O145) is problematic (Bosilevac et al., 2013). Compounding targeting the *stx* and *eae* genes is the fact that there are a number of genetic variants of these genes. Great difficulty must be overcome in designing primers for all variants. The *stx*2f variant was not believed to be associated with STEC strains that cause illness in humans; however, STEC carrying this subtype have been increasingly associated with human infections (Friesema et al., 2014). In addition, certain *eae* subtypes have been found to be associated with certain STEC serogroups, and PCR assays targeting specific *eae* variants have been designed (Bibbal et al., 2014)

Antibody-based or nucleic acid-based multiplexed detection systems that can be used to detect more than one pathogen simultaneously increase the speed of detection and result in cost savings; results of these methods are comparable to those of culture-based assays (Zhao et al., 2016; Zhou et al., 2013). Finally, if confirmation and additional testing of isolates is required, development of improved selective and differential agars, particularly for non-O157 STEC is essential (Kalchayanand et al., 2013a).

2 Rapid methods for detecting pathogens in beef

As opposed to traditional methods for detection of food-borne pathogens that rely on growth in culture media followed by isolation and confirmatory tests, rapid methods are various types of assays that accelerate analysis, are of higher throughput, may be automated, are relatively simpler to use and generally do not require elaborate instrumentation. However, although the method itself may be rapid, an enrichment step is still required to allow for growth of the target pathogen to detectable levels. A major goal in testing for food-borne pathogens is to develop better techniques to recover and concentrate pathogens and reduce or eliminate the enrichment step to actually achieve rapid, real-time detection.

Rapid methods generally include antibody-based and nucleic acid-based approaches, and these types of assays have been commonly used by the food industry. Biosensor-based methods consist of devices that have a bioreceptor that can be an antibody, enzyme, nucleic acid, or cells and a transducer that converts the interactions into a measureable signal. Examples are optical, electrochemical or mass-based biosensors, surface plasmon resonance, multi-analyte array/evanescent wave, quartz crystal microbalance or surface-enhanced Raman scattering biosensors, among others (Law et al., 2015; Dwivedi and Jaykus, 2016). Wiedmann et al. (2014) published an in-depth review of rapid methods and criteria used by the food industry for selection of specific methods. The AOAC Research Institute certifies rapid test kits through validation programmes involving independent third-party review testing of kit performance. Kits that perform according to manufacturer's claims are given performance tested methods status.

Immunological methods involve the use of monoclonal or polyclonal antibodies that bind to an antigen of the bacterium or to a toxin. Types of immunoassays include enzyme-linked immunosorbent assays (ELISA) or enzyme immunoassays, fluoroimmunoassays, radioimmunoassays, chemiluminescence immunoassays, lateral flow immunoassays and others, with a difference being the labels linked to the antibodies (radioactive, enzyme, chemiluminescent or fluorescent). The choice of antibodies used in immunoassays influences assay sensitivity and specificity. Immunoassays can also be designed to detect multiple analytes from a single sample, such as *E. coli* O57:H7 cells, as well as the Shiga toxins (Gehring et al., 2015). An immunoassay described by Shen et al. (2014) used nanoparticles to amplify the detection signal. The detection limit for the functional nanoparticle-enhanced ELISA (FNP-ELISA) was higher (68 CFU/ml in PBS and 6.8×10^2–6.8×10^3 CFU/ml in food samples) than the detection limit typical of other types of immunoassays. In Table 1, several antibody-based commercial detection systems for STEC detection and identification are listed.

Nucleic acid-based methods rely on detection of specific DNA or RNA sequences of the target organism. The most commonly used DNA-based methods utilize the PCR that involves amplification of specific sequences of the pathogen and detection of the ethidium bromide (or other stain)-stained PCR products by agarose gel electrophoresis (Law et al., 2015; Dwivedi and Jaykus, 2016). Real-time PCR assays do not rely on visualizing products on agarose gels, but instead a fluorescent signal is produced during the PCR analysis, with the intensity of the signal proportional to the amount of amplicons. Multiplex PCR assays allow amplification and detection of more than one target sequence simultaneously; therefore, they can be used for detection of one or more pathogens in a single sample or can be employed to target more than one gene/sequence of a single organism. A number

Table 1 Partial list of commercially available kits and reagents for detection, isolation and identification of STEC and *Salmonella*

Target organism	Assay format	Kit name and manufacturer
Salmonella	TMA	Atlas *Salmonella* SEN Detection Assay; Roka Bioscience
Salmonella	Real-time PCR	BAX Real-Time PCR for *Salmonella*; DuPont
Salmonella	Real-time PCR	GeneDisc *Salmonella* spp.; Pall GeneSystems
Salmonella	Real-time PCR	iQ-Check PCR; Bio-Rad
Salmonella	Real-time PCR	MicroSEQ *Salmonella*; Applied Biosystems
Salmonella	Real-time PCR	Assurance GDS *Salmonella*; Biocontrol Systems
Salmonella	Latex agglutination	RapidChek SELECT *Salmonella*; SDIX
Salmonella	Latex agglutination	*Salmonella* Latex test; Oxoid
Salmonella	Plating medium	XLT-4; Hardy Diagnostics
Salmonella	Plating medium	BBLTM CHROMagar; BD Diagnostics
Salmonella	Plating medium	RAPID *Salmonella*; Bio-Rad
Salmonella	Enrichment/isolation	S.P.R.I.N.T. *Salmonella*; Oxoid
Salmonella	IMS	Dynabeads anti-*Salmonella*; Thermo Fisher Scientific
Salmonella	ELISA	Bioline *Salmonella* ELISA Test Kit; Hardy Diagnostics
Salmonella	ELISA	RIDASCREEN *Salmonella*; R-Biopharm
Salmonella	Lateral flow	Lateral Flow System for *Salmonella* AD; DuPont
STEC	TMA	Atlas *E. coli* O157:H7 EG2; Atlas STEC EG2 Combo; Roka Bioscience
STEC	Real-time PCR	BioControl Assurance Genetic Detection System MPX top 7 STEC; BioControl Systems
STEC	Real-time PCR	DuPont BAX System STEC Suite; DuPont
STEC	Real-time PCR	Pall GeneDisc System for STEC and EHEC; Pall GeneSystems
STEC	Real-time PCR	foodproof STEC Screening LyoKit; Biotecon Diagnostics
STEC	Real-time PCR	iQ-Check STEC (virX, SerO); Bio-Rad
STEC	PCR with SNP identification	NeoSEEK STEC Detection and Identification; Neogen
Non-O157 STEC	Plating medium	R&F non-O157 STEC chromogenic agar; R&F Laboratories
STEC	Plating medium	CHROMagar STEC; CHROMagar
STEC O157	Plating medium	RAPID *E. coli* O157:H7; Bio-Rad
STEC	Plating medium	R&F Non-O157 STEC Plating Medium; R&F Laboratories
STEC O157	Plating medium	Fluorocult O157:H7; Merck-Millipore
STEC O157	Plating medium	Rainbow Agar O157; Biolog
STEC O157	ELISA	3M Tecra *E. coli* O157; 3M
STEC O157	ELISA	Assurance EIA EHEC; BioControl Systems
Shiga toxins	Lateral flow	ImmunoCardSTAT; Meridian
Shiga toxins	Lateral flow	SHIGA TOXIN QUIK CHECK; Alere
STEC	Latex agglutination	*E. coli* O157/STEC Latex ID; Thermo Fisher Scientific
STEC	Latex agglutination	*E. coli* Latex Agglutination tests; Abraxis
STEC	IMS	STEC (various serogroups) Latex Agglutination Beads; Abraxis
STEC	IMS	Dynabeads O157 (and other STEC); Thermo Fisher Scientific

TMA = isothermal transcription-mediated amplification; SNP = single nucleotide polymorphism of multiplex-based PCR methods for detection of STEC have been developed (Fratamico and Bagi, 2012; Verstraete et al., 2012; Fratamico et al., 2014).

Unlike PCR, isothermal nucleic acid amplification methods do not require a temperature cycling process. These methods include nucleic acid-based sequence amplification (NASBA), loop-mediated isothermal amplification (LAMP), strand displacement amplification and others (Yan et al., 2014; Law et al., 2015). NASBA is a method that detects RNA targets and has been used to detect *E. coli* in drinking water and *Salmonella* in various types of foods, including fresh meats (Min and Baeumner, 2002; D'Souza and Jaykus, 2003). LAMP is a rapid and very specific method as it employs various inner and outer primers for amplification. This technique has been applied for detection of *Salmonella*, *Shigella* and STEC in various foods (Shao et al., 2011; Wang et al., 2014). Several commercial PCR-based assays for detection of STEC and *Salmonella* are listed in Table 1. In 2010, the USDA Food Safety and Inspection Service (FSIS) released a guidance document for test kit manufacturers for evaluating kit performance (http://www.fsis.usda.gov/shared/PDF/Validation_Studies_Pathogen_Detection_Methods.pdf). The document provides information on how to design a validation study to demonstrate equivalence of a test kit with FSIS reference methods.

3 Advantages and limitations of rapid methods

Although there are advantages of using rapid methods in food testing, a number of challenges also exist. One of the several advantages of rapid methods is that they decrease the time to obtain results; therefore, they are very useful as screening methods targeting specific pathogens in large numbers of food samples. However, positive results are often regarded as presumptive and must be confirmed by traditional methods. Nucleic acid-based methods can generally be designed to be more sensitive and specific than immunologic-based methods. The antibody used in immunoassays is the major determinant of the specificity and sensitivity of the assay, and cross-reactivity of the antibodies with different bacteria is often difficult to avoid with polyclonal antisera. Enrichment of samples is required before testing by rapid methods, although enrichment times can be reduced depending on the sensitivity of the method. It may be practical for food industry and regulatory testing to perform enrichment of the same beef sample for *Salmonella* and STEC. The FSIS announced in the Federal Register (79 FR 32436; https://www.federalregister.gov/articles/2014/06/05/2014-13064/changes-to-salmonella-verification-sampling-program-analysis-of-raw-beef-for-shiga-toxin-producing) that testing for *Salmonella* would be performed on the same beef product that is analyzed for STEC. Modified tryptic soy broth (mTSB) without novobiocin was selected as the enrichment medium suitable for both *Salmonella* and STEC. mTSB with 2 mg novobiocin per litre was used for enrichment of naturally contaminated ground beef samples to determine the presence of *Salmonella* and STEC using DuPont™ BAX® PCR-based assays (Wasilenko et al., 2014). Higher levels of novobiocin have been found to inhibit or slow the growth of some STEC strains (Vimont et al., 2007).

Multiplex PCR typically can be less sensitive than PCR assays in which only one target is amplified, and multiplex assays require more optimization of reaction components and conditions. Another challenge associated with multiplex PCR assays performed on complex samples containing many types of bacteria is that it cannot easily be determined if positive results indicate that all of the amplicons are from the single target organism or if the gene targets are carried by multiple organisms. Multiplex assays can be used to detect more than one pathogen simultaneously if targeting specific genes or sequences for each pathogen (Botkin et al., 2012).

As foods are complex matrices that may contain very low numbers of the target organism, high levels of background flora and many components that are not compatible with rapid analytical methods, an enrichment step and sample clean-up procedures are necessary before testing. Inhibitory components may need to be removed and the target organism recovered and concentrated. Many types of nucleic acid extraction kits are commercially available, and Dwivedi and Jaykus (2016) describe some approaches for pre-analytical sample processing, including immunomagnetic separation (IMS), filtration, centrifugation and others.

If the sample contains DNA from a large amount of dead cells, a positive signal can occur. Therefore, false-positive results due to amplification of DNA from dead cells is a limitation of PCR assays. Methods have been developed to overcome this limitation, which are based on the use of DNA-binding dyes, such as ethidium monoazide or propidium monoazide that prevent amplification of DNA from dead cells.

4 Detecting antibiotic resistance in pathogens

AMR bacteria may be transmitted from cattle to humans by one of three general routes: beef products, environmental (manure, runoff, dust, etc.) or direct contact (animal handling, etc.) (CDC, 2013; FDA, 2013a; Rinsky et al., 2013; Durso et al., 2014; Neyra et al., 2014; Nadimpalli et al., 2015). For beef products, eminent concerns about AMR in food-borne pathogens are limited to non-typhoid *Salmonella enterica* (hereafter '*Salmonella*'). AMR in STECs is generally not considered a high-priority concern since antimicrobials are typically not employed for treatment of STEC illnesses (Wong et al., 2000; Tarr et al., 2005; Smith et al., 2012), and 91% of human *E. coli* O157 isolates tested by the United States' Centers for Disease Control and Prevention between 2004 and 2013 were pan-susceptible (CDC, 2015). Genes encoding resistance to antimicrobials ('antimicrobial resistance genes', hereafter 'ARGs') are often contained within mobile genetic elements harboured in non-pathogenic 'commensal' bacteria, and horizontal transfer of these elements can occur among unrelated bacteria (Shoemaker et al., 2001; Djordjevic et al., 2013). The occurrence of AMR in gram-negative and gram-positive commensal bacteria is regularly assessed by determining the antimicrobial susceptibilities of *E. coli* and *Enterococcus* spp. isolates, respectively. Increasingly, the potential for horizontal transfer of ARGs is accounted for by determining the occurrence of specific ARGs rather than the occurrence of specific antimicrobial-resistant bacteria.

In the United States, the National Antimicrobial Resistance Monitoring System (NARMS) surveys AMR in beef production by culturing for four organisms: *Campylobacter* spp., *Enterococcus* spp., *E. coli* and *Salmonella*. NARMS obtains samples from cattle caecal contents, carcasses and ground beef, but the organisms cultured vary by the sample type. *Campylobacter* spp. are cultured from caecal contents, *Enterococcus* spp. and *E. coli* are cultured from caecal contents and ground beef, and *Salmonella* is cultured from all three sample types (FDA, 2013a,b; USDA FSIS, 2013). NARMS methods vary by organism and sample type, but all are designed to yield one isolate of the cultured organism per sample (if present) regardless of antimicrobial susceptibility. Then for each isolate, minimum inhibitory concentrations (MICs) for 9–16 antimicrobials (depending on the organism) are obtained by performing antimicrobial susceptibility tests (ASTs) using a semi-automatic broth microdilution system. NARMS methods are ideal for national epidemiological tracking since susceptibility data for several antimicrobial classes are obtained from every sample from which the organism is recovered.

Sub-populations of *Enterococcus* spp., *E. coli* and *Salmonella* resistant to antimicrobials critically important to human medicine have been detected in cattle production and processing environments (Schmidt et al., 2013, 2015; Agga et al., 2015; Beukers et al., 2015). Erythromycin-resistant (ERY^r) *Enterococcus* spp. and tylosin-resistant (TYL^r) *Enterococcus* spp. are important, as erythromycin is a macrolide-class antimicrobial important for the treatment of human *Enterococcus* spp. infections (Roberts, 2008), while tylosin is a macrolide-class antimicrobial administered in-feed to approximately 71% of US feedlot cattle to reduce the incidence of liver abscesses (Pagel and Gautier, 2012; USDA, 2013). Concerns have been raised that use of tylosin in cattle feed may increase the occurrence of AMR human infections, specifically, *Enterococcus* spp. (Marshall and Levy, 2011). ERY^r and TYL^r *Enterococcus* spp. sub-populations within larger *Enterococcus* spp. populations have been recovered from cattle production environments (Agga et al., 2015; Beukers et al., 2015).

Non-pathogenic, commensal *E. coli* are thought to be 'reservoirs' of third-generation cephalosorins mediated by proteins expressed by the bla_{CMY2} or $bla_{CTX\text{-}M}$ genes and harboured in mobile genetic elements, which may be transferred to pathogenic bacteria (Winokur et al., 2001; Kluytmans et al., 2013). The folate synthesis inhibitor combination trimethoprim-sulphamethoxazole (co-trimoxazole) is the preferred therapy for urinary tract infections caused by extraintestinal pathogenic *E. coli* (ExPEC) (Russo and Johnson, 2003; Gupta et al., 2011). Concerns have been raised that co-trimoxazole-resistant (COT^r) ExPEC have possible food-borne origins that include beef products (Ramchandani et al., 2005; Manges et al., 2007; Xia et al., 2011). Third-generation cephalosporin-resistant ($3GC^r$) *E. coli* and COT^r *E. coli* sub-populations are present in cattle production and processing environments (Schmidt et al., 2013, 2015; Agga et al., 2015).

Quinolone-resistant (QNL^r) *Salmonella* and third-generation cephalosporin-resistant ($3GC^r$) *Salmonella* are important, as quinolones (ciprofloxacin, levofloxacin) and third-generation cephalosporins (cefotaxime, ceftriaxone) are the preferred therapies for adults and children with serious *Salmonella* infections, respectively (Su et al., 2004; Bradley et al., 2011; Chen et al., 2013). $3GC^r$ *Salmonella* and QNL^r *Salmonella* exist as sub-populations within larger susceptible *Salmonella* populations in cattle production and processing facilities (Agga et al., 2015; Schmidt et al., 2015).

ERY^r *Enterococcus* spp., $3GC^r$ *E. coli*, COT^r *E. coli*, $3GC^r$ *Salmonella* and nalidixic acid-resistant (NAL^r) *Salmonella* sub-populations can be detected and enumerated using a series of plates with and without supplemental antimicrobials in combination with enrichments (Agga et al., 2015; Schmidt et al., 2015). First, as shown in panels A and B of Fig. 1, samples or appropriate dilutions are directly plated on the following agar plates: Slanetz and Bartley medium agar (SBM, Thermo-Fisher, St. Louis, MO), SBM supplemented with 8 mg/L erythromycin plates (SBM+ERY), CHROMagar *E. coli* agar (CEC, DRG International, Mountainside, NJ), CEC supplemented with 2 mg/L cefotaxime (CEC+CTX), CEC supplemented with 4 mg/L trimethoprim and 76 mg/L sulphamethoxazole (CEC+COT), xylose lysine deoxycholate agar (XLD, Thermo Fisher) supplemented with 4.6 mL/L tergitol, 15 mg/L novobiocin and 5 mg/L cefesulodin (XLDtnc; tergitol, novobiocin and cefesulodin reduce background growth) (Brichta-Harhay et al., 2007), XLD supplemented with 2 mg/L

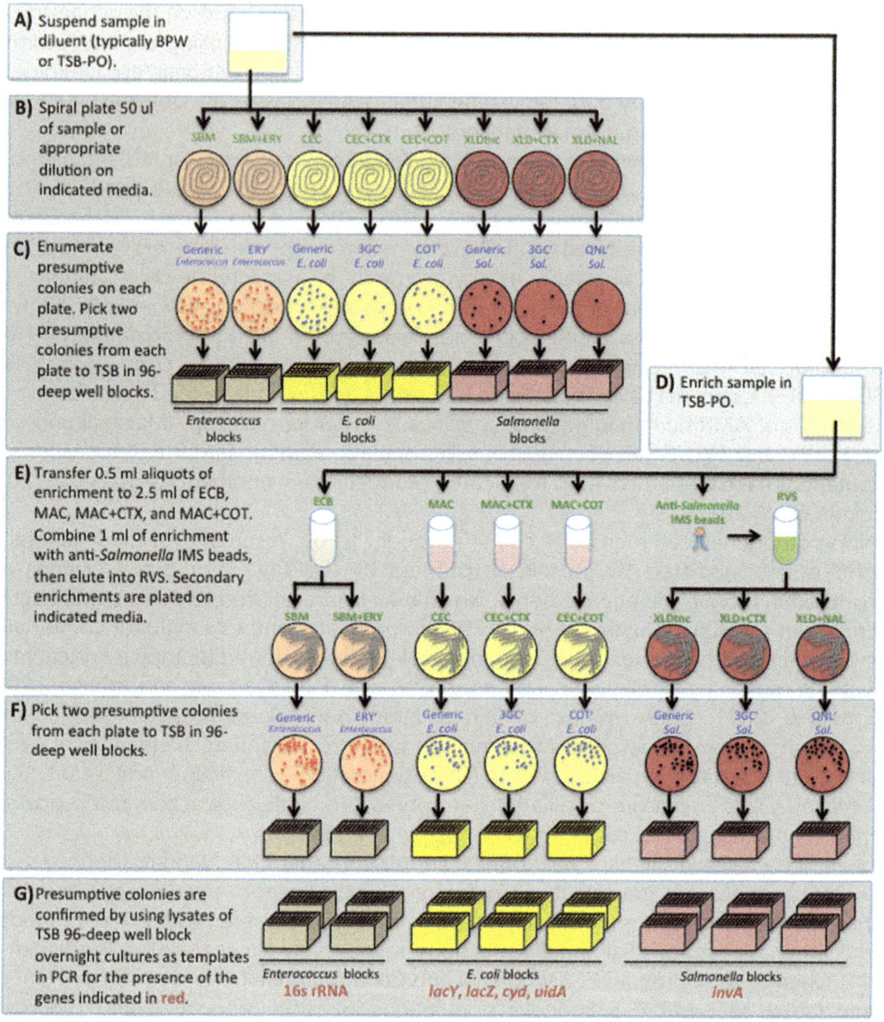

Figure 1 Flow diagram of a method for selective enumeration and detection of *Enterococcus* spp., *E. coli* and *Salmonella* resistant to specific important antimicrobials. Panel A) Samples are suspended in the appropriate volume of diluent. Faecal swabs are suspended in 5 ml phosphate-buffered tryptic soy broth (TSB-PO, 30 g/L TSB, 2.31 g/L KH_2PO_4, 12.54 g/L K_2HPO_4 per litre, final pH 7.2). Hide sponge swabs are suspended in 20 ml buffered peptone water. Faecal grab samples (10 g) and soil samples (10 g) are suspended in 90 ml of TSB-PO. Samples of cattle wastewater retention ponds are not diluted. Panel B) Using a spiral plater, 50 μl sample or appropriate dilutions of sample are plated onto the following media: Slanetz and Bartley medium agar (SBM), SBM supplemented with 8 mg/L erythromycin (SBM+ERY), CHROMagar *E. coli* agar (CEC), CEC supplemented with 2 mg/L cefotaxime (CEC+CTX), CEC supplemented with 4 mg/L trimethoprim and 76 mg/L sulphamethoxazole (CEC+COT), xylose lysine deoxycholate agar (XLD) supplemented with 4.6 ml/L tergitol, 15 mg/L novobiocin and 5 mg/L cefesulodin (XLDtnc), XLD supplemented with 2 mg/L cefotaxime (XLD+CTX), and XLD supplemented with 32 mg/L nalidixic acid (XLD+NAL). SBM and SBM+ERY plates are

Figure 1 *(Continued)*

incubated at 35°C for 4 h then 44°C for 48 h. CEC, CEC+CTX and CEC+COT plates are incubated at 37°C for 16 to 24 h. XLDtnc, XLD+CTX and XLD+NAL plates are incubated at 37°C for 16 to 24 h then held at 25°C for up to 72 h to allow H_2S production and black colour development. Panel C) Red colonies on SBM plates are enumerated as presumptive generic *Enterococcus* spp. Red colonies on SBM+ERY plates are enumerated as presumptive ERY^r *Enterococcus* spp. Blue colonies on CEC plates are enumerated as presumptive generic *E. coli*. Blue colonies on CEC+CTX plates are enumerated as presumptive 3GC^r *E. coli*. Blue colonies on CEC+COT plates are enumerated as presumptive COT^r *E. coli*. Black colonies on XLDtnc plates are enumerated as presumptive generic *Salmonella*. Black colonies on XLD+CTX plates are enumerated as presumptive 3GC^r *Salmonella*. Black colonies on XLD+NAL plates are enumerated as presumptive QNL^r *Salmonella*. Two presumptive colonies from each plate are picked and inoculated into TSB in 96 deep well blocks. Blocks are incubated at 37°C overnight. Panel D) Hide sponge samples (20 ml) and cattle wastewater retention pond samples (20 ml) are combined with 80 ml TSB-PO. Faecal swab, faecal grab and soil samples are not further diluted. Faecal swab and faecal grab samples are enriched by incubating at 42°C for 8 h. Hide swab samples, soil samples and cattle wastewater retention pond samples are enriched by incubating at 25°C for 2 h and 42°C for 6 h. Panel E) One-half ml aliquots of TSB-PO enrichments are transferred to the following secondary enrichments: 2.5 ml enterococcal broth (ECB), 2.5 ml MacConkey broth (MCB), 2.5 ml MCB supplemented with 2.4 mg/L cefotaxime (MCB+CTX) and MCB supplemented with 4.8 mg/L trimethoprim and 91.2 mg/L sulphamethoxazole (MCB+COT). One ml TSB-PO enrichment is mixed with 20 μl *Salmonella*-specific immunomagnetic separation (IMS) beads. The *Salmonella*–IMS bead complex is eluted into 3 ml of Rappaport-Vassiliadis soy a peptone broth (RVS). ECB is incubated overnight at 37°C. MCB, MCB+CTX, MCB+COT and RVS are incubated overnight at 42°C. ECB secondary enrichments are struck onto SBM and SBM+ERY plates. MCB, MCB+CTX and MCB+COT secondary enrichments are struck onto CEC, CEC+CTX and CEC+COT plates, respectively. RVS secondary enrichments are struck onto XLDtnc, XLD+CTX and XLD+NAL plates. SBM and SBM+ERY plates are incubated at 35°C for 4 h, and then 44°C for 48 h. CEC, CEC+CTX and CEC+COT plates are incubated at 37°C for 16 to 24 h. XLDtnc, XLD+CTX, and XLD+NAL plates are incubated at 37°C for 16 to 24 h, and then held at 25°C for up to 72 h to allow H_2S production and black colour development. Panel F) Two presumptive colonies from each plate are picked and inoculated into TSB in 96 deep well blocks. Blocks are incubated at 37°C overnight. Red colonies on SBM plates are presumptive generic *Enterococcus* spp. Red colonies on SBM+ERY plates are presumptive ERY^r *Enterococcus* spp. Blue colonies on CEC plates are presumptive generic *E. coli*. Blue colonies on CEC+CTX plates are presumptive 3GC^r *E. coli*. Blue colonies on CEC+COT plates are presumptive COT^r *E. coli*. Black colonies on XLDtnc plates are presumptive generic *Salmonella*. Black colonies on XLD+CTX plates are presumptive 3GC^r *Salmonella*. Black colonies on XLD+NAL plates are presumptive QNL^r *Salmonella*. Panel G) Lysates are generated from aliquots of TSB 96 deep well block overnight cultures of presumptive colonies. Lysates of presumptive *Enterococcus* spp. are confirmed by PCR for the presence of 16s rRNA gene specific for *Enterococcus* spp. (35). Lysates of presumptive *E. coli* are confirmed by multiplex PCR for the presence of *lacY, lacZ, cyd* and *uidA* genes (36). Lysates of presumptive *Salmonella* are confirmed by PCR for the *invA* gene (Rahn et al., 1992; Nucera et al., 2006).

cefotaxime (XLD+CTX) and XLD supplemented with 32 mg/L nalidixic acid (XLD+NAL). Secondly, as shown in panel C of Fig. 1, presumptive colonies are enumerated, and from each plate, two presumptive colonies are inoculated into 1 ml tryptic soy broth (TSB) in a 96 deep well block. Thirdly, as shown in panel D of Fig. 1, samples are non-specifically enriched in a phosphate-buffered TSB (30 g/L TSB, 2.31 g/L KH_2PO_4, 12.54 g/L K_2HPO_4, final pH 7.2). Fourthly, as shown in panel E of Fig. 1, the enrichment is sub-cultured to the following secondary enrichment broths: Enterococcal broth (ECB), MacConkey broth (MCB), MCB supplemented with 2 mg/L (final concentration) cefotaxime (MCB+CTX), MCB supplemented with 4 mg/L trimethoprim and 76 mg/L (final concentrations) sulphamethoxazole (MCB+COT), and Rappaport-Vassiliadis soy peptone broth (RVS; Remel) following IMS using *Salmonella*-specific beads.

The secondary enrichments are plated as follows: ECB is swabbed onto SBM and SBM+ERY, MCB is swabbed onto CEC, MCB+CTX is swabbed onto CEC+CTX, MCB+COT is swabbed onto CEC+COT and RVS is swabbed onto XLDtnc, XLD+CTX and XLD+NAL. Fifthly, as shown in panel F of Fig. 1, two presumptive colonies from each plate are inoculated into 1 ml TSB in a 96 deep well block. Sixthly, as shown in panel G of Fig. 1, from each overnight TSB culture, an aliquot is removed and a lysate prepared. Lysates are then used as templates to confirm the presumptive enumeration and prevalence. Presumptive *Enterococcus* spp. are confirmed by PCR for the 16s rRNA gene specific for *Enterococcus* spp. (Deasy et al., 2000). Presumptive *E. coli* are confirmed by multiplex PCR for the presence of *lacY, lacZ, cyd* and *uidA* genes (Horakova et al., 2008). Presumptive *Salmonella* are confirmed by PCR for the *invA* gene (Rahn et al., 1992; Nucera et al., 2006).

This 'selective enumeration and detection' method has been used to identify ERY^r *Enterococcus* spp., $3GC^r$ *E. coli*, COT^r *E. coli*, $3GC^r$ *Salmonella* and NAL^r *Salmonella* sub-populations within generic *E. coli* populations in cattle, swine and human wastes (Agga et al., 2015; Schmidt et al., 2015). Furthermore, the 'selective enumeration and detection' method has been demonstrated to improve detection of $3GC^r$ *E. coli*, COT^r *E. coli*, $3GC^r$ *Salmonella* and NAL^r *Salmonella* sub-populations compared with methods that rely on isolation of a single 'generic' isolate, such as NARMS methods (Schmidt et al., 2015). The primary advantage of the 'selective enumeration and detection' method is that sub-populations resistant to specific important antimicrobials are detected and quantified, and it is useful for field studies focused on the occurrence and microbial ecology of these critically important resistances. Weaknesses of the 'selective enumeration and detection' method are that additional plates supplemented with the appropriate antimicrobial are required for each resistance to be studied and that MICs are not obtained for resistant isolates. However, this weakness can be somewhat overcome by streaking for isolation, confirming generic and resistant colonies, and then performing ASTs on these isolates (Schmidt et al., 2015).

Concerns regarding specific antimicrobial-resistant organisms that may be present in cattle production and processing environments extend beyond these organisms and include QNL^r and azithromycin-resistant *Campylobacter* (Inglis et al., 2006; Luangtongkum et al., 2009; CDC, 2013), carbapenem-resistant Enterobacteriaceae (*Klebsiella* spp. and *E. coli*) (CDC, 2013; Webb et al., 2016) and methicillin-resistant *Staphylococcus aureus* (MRSA) (Graveland et al., 2011; CDC, 2013; Vandendriessche et al., 2013). To our knowledge, 'selective enumeration and detection' methods for these organisms have not been published.

Carriage of ARGs in non-pathogenic bacteria are a concern as ARGs can be contained within mobile genetic elements that facilitate transfer of AMR to, theoretically, any bacterial species (Allen et al., 2010). As ARGs may be vastly distributed among diverse bacterial species, culture-independent methods may also be applied to AMR in cattle production and processing environments. Culture-independent methods can be placed into three general classes: quantitative PCR (qPCR) of specific ARGs, commercial arrays of specific ARGs and metagenomic sequencing. All culture-independent studies begin with isolation of metagenomic DNA (mDNA) from a sample and often include a bead-beating step. Isolation of mDNA from cattle production environments is complicated by the presence of PCR inhibitors in soil (humic acids) and faeces (bile salts); however, several commercially available genomic DNA isolation protocols remove these inhibitors (Schrader et al., 2012). Once prepared, mDNA can be stored indefinitely at –20°C.

qPCR methods have been used to measure levels of the following specific genes in cattle production environments: *aadA*, *aph* and *aac*, aminoglycoside resistance genes (Chen et al., 2015); $bla_{CTX\text{-}M}$ and $bla_{CMY\text{-}2}$, third-generation cephalosporin resistance genes (Alali et al., 2009; Boyer et al., 2012; Kanwar et al., 2014); *ermA*, *ermB*, *ermC*, *ermF*, *ermT* and *ermX*, macrolide resistance genes (Chen et al., 2007; Alexander et al., 2011; Chen et al., 2015); *mecA*, methicillin resistance gene (Velasco et al., 2014); *sul1*, *sul2* and *sul3*, sulfonamide resistance genes (Alexander et al., 2011; Chen et al., 2015); and *tetA*, *tetB*, *tetC*, *tetE*, *tetG*, *tetL*, *tetM*, *tetO*, *tetT* and *tetW*, tetracycline resistance genes (Peak et al., 2007; Alexander et al., 2011; Kanwar et al., 2014; Chen et al., 2015). The primer sequences used in these studies are listed in Table 2. A commercially available array (Microbial DNA qPCR array BAID-1901Z, QIAGEN, Valencia, CA) has been used to determine the presence and concentrations of 84 ARGs in cattle feedlot wastewater (Agga et al., 2015).

Metagenomic sequencing (hereafter 'metagenomics') has been used to detect ARGs in cattle production environments (Durso et al., 2011; Wichmann et al., 2014; Chambers et al., 2015). Metagenomic studies consist of isolation of mDNA as described above, the construction of libraries from the mDNA which allow high-throughput 'next-generation' sequencing (NGS), the removal of low quality sequence information (or data) with a software tool such as Trimmomatic (Bolger et al., 2014), removal of non-bacterial host DNA (i.e. bovine DNA for studies of cattle production) using an alignment tool such as the Burrows-Wheeler Alignment tool (Li and Durbin, 2009), then the identification of ARGs by querying curated databases such as the 'Antibiotic Resistance Genes Database' (Liu and Pop, 2009) and the 'Comprehensive Antibiotic Resistance Database' (McArthur et al., 2013).

The strengths shared by all three culture-independent methods include the ability to detect AMR genes without culture bias and the ability to examine archived samples (either mDNA or sample matrix stored at –20°C to –80°C). Weaknesses shared by all three culture-independent methods include less sensitivity compared with culture methods (culture-based enrichments can use 10 g or more of sample, most genomic DNA preparation protocols use 100–250 mg of sample), inability to connect ARGs with host mobile genetic elements or organisms, unknown expression and functionality of detected ARGs and unknown viability of the host organism. A major advantage specific to metagenomics is the elimination of the need to identify specific ARG targets. A drawback of metagenomics is that ARGs are only a small proportion of metagenomic sequences; 0.05–5.6% of sequences obtained from faecal and soil samples were ARGs (Nesme et al., 2014).

The scientific study of the factors influencing AMR occurrence in cattle production and processing environments is in its infancy. No single method, culture or culture-independent will fully illuminate these factors. Well-designed studies on cattle production and processing environments using a combination of culture-dependent and culture-independent

Table 2 Antibiotic resistance gene targets and PCR primer sequences

Class of antibiotic	Resistance gene	Primer name	Primer sequences (5′ to 3′)	Reference
Aminoglycoside	*aadA*	aadA-F	ATGGGGATGATGTTAAGGCT	Chen et al., 2015
		aadA-R	TCACTTCCACCTTCCACTCA	
Aminoglycoside	*aph*	aph-F	ATTCAACGGGAAACGTCTTG	Chen et al., 2015
		aph-R	ACGCTACCTTTGCCATGTTT	
Aminoglycoside	*aac*	aac-F	GGAAGAAGAAACCGATCCAG	Chen et al., 2015
		aac-R	CAACTCAACCAGAGCTCGAA	
Third-generation cephalosporin	bla_{CMY2}	CMY2-f	CAGCATCTCCCAGCCTAATC	Boyer and Singer, 2012
		CMY2-r	GAAGCCCGTACACGTTTCTC	
Third-generation cephalosporin	bla_{CMY2}	675F	AGGGAAGCCCGTACACGTT	Alali et al., 2009
		738R	GCTGGATTTCACGCCATAGG	
Third-generation cephalosporin	$bla_{CTX\text{-}M}$	CTX-M(F)	ATGTGCAGYACCAGTAA	Kanwar et al., 2014
		CTX-M(R)	CCGCTGCCGGTYTTATC	
Macrolide	*ermA*	erm(A)-106	GAAATYGGRTCAGGAAAAGG	Chen et al., 2007
		erm(A)-437r	AAYAGYAAACCYAAAGCTC	
Macrolide	*ermB*	erm(B)-91f	GATACCGTTTACGAAATTGG	Chen et al., 2007
		erm(B)-454r	GAATCGAGACTTGAGTGTGC	
Macrolide	*ermB*	ermB-F	AGGGTTGCTCTTGCACACTC	Chen et al., 2015
		ermB-R	CTGTGGTATGGCGGGTAAGT	
Macrolide	*ermC*	erm(C)-43f	TCAAAACATAATATAGATAAA	Chen et al., 2007
		erm(C)-684r	GCTAATATTGTTTAAATCGTCAAT	
Macrolide	*ermC*	ermC-F	GAAATCGGCTCAGGAAAAGG	Chen et al., 2015
		ermC-R	TAGCAAACCCGTATTCCACG	
Macrolide	*ermF*	erm(F)-189f	CGACACAGCTTTGGTTGAAC	Chen et al., 2007
		erm(F)-497r	GGACCTACCTCATAGACAAG	
Macrolide	*ermT*	erm(T)-52f	CATATAAATGAAATTTTGAG	Chen et al., 2007
	erm(T)-420r	erm(T)-420r	ACGATTTGTATTTAGCAACC	
Macrolide	*ermX*	erm(X)-112f	GAGATCGGRCCAGGAAGC	Chen et al., 2007
		erm(X)-599r	GTGTGCACCATCGCCTGA	
Methicillin	*mecA*	mecA For	GGCAATATTACCGCACCTCA	Velasco et al., 2014
		mecA Rev	GTCTGCCACTTTCTCCTTGT	
		mecA Probe	AGATCTTATGCAAACTTAATTGGCAAATCC	

Class of antibiotic	Resistance gene	Primer name	Primer sequences (5′ to 3′)	Reference
Sulphonamide	*sul1*	sul1-F	TGTCGAACCTTCAAAAGCTG	Chen et al., 2015
		sul1-R	TGGACCCAGATCCTTTACAG	
Sulphonamide	*sul1*	sulI-L	GTGACGGTGTTCGGCATTCT	Lanz et al., 2003
		sulI-R	TCCGAGAAGGTGATTGCGCT	
Sulphonamide	*sul2*	sul2-F	ATCTGCCAAACTCGTCGTTA	Chen et al., 2015
		sul2-R	CAATGTGATCCATGATGTCG	
Sulphonamide	*sul2*	sulII-L	CGGCATCGTCAACATAACCT	Lanz et al., 2003
		sulII-R	TGTGCGGATGAAGTCAGCTC	
Sulphonamide	*sul3*	sul3-F	AGGCTTGGCAAAGTCAGATT	Chen et al., 2015
		sul3-R	CACCAGCCTCAACTAAAGCA	
Tetracycline	*tetA*	tet(A)-F	GCTACATCCTGCTTGCCTTC	Ng et al., 2001
		tet(A)-R	CATAGATCGCCGTGAAGAGG	
Tetracycline	*tetB*	tet(B)-F	TTGGTTAGGGGCAAGTTTTG	Ng et al., 2001
		tet(B)-R	GTAATGGGCCAATAACACCG	
Tetracycline		tet(B)-FW	ACACTCAGTATTCCAAGCCTTTG	
	tetB	tet(B)-RV	GATAGACATCACTCCCTGTAATGC	Peak et al., 2007
		tet(B)-probe	AAAGCGATCCCACCACCAGCCAAT	
Tetracycline	*tetC*	tet(C)-F	CTTGAGAGCCTTCAACCCAG	Ng et al., 2001
		tet(C)-R	ATGGTCGTCATCTACCTGCC	
Tetracycline	*tetC*	tetC-F	TGCAACTCGTAGGACAGGTG	Chen et al., 2015
		tetC-R	ACCAGTGACGAAGGCTTGAG	
Tetracycline	*tetE*	tetE-F	ACTGTGATGATGGCACTGGT	Chen et al., 2015
		tetE-R	CTGCAACATTAGCCTTTCCA	
Tetracycline	*tetG*	tetG-F	CTCGGTGGTATCTCTGCTCA	Chen et al., 2015
		tetG-R	CAGAACGAATGGTTTGATGC	
Tetracycline		tet(L)-FW	GGTTTTGAACGTCTCATTACCTGAT	
	tetL	tet(L)-RV	CCAATGGAAAAGGTTAACATAAAGG	Peak et al., 2007
		tet(L)-probe	CCACCTGCGAGTACAAACTGGGTGAAC	
Tetracycline	*tetL*	tet(L)-F	TCGTTAGCGTGCTGTCATTC	Ng et al., 2001
		tet(L)-R	GTATCCCACCAATGTAGCCG	

Table 2 (Continued)

Class of antibiotic	Resistance gene	Primer name	Primer sequences (5′ to 3′)	Reference
Tetracycline	*tetM*	tet(M)-F	GTGGACAAAGGTACAACGAG	Ng et al., 2001
		tet(M)-R	CGGTAAAGTTCGTCACACAC	
Tetracycline	*tetM*	tetM-F	ACAGAAAGCTTATTATATAAC	Chen et al., 2015
		tetM-R	GGCGTGTCTATGATGTTCAC	
Tetracycline	*tetO*	tetO-F	ACGGARAGTTTATTGTATACC	Chen et al., 2015
		tetO-R	TGGCGTATCTATAATGTTGAC	
Tetracycline	*tetQ*	tetQ-F	AGAATCTGCTGTTTGCCAGTG	Chen et al., 2015
		tetQ-R	CGGAGTGTCAATGATATTGCA	
Tetracycline	*tetT*	tetT-F	AGAGTCGACAATGCTACAACGA	Chen et al., 2015
		tetT-R	TGAAACTGTAGACGCCCGTA	
Tetracycline	*tetW*	tetW-F	GAGAGCCTGCTATATGCCAGC	Chen et al., 2015
		tetW-R	GGGCGTATCCACAATGTTAAC	
Tetracycline	*tetW*	tet(W)-FW	GCAGAGCGTGGTTCAGTCT	Peak et al., 2007
		tet(W)-RV	GACACCGTCTGCTTGATGATAAT	
		tet(W)-probe	TTCGGGATAAGCTCTCCGCCGA	

methods with proper controls incorporated into the study are needed to advance the understanding of and to identify the factors responsible for AMR occurrence.

5 Summary and future trends

Food safety regulations and food production have relied on microbiological testing as an essential element for ensuring the microbiological safety of the food supply. A limitation of traditional culture-based testing methods is the relatively long period of time required to obtain results. The increasing emphasis on farm-to-fork approaches to food safety necessitates the availability of rapid methods to provide timely information regarding the presence of pathogens in food. Many types of rapid methods, including nucleic acid- and immunologic-based methods, have been developed and are commercially available for detection of pathogens, including STEC and *Salmonella*, which are important pathogens associated with beef. Nucleic acid-based methods, particularly the PCR, are commonly used as they have high sensitivity and specificity, can be automated, allow detection of multiple pathogens and provide reliable results. However, sample processing is necessary to remove inhibitors of the PCR assay or other nucleic acid-based methods. Immunologic-based assays can also be designed to be specific, can be automated, are easy to perform and allow detection of bacterial toxins. However, sensitivity may be lower than that of nucleic acid-based assays, and cross-reactivity of the antibodies with closely related antigens can occur. Improvements in rapid methods will continue to be made, with the goal of being able to use these methods to obtain real-time or near real-time results during food processing.

Increasingly more evidence is demonstrating the link between the use of antibiotics in food animals, including cattle, the development of resistant microorganisms in the animals the spread of the pathogens to humans. Methods that can be employed for detection of resistance to antibiotics of importance in human medicine in pathogens found in cattle and beef, including *E. coli*, *Salmonella*, *Campylobacter*, MRSA and other bacteria, include 'selective enumeration and detection methods' and culture-independent methods, including qPCR of specific ARGs, commercial arrays of specific ARGs and metagenomic sequencing.

The introduction of NGS technologies has made it possible to rapidly obtain whole-genome sequence (WGS) data of bacteria at relatively low cost. An overview of NGS technologies and bioinformatics software is presented in a review by Franz et al. (2014). They present benefits, pitfalls, needs and recommendations related to WGS for characterization of bacterial pathogens, specifically STEC, for clinical, food safety and other applications. Bacterial WGS has demonstrated its utilization for epidemiological investigations (Jenkins et al., 2015) and the potential for complete characterization of bacteria, allowing unambiguous identification of the pathogen, determination of information regarding serotype, resistome (antibiotic resistance gene profile) and the presence of specific virulence genes, prophages and plasmids. In conjunction with robust bioinformatic pipelines, WGS is useful for distinguishing strains to the source level, including specific farms, facilities and geographic locations. WGS was found to be superior to PFGE in distinguishing *Salmonella* isolates related to an outbreak linked to spicy tuna sushi rolls from unrelated historical isolates (Allard et al., 2016). A report from Jenkins et al. (2015) demonstrated the value of WGS for outbreak investigation showing that two simultaneous O157:H7 outbreaks were linked to watercress; however, the sources of contamination differed.

Comparative analyses of WGS data allow identification of genetic markers that can be used in the development of pathogen-specific PCR-based assays. A WGS bioinformatic pipeline approach known as GeneSippr was used to identify STEC colonies based on identifying the presence of specific marker genes by mapping sequence reads to reference libraries (Lambert et al., 2015). Joensen et al. (2015) developed SeroTypeFinder, a publicly available web tool hosted by the Center for Genomic Epidemiology, which enables WGS-based serotyping of *E. coli*. Typing is based on O-group-specific *wzx*, *wzy*, *wzm* and *wzt*, as well as several flagellin genes. Similar to SeroTypeFinder is the VirulenceFinder tool (Joensen et al., 2014). The authors describe sequencing of 46 suspected STEC isolates using the Ion Torrent PGM sequencer and analysis of the WGS data through VirulenceFinder to identify virulence genes carried by the strains. VirulenceFinder can also assign Shiga toxin subtypes. A network of state, federal, hospital and other laboratories was created by the Food and Drug Administration (FDA) Center for Food Safety and Applied Nutrition known as the GenomeTrakr database (http://www.fda.gov/Food/FoodScienceResearch/WholeGenomeSequencingProgramWGS/ucm363134.htm), collecting WGS data from food-borne pathogens and uploading them to a publically accessible database that is managed by the National Center for Biotechnology Information. This integrated network using WGS assists public health and regulatory agencies in traceability efforts during outbreak investigations (Allard et al., 2016).

Culture-independent methods based on metagenomics, used for microbial ecology and community analyses, can potentially allow detection of all microorganisms in a sample in addition to the target pathogen(s) through the use of NGS (Miller et al., 2013). With this approach and using appropriate bioinformatic tools, the WGS of the organisms of interest can be determined, providing information on virulence genes, SNPs, antibiotic resistance genes and so on carried by the organisms. Using a metagenomic approach, Loman et al. (2013) analyzed faecal specimens from patients who developed diarrhoea during the STEC O104:H4 outbreak in Germany in 2011. They were able to obtain the draft genome of the outbreak strain from the faecal samples, as well as the sequences of *C. difficile, C. jejuni, C. concisus* and *S. enterica*.These types of technologies are being explored for detection and analysis of pathogens in food samples, and culture-independent approaches that will provide a wealth of information will soon become a reality.

6 Where to look for further information

There are several books and comprehensive review articles that cover pathogens of importance in beef and their detection technologies. Several of these are the following:

Advances in Microbial Food Safety. 2013. In J. Sofos (ed.), vol. 1, Woodhead Publishing in Food Science, Technology, and Nutrition, Cambridge, UK.

Advances in Microbial Food Safety. 2015. In J. Sofos (ed.), vol. 2, Woodhead Publishing in Food Science, Technology, and Nutrition, Cambridge, UK.

Dwivedi, H. P. and Jaykus, L.-A. 2011. Detection of pathogens in foods: the current state-of-the-art and future directions. *Crit. Rev. Microbiol.* 37:40–63.

Franz, E., Delaquis, P., Morabito, S., Beutin, L., Gobius, K., Rasko, D. A., Bono, J, French, N., Osek, J., Lindstedt, B. A., Muniesa, M., Manning, S., LeJeune, J., Callaway, T., Beatson, S., Eppinger, M., Dallman, T., Forbes, K. J., Aarts, H., Pearl, D. L., Gannon, V. P., Laing, C. R., and Strachan, N. J. 2014. Exploiting the explosion of information associated with whole genome sequencing to tackle Shiga toxin-producing *Escherichia coli* (STEC) in global food production systems. *Int. J. Food Microbiol.* 187:57–72.

Improving the Safety of Fresh Meat. 2005. In J. Sofos (ed.), Woodhead Publishing in Food Science, Technology, and Nutrition, Cambridge, UK.

Law, J. W.-F., Ab Mutalib, N.-S., Chan, K.-G. and Lee, L.-H. 2014. Rapid methods for the detection of foodborne bacterial pathogens: principles, applications, advantages and limitations. *Front. Microbiol.* 5:770. doi: 10.3389/fmicb.2014.00770.

Lee, K.-M., Runyon, M., Herrman, T. J., Phillips, R. and Hsieh, J. 2015. Review of *Salmonella* detection and identification methods: aspects of rapid emergency response and food safety. *Food Control.* 47:264–76.

Rapid Detection, Characterization, and Enumeration of Foodborne Pathogens. 2011. In J. Hoorfar (ed.). ASM Press, Washington, DC.

Wang, Y. and Salazar, J. K. 2016. Culture-independent rapid detection methods for bacterial pathogens and toxins in food matrices. *Comp. Rev. Food Sci. Technol.* 15:183–205.

Wiedmann, M., Wang, S., Post, L. and Nightingale, K. 2014. Assessment of criteria and approaches for rapid detection methods to be used in the food industry. *J. Food Prot.* 77:670–90.

Websites:

USDA FSIS website: http://www.fsis.usda.gov/wps/portal/fsis/home
The USDA FSIS Microbiology Laboratory Guidebook: http://www.fsis.usda.gov/wps/portal/fsis/topics/science/laboratories-and-procedures/guidebooks-and-methods/microbiology-laboratory-guidebook/microbiology-laboratory-guidebook
U.S. Food and Drug Administration (FDA), Bad Bug Book (2nd edition): http://www.fda.gov/Food/FoodborneIllnessContaminants/CausesOfIllnessBadBugBook/
U.S. FDA, Microbiological Methods and Bacteriological Analytical Manual (BAM): http://www.fda.gov/Food/FoodScienceResearch/LaboratoryMethods/ucm114664.htm

Professional societies and organizations and journals:

The International Association for Food Protection (IAFP; https://www.foodprotection.org/) and the scientific journal of IAFP, The Journal of Food Protection.
The Institute of Food Technologists (IFT; https://www.ift.org/) and the scientific journal of IFT, The Journal of Food Science.
Beef Industry Food Safety Council (BIFSCo) https://www.bifsco.org/
National Cattlemen's Beef Association:
https://www.beef.org/; http://www.beefusa.org/

7 References

Agga, G. E., Arthur, T. M., Durso, L. M., Harhay, D. M. and Schmidt, J. W. 2015. Antimicrobial-resistant bacterial populations and antimicrobial resistance genes obtained from environments impacted by livestock and municipal waste. *PLoS ONE* 10:e0132586.
Alali, W. Q., Scott, H. M., Norby, B., Gebreyes, W. and Loneragan, G. H. 2009. Quantification of the *bla*(CMY-2) in feces from beef feedlot cattle administered three different doses of ceftiofur in a longitudinal controlled field trial. *Foodborne Pathog. Dis.* 6:917–24.
Alexander, T. W., Yanke, J. L., Reuter, T., Topp, E., Read, R. R., Selinger, B. L. and McAllister, T. A. 2011. Longitudinal characterization of antimicrobial resistance genes in feces shed from cattle fed different subtherapeutic antibiotics. *BMC Microbiol.* 11:19.
Allard, M. W., Strain, E., Melka, D., Bunning, K., Muser, S. M., Brown, E. W. and Timme, R. 2016. The practical value of food pathogen traceability through building a whole-genome sequencing network database. *J. Clin. Microbiol.*, 54:1975–83.
Allen, H. K., Donato, J., Wang, H. H., Cloud-Hansen, K. A., Davies, J. and Handelsman, J. 2010. Call of the wild: antibiotic resistance genes in natural environments. *Nat. Rev. Microbiol.* 8:251–9.
Beukers, A. G., Zaheer, R., Cook, S. R., Stanford, K., Chaves, A. V., Ward, M. P. and McAllister, T. A. 2015. Effect of in-feed administration and withdrawal of tylosin phosphate on antibiotic resistance in enterococci isolated from feedlot steers. *Front. Microbiol.* 6:483.
Bibbal, D., Loukiadis, E., Kérourédan, M., Peytavin de Garam, C., Ferré, F., Cartier, P., Gay, E., Oswald, E., Auvray, F. and Brugère, H. 2014. Intimin gene (*eae*) subtype-based real-time PCR strategy for specific detection of Shiga toxin-producing *Escherichia coli* serotypes O157:H7, O26:H11, O103:H2, O111:H8, and O145:H28 in cattle feces. *Appl. Environ. Microbiol.* 80:1177–84.
Bolger, A. M., Lohse, M. and Usadel, B. 2014. Trimmomatic: a flexible trimmer for Illumina sequence data. *Bioinformatics* 30:2114–20.

Bosilevac, J. M. 2013. When STEC are our target, Where do you aim? A review of the current state of STEC testing. *Food Safety Mag*. April–May:30–35.

Bosilevac, J. M., Kalchayanand, N., Schmidt, J. W., Shackelford, S. D., Wheeler, T. L. and Koohmaraie, M. 2010. Inoculation of beef with low concentrations of *Escherichia coli* O157:H7 and examination of factors that interfere with its detection by culture isolation and rapid methods. *J. Food Prot*. 73:2180–8.

Botkin, D. J., Galli, L., Sankarapani, V., Soler, M. Rivas, M. and Torres, A. G. 2012. Development of a multiplex PCR assay for detection of Shiga toxin-producing *Escherichia coli*, enterohemorrhagic *E. coli*, and enteropathogenic *E. coli* strains. *Front. Cell. Infect. Microbiol*. 2:8. doi: 10.3389/fcimb.2012.00008.

Boyer, T. C. and Singer, R. S. 2012. Quantitative measurement of blaCMY-2 in a longitudinal observational study of dairy cattle treated with ceftiofur. *Foodborne Pathog. Dis*. 9:1022–7.

Bradley, J. S., Jackson, M. A. and Committee on Infectious. 2011. The use of systemic and topical fluoroquinolones. *American Academy of Pediatrics* 128:e1034–45.

Brichta-Harhay, D. M., Arthur, T. M., Bosilevac, J. M., Guerini, M. N., Kalchayanand, N. and Koohmaraie, M. 2007. Enumeration of *Salmonella* and *Escherichia coli* O157:H7 in ground beef, cattle carcass, hide and faecal samples using direct plating methods. *J. Appl. Microbiol*. 103:1657–68.

Brichta-Harhay, D. M., Arthur, T. M., Bosilevac, J. M., Kalchayanand, N., Schmidt, J. W., Wang, R., Shackelford, S. D., Loneragan, G. H. and Wheeler, T. L. 2012. Microbiological analysis of bovine lymph nodes for the detection of *Salmonella enterica*. *J Food Prot*. 75:854–8.

CDC. 2013. Antibiotic Resistance Threats in the United States, 2013. Centers for Disease Control and Prevention. Available: http://www.cdc.gov/drugresistance/threat-report-2013/pdf/ar-threats-2013-508.pdf. Accessed 6 July 2016.

CDC. 2015. National Antimicrobial Resistance Monitoring System for Enteric Bacteria (NARMS): Human Isolates Final Report, 2013. Atlanta, Georgia: U.S. Department of Health and Human Services, CDC. Available: http://www.cdc.gov/narms/pdf/2013-annual-report-narms-508c.pdf. Accessed 6 July 2016.

Chambers, L., Yang, Y., Littier, H., Ray, R., Zhang, T., Pruden, A., Strickland, M. and Knowlton, K. 2015. Metagenomic analysis of antibiotic resistance genes in dairy cow feces following therapeutic administration of third generation cephalosporin. *PLoS ONE* 10:e0133764.

Chen, B., Hao, L., Guo, X., Wang, N. and Ye, B. 2015. Prevalence of antibiotic resistance genes of wastewater and surface water in livestock farms of Jiangsu Province. *China. Environ. Sci. Pollut. Res. Int*. 22:13950–9.

Chen, H. M., Wang, Y., Su, L. H. and Chiu, C. H. 2013. Nontyphoid *Salmonella* infection: microbiology, clinical features, and antimicrobial therapy. *Pediatr. Neonatol*. 54:147–52.

Chen, J., Yu, Z., Michel Jr., F. C., Wittum, T. and Morrison, M. 2007. Development and application of real-time PCR assays for quantification of erm genes conferring resistance to macrolides-lincosamides-streptogramin B in livestock manure and manure management systems. *Appl. Environ. Microbiol*. 73:4407–16.

Deasy, B. M., Rea, M. C., Fitzgerald, G. F., Cogan, T. M. and Beresford, T. P. 2000. A rapid PCR based method to distinguish between *Lactococcus* and *Enterococcus*. *Syst. Appl. Microbiol*. 23:510–22.

Djordjevic, S. P., Stokes, H. W. and Roy Chowdhury, P. 2013. Mobile elements, zoonotic pathogens and commensal bacteria: conduits for the delivery of resistance genes into humans, production animals and soil microbiota. *Front. Microbiol*. 4:86.

D'Souza, D. H. and Jaykus, L. A. 2003. Nucleic acid sequence based amplification for the rapid and sensitive detection of *Salmonella* enterica from foods. *J. Appl. Microbiol*. 95:1343–50.

Durso, L. M. and Cook, K. L. 2014. Impacts of antibiotic use in agriculture: what are the benefits and risks? *Curr. Opin. Microbiol*. 19C:37–44.

Durso, L. M., Harhay, G. P., Bono, J. L. and Smith, T. P. 2011. Virulence-associated and antibiotic resistance genes of microbial populations in cattle feces analyzed using a metagenomic approach. *J. Microbiol. Methods*. 84:278–82.

Dwivedi, H. P. and Jaykus, L.-A. 2011. Detection of patohgens in foods: the current state-of-the-art and future directions. *Crit. Rev. Microbiol.* 37:40–63.

FDA. 2013a. National Antimicrobial Resistance Monitoring System – Enteric Bacteria (NARMS): 2011 Executive Report. Rockville, MD: U.S. Department of Health and Human Services, Food and Drug Administration, 2013. Available: http://www.fda.gov/downloads/AnimalVeterinary/SafetyHealth/AntimicrobialResistance/NationalAntimicrobialResistanceMonitoringSystem/UCM407962.pdf. Accessed 6 July 2016.

FDA. 2013b. NARMS retail meat annual report, 2011. U.S. Food and Drug Administration, Washington, DC. Available: http://www.fda.gov/downloads/AnimalVeterinary/SafetyHealth/AntimicrobialResistance/NationalAntimicrobialResistanceMonitoringSystem/UCM334834.pdf. Accessed 6 July 2016.

Franz, E., Delaquis, P., Morabito, S., Beutin, L., Gobius, K., Rasko, D. A., Bono, J, French, N., Osek, J., Lindstedt, B. A., Muniesa, M., Manning, S., LeJeune, J., Callaway, T., Beatson, S., Eppinger, M., Dallman, T., Forbes, K. J., Aarts, H., Pearl, D. L., Gannon, V. P., Laing, C. R. and Strachan, N. J. 2014. Exploiting the explosion of information associated with whole genome sequencing to tackle Shiga toxin-producing *Escherichia coli* (STEC) in global food production systems. *Int. J. Food Microbiol.* 187:57–72.

Fratamico, P. M. and Bagi, L. K. 2012. Detection of Shiga toxin-producing *Escherichia coli* in ground beef using the GeneDisc real-time PCR system. *Front. Cell. Inf. Microbiol.* 2:152. doi: 10.3389/fcimb.2012.00152

Fratamico, P. M., Wasilenko, J. L., Garman, B., DeMarco, D. R., Varkey, S., Jensen, M., Rhoden, K. and Tice, G. 2014. Evaluation of a multiplex real-time PCR method for detecting Shiga toxin-producing *Escherichia coli* in beef and comparison to the FSIS Microbiology Laboratory Guidebook method. *J. Food Prot.* 77:180–8.

Friesema, I., van der Zwaluw, K., Schuurman, T., Kooistra-Smid, M., Franz, E., van Duynhoven, Y. and van Pelt, W. 2014. Emergence of Escherichia coli encoding Shiga toxin 2f in human Shiga toxin-producing *E. coli* (STEC) infections in the Netherlands, January 2008 to December 2011. *Euro. Surveill.* 19(17):26–32.

Gehring, A. G., Brewster, J. D., He, Y., Irwin, P. L., Paoli, G. C., Simons, T., Tu, S. I. and Uknalis, J. 2015. Antibody microarray for *E. coli* O157:H7 and Shiga toxin in microtiter plates. *Sensors (Basel)* 15:30429–42.

Graveland, H., Wagenaar, J. A., Bergs, K., Heesterbeek, H. and Heederik, D. 2011. Persistence of livestock associated MRSA CC398 in humans is dependent on intensity of animal contact. *PLoS ONE* 6:e16830.

Gupta, K., Hooton, T. M., Naber, K. G.,Wullt, B., Colgan, R., Miller, L. G., Moran, G. J., Nicolle, L. E., Raz, R., Schaeffer, A. J., Soper, D. E.; Infectious Diseases Society of America and the European Society for Microbiology and Infectious Diseases. 2011. International clinical practice guidelines for the treatment of acute uncomplicated cystitis and pyelonephritis in women: A 2010 update by the Infectious Diseases Society of America and the European Society for Microbiology and Infectious Diseases. *Clin. Infect. Dis.* 52:e103–20.

Horakova, K., Mlejnkova, H. and Mlejnek, P. 2008. Specific detection of *Escherichia coli* isolated from water samples using polymerase chain reaction targeting four genes: cytochrome *bd* complex, lactose permease, beta-D-glucuronidase, and beta-D-galactosidase. *J. Appl. Microbiol.* 105:970–6.

Inglis, G. D., Morck, D. W., McAllister, T. A., Entz, T., Olson, M. E., Yanke, L. J. and Read, R. R. 2006. Temporal prevalence of antimicrobial resistance in *Campylobacter* spp. from beef cattle in Alberta feedlots. *Appl. Environ. Microbiol.* 72:4088–95.

Jenkins, C., Dallman, T. J., Launders, N., Willis, C., Byrne, L., Jorgensen, F., Eppinger, M., Adak, G. K., Aird, H., Elviss, N., Grant, K. A., Morgan, D. and McLauchlin, J. 2015. Public health investigation of two outbreaks of Shiga toxin-producing *Escherichia coli* O157 associated with consumption of watercress. *Appl. Environ. Microbiol.* 81:3946–52.

Joensen, K. G., Scheutz, F., Lund, O., Hasman, H., Kaas, R. S., Nielsen, E. M. and Aarestrup, F. M. 2014. Real-time whole-genome sequencing for routine typing, surveillance, and outbreak investigation of verotoxigenic Escherichia coli. *J. Clin. Microbiol.* 52:1501–10.

Joensen, K. G., Tetzschner, A. M. M., Iguchi, A., Aarestrup, F. M. and Scheutz, F. 2015. Rapid and easy *in silico* serotyping of *Escherichia coli* isolates by use of whole-genome sequencing. *J. Clin. Microbiol.* 53:2410–26.

Kalchayanand, N., Arthur, T. M., Bosilevac, J. M., Wells, J. E. and Wheeler, T. L. 2013a. Chromogenic agar medium for detection and isolation of *Escherichia coli* serogroups O26, O45, O103, O111, O121, and O145 from fresh beef and cattle feces. *J. Food Prot.* 76:192–9.

Kanwar, N., Scott, H. M., Norby, B., Loneragan, G. H., Vinasco, J., Cottell, J. L., Chalmers, G., Chengappa, M. M., Bai, J. and Boerlin, P. 2014. Impact of treatment strategies on cephalosporin and tetracycline resistance gene quantities in the bovine fecal metagenome. *Sci. Rep.* 4:5100.

Kluytmans, J. A., Overdevest, I. T., Willemsen, I., Kluytmans-van den Bergh, M. F., van der Zwaluw, K., Heck, M., Rijnsburger, M., Vandenbroucke-Grauls, C. M., Savelkoul, P. H., Johnston, B. D., Gordon, D. and Johnson, J. R. 2013. Extended-spectrum beta-lactamase-producing *Escherichia coli* from retail chicken meat and humans: comparison of strains, plasmids, resistance genes, and virulence factors. *Clin. Infect. Dis.* 56:478–87.

Lambert, D., Carrillo, C. D., Koziol, A. G., Manninger, P. and Blais, B. W. 2015. GeneSippr: a rapid whole-genome approach for the identification and characterization of foodborne pathogens such as priority Shiga toxigenic *Escherichia coli. PLoS ONE* 10:e0122928. doi: 10.1371/journal.pone.0122928.

Lanz, R., Kuhnert, P. and Boerlin, P. 2003. Antimicrobial resistance and resistance gene determinants in clinical *Escherichia coli* from different animal species in Switzerland. *Vet. Microbiol.* 91:73-84.

Law, J. W., Mutalib, A. B., Chan, N. S. and Lee, L. H. 2015. Rapid methods for the detection of foodborne bacterial pathogens: principles, applications, advantages and limitations. *Front. Microbiol.* 5:770. doi: 10.3389/fmicb.2014.00770.

Li, H. and Durbin, R. 2009. Fast and accurate short read alignment with Burrows-Wheeler transform. *Bioinformatics* 25:1754–60.

Liu, B. and Pop, M. 2009. ARDB–Antibiotic Resistance Genes Database. *Nucleic Acids Res.* 37:D443–7.

Loman, N. J., Constantinidou, C., Christner, M., Rohde, H., Chan, J. Z., Quick, J., Weir, J. C., Quince, C., Smith, G. P., Betley, J. R., Aepfelbacher, M. and Pallen, M. J. 2013. A culture-independent sequence-based metagenomics approach to the investigation of an outbreak of Shiga-toxigenic *Escherichia coli* O104:H4. *JAMA* 309:1502–10.

Luangtongkum, T., Jeon, B., Han, J., Plummer, P., Logue, C. M. and Zhang, Q. 2009. Antibiotic resistance in *Campylobacter*: emergence, transmission and persistence. *Future Microbiol.* 4:189–200.

Luedtke, B. E., Bono, J. L., Bosilevac, J. M. 2014. Evaluation of real time PCR assays for the detection and enumeration of enterohemorrhagic *Escherichia coli* directly from cattle feces. *J. Microbiol. Methods.* 105:72–9.

Manges, A. R., Smith, S. P., Lau, B. J., Nuval, C. J., Eisenberg, J. N., Dietrich, P. S. and Riley, L. W. 2007. Retail meat consumption and the acquisition of antimicrobial resistant *Escherichia coli* causing urinary tract infections: a case-control study. *Foodborne Pathog. Dis.* 4:419–31.

Marshall, B. M. and Levy, S. B. 2011. Food animals and antimicrobials: impacts on human health. *Clin. Microbiol. Rev.* 24:718–33.

McArthur, A. G., Waglechner, N., Nizam, F., Yan, A., Azad, M. A., Baylay, A. J., Bhullar, K., Canova, M. J., De Pascale, G., Ejim, L., Kalan, L., King, A. M., Koteva, K., Morar, M., Mulvey, M. R., O'Brien, J. S., Pawlowski, A. C., Piddock, L. J., Spanogiannopoulos, P., Sutherland, A. D., Tang, I., Taylor, P. L., Thaker, M., Wang, W., Yan, M., Yu, T. and Wright, G. D. 2013. The comprehensive antibiotic resistance database. *Antimicrob. Agents Chemother.* 57:3348–57.

Miller, R. R., Montoya, V., Gardy, J. L., Patrick, D. M. and Tang, P. 2013. Metagenomics for pathogen detection in public health. *Genome Med.* 5:81. doi: 10.1186/gm485.

Min, J. and Baeumner, A. J. 2002. Highly sensitive and specific detection of viable *Escherichia coli* in drinking water. *Anal. Biochem.* 303:186–93.

Nadimpalli, M., Rinsky, J. L., Wing, S., Hall, D., Stewart, J., Larsen, J., Nachman, K. E., Love, D. C., Pierce, E., Pisanic, N., Strelitz, J., Harduar-Morano, L. and Heaney, C. D. 2015. Persistence of livestock-associated antibiotic-resistant *Staphylococcus aureus* among industrial hog operation workers in North Carolina over 14 days. *Occup. Environ. Med.* 72:90–9.

Nesme, J., Cecillon, S., Delmont, T. O., Monier, J. M., Vogel, T. M. and Simonet, P. 2014. Large-scale metagenomic-based study of antibiotic resistance in the environment. *Curr. Biol.* 24:1096–100.

Neyra, R. C., Frisancho, J. A., Rinsky, J. L., Resnick, C., Carroll, K. C., Rule, A. M., Ross, T., You, Y., Price, L. B. and Silbergeld, E. K. 2014. Multidrug-resistant and methicillin-resistant *Staphylococcus aureus* (MRSA) in hog slaughter and processing plant workers and their community in North Carolina (USA). *Environ. Health Perspect.* 122:471–7.

Nucera, D. M., Maddox, C. W., Hoien-Dalen, P. and Weigel, R. M. 2006. Comparison of API 20E and *invA* PCR for identification of *Salmonella enterica* isolates from swine production units. *J. Clin. Microbiol.* 44:3388–90.

Pagel, S. W. and Gautier, P. 2012. Use of antimicrobial agents in livestock. *Rev. Sci. Tech.* 31:145–88.

Peak, N., Knapp, C. W., Yang, R. K., Hanfelt, M. M., Smith, M. S., Aga, D. S. and Graham, D. W. 2007. Abundance of six tetracycline resistance genes in wastewater lagoons at cattle feedlots with different antibiotic use strategies. *Environ. Microbiol.* 9:143–51.

Presad, A. K. and Lejeune, J. T. 2014. Animal reservoirs of Shiga toxin-producing *Escherichia coli*. *Microbiol Spectr.* 2(4):EHEC-0027-2014. doi: 10.1128/microbiolspec.EHEC-0027-2014.

Rahn, K., De Grandis, S. A., Clarke, R. C., McEwen, S. A., Galan, J. E., Ginocchio, C., Curtiss, R. and Gyles, C. L. 1992. Amplification of an *invA* gene sequence of *Salmonella typhimurium* by polymerase chain reaction as a specific method of detection of *Salmonella*. *Mol. Cell. Probes* 6:271–9.

Ramchandani, M., Manges, A. R., DebRoy, C., Smith, S. P., Johnson, J. R. and Riley, L. W. 2005. Possible animal origin of human-associated, multidrug-resistant, uropathogenic *Escherichia coli*. *Clin. Infect. Dis.* 40:251–7.

Rinsky, J. L., Nadimpalli, M., Wing, S., Hall, D., Baron, D., Price, L. B., Larsen, J., Stegger, M., Stewart, J. and Heaney, C. D. 2013. Livestock-associated methicillin and multidrug resistant *Staphylococcus aureus* is present among industrial, not antibiotic-free livestock operation workers in North Carolina. *PLoS ONE* 8:e67641.

Roberts, M. C. 2008. Update on macrolide-lincosamide-streptogramin, ketolide, and oxazolidinone resistance genes. *FEMS Microbiol. Lett.* 282:147–59.

Russo, T. A. and Johnson, J. R. 2003. Medical and economic impact of extraintestinal infections due to *Escherichia coli*: focus on an increasingly important endemic problem. *Microbes Infect.* 5:449–56.

Schmidt, J. W., Agga, G. E., Bosilevac, J. M., Brichta-Harhay, D. M., Shackelford, S. D., Wang, R., Wheeler, T. L. and Arthur, T. M. 2015. Occurrence of antimicrobial-resistant *Escherichia coli* and *Salmonella enterica* in the beef cattle production and processing continuum. *Appl. Environ. Microbiol.* 81:713–25.

Schmidt, J. W., Griffin, D., Kuehn, L. A. and Brichta-Harhay, D. M. 2013. Influence of therapeutic ceftiofur treatments of feedlot cattle on fecal and hide prevalences of commensal *Escherichia coli* resistant to expanded-spectrum cephalosporins, and molecular characterization of resistant isolates. *Appl. Environ. Microbiol.* 79:2273–83.

Schrader, C., Schielke, A., Ellerbroek, L. and Johne, R. 2012. PCR inhibitors - occurrence, properties and removal. *J. Appl. Microbiol.* 113:1014–26.

Shao, Y., Zhu, S., Jin, C. and Chen, F. 2011. Development of multiplex loop-mediated isothermal amplifcation-RFLP (mLAMP-RFLP) to detect *Salmonella* spp. and *Shigella* spp. in milk. *Int. J. Food Microbiol.* 148:75–9.

Shen, Z., Hou, N., Jin, M., Qiu, Z., Wang, J., Zhang, B., Wang, X., Wang, J., Zhou, D. and Li, J. 2014. A novel enzyme-linked immunosorbent assay for detection of Escherichia coli O157:H7 using immunomagnetic and beacon gold nanoparticles. *Gut Pathog.* 6:14. doi: 10.1186/1757-4749-6-14.

Shoemaker, N. B., Vlamakis, H., Hayes, K. and Salyers, A. A. 2001. Evidence for extensive resistance gene transfer among *Bacteroides* spp. and among *Bacteroides* and other genera in the human colon. *Appl. Environ. Microbiol.* 67:561–8.

Smith, K. E., Wilker, P. R., Reiter, P. L., Hedican, E. B., Bender, J. B. and Hedberg, C. W. 2012. Antibiotic treatment of *Escherichia coli* O157 infection and the risk of hemolytic uremic syndrome, Minnesota. *Pediatr. Infect. Dis. J.* 31:37–41.

Stevens, M. P., Humphrey, T. J. and Maskell, D. J. 2009. Molecular insights into farm animal and zoonotic *Salmonella* infections. *Philos. Trans. R. Soc. Lond. B. Biol. Sci.* 364:2709–23.

Su, L. H., Chiu, C. H., Chu, C. and Ou, J. T. 2004. Antimicrobial resistance in nontyphoid *Salmonella* serotypes: a global challenge. *Clin. Infect. Dis.* 39:546–51.

Tarr, P. I., Gordon, C. A. and Chandler, W. L. 2005. Shiga toxin-producing *Escherichia coli* and haemolytic uraemic syndrome. *Lancet* 365:1073–86.

USDA. 2013. Feedlot 2011, Part IV: Health and Health Management on U.S. Feedlots with a Capacity of 1,000 or More Head. National Animal Health Monitoring System, Animal and Plant Health Inspection Service, U.S. Dept. of Agriculture. Available: http://www.aphis.usda.gov/animal_health/nahms/feedlot/downloads/feedlot2011/Feed11_dr_PartIV.pdf. Accessed 6 July 2016.

USDA FSIS. 2013. Notice 13-13: FSIS sampling for the national antimicrobial resistance monitoring system. In F. S. A. I. Service (ed.), Washington, DC. http://www.fda.gov/downloads/AnimalVeterinary/SafetyHealth/AntimicrobialResistance/NationalAntimicrobialResistanceMonitoringSystem/UCM334834.pdf. Accessed 6 July 2016.

Vandendriessche, S., Vanderhaeghen, W., Soares, F. V., Hallin, M., Catry, B., Hermans, K., Butaye, P., Haesebrouck, F., Struelens, M. J. and Denis, O. 2013. Prevalence, risk factors and genetic diversity of methicillin-resistant *Staphylococcus aureus* carried by humans and animals across livestock production sectors. *J. Antimicrob. Chemother.* 68:1510–16.

Velasco, V., Sherwood, J. S., Rojas-Garcia, P. P. and Logue, C. M. 2014. Multiplex real-time PCR for detection of *Staphylococcus aureus*, mecA and Panton-Valentine Leukocidin (PVL) genes from selective enrichments from animals and retail meat. *PLoS ONE* 9:e97617.

Verstraete, K., Robyn, J., Del-Favero, J., De Rijk, P., Joris, M. A., Herman, L., Heyndrickx, M., De Zutter, L. and De Reu, K. 2012. Evaluation of a multiplex-PCR detection in combination with an isolation method for STEC O26, O103, O111, O145 and sorbitol fermenting O157 in food. *Food Microbiol.* 29:49–55.

Vimont, A., Delignette-Muller, M.-L. and Vernozy-Rozand, C. 2007. Supplementation of enrichment broths by novobiocin for detecting Shiga toxin-producing *Escherichia coli* from food: a controversial use. *Lett. Appl. Microbiol.* 44:326–31.

Wang, F., Yang, Q., Qu, Y., Meng, J. and Ge, B. 2014. Evaluation of a loop-mediated isothermal amplification suite for the rapid, reliable, and robust detection of Shiga toxin-producing *Escherichia coli* in produce. *Appl. Environ. Microbiol.* 80:2516–25.

Wasilenko, J. L., Fratamico, P. M., Sommers, C., DeMarco, D. R., Varkey, S., Rhoden, K. and Tice, G. 2014. Detection of Shiga toxin-producing *Escherichia coli* (STEC) O157:H7, O26, O45, O103, O111, O121, and O145, and *Salmonella* in naturally-contaminated ground beef using the DuPont BAX® System. *Front. Cell. Infect. Microbiol.* 18 June. 4:81. doi: 10.3389/fcimb.2014.00081.

Webb, H. E., Bugarel, M., den Bakker, H. C., Nightingale, K. K., Granier, S. A., Scott, H. M. and Loneragan, G. H. 2016. Carbapenem-resistant bacteria recovered from faeces of dairy cattle in the high plains region of the USA. *PLoS ONE* 11:e0147363.

Wichmann, F., Udikovic-Kolic, N., Andrew, S. and Handelsman, J. 2014. Diverse antibiotic resistance genes in dairy cow manure. *mBio* 5:e01017.

Wiedmann, M., Wang, S., Post, L. and Nightingale, K. 2014. Assessment criteria and approaches for rapid detection methods to be used in the food industry. *J. Food Prot.* 77:670–90.

Winokur, P. L., Vonstein, D. L., Hoffman, L. J., Uhlenhopp, E. K. and Doern, G. V. 2001. Evidence for transfer of CMY-2 AmpC beta-lactamase plasmids between *Escherichia coli* and *Salmonella* isolates from food animals and humans. *Antimicrob. Agents Chemother.* 45:2716–22.

Wong, C. S., Jelacic, S., Habeeb, R. L., Watkins, S. L. and Tarr, P. I. 2000. The risk of the hemolytic-uremic syndrome after antibiotic treatment of *Escherichia coli* O157:H7 infections. *N. Engl. J. Med.* 342:1930–6.

Xia, X., Meng, J., Zhao, S., Bodeis-Jones, S., Gaines, S. A., Ayers, S. L. and McDermott, P. F. 2011. Identification and antimicrobial resistance of extraintestinal pathogenic *Escherichia coli* from retail meats. *J. Food Prot.* 74:38–44.

Yan, L., Zhou, J., Zheng, Y., Gamson, A. S., Roembke, B. T., Nakayama, S. and Sintim, H. O. 2014. Isothermal amplified detection of DNA and RNA. *Mol. BioSyst.* 10:970–1003.

Zhao, Y., Wang, H., Zhang, P., Sun, C., Wang, X., Wang. X., Yang, R., Wang, C., Zhou, L. 2016. Rapid multiplex detection of 10 foodborne pathogens with an up-converting phosphor technology-based 10-channel lateral flow assay. See comment in PubMed Commons below*Sci. Rep.* 17(6):21342. doi: 10.1038/srep21342.

Zhou, B., Xiao, J., Liu, S., Yang, J., Wang, Y., Nie, F., Zhou, Q., Li, Y. and Zhao, G. 2013. Simultaneous detection of six food-borne pathogens by multiplex PCR with GEXP analyzer. *Food Cont.* 32, 198–204.

Chapter 3

Methods for detecting pathogens in the beef food chain: detecting particular pathogens

Pina M. Fratamico, Joseph M. Bosilevac and John W. Schmidt, United States Department of Agriculture, USA

1 Introduction

Pathogens of importance in beef include Shiga toxin-producing *Escherichia coli* (STEC), especially STEC that cause haemorrhagic colitis and haemolytic uraemic syndrome known as enterohaemorrhagic *E. coli* (EHEC). Beef and dairy cattle are an important reservoir of STEC and EHEC belonging to a variety of serotypes (Presad and Lejeune, 2014). *Salmonella* are also carried by beef cattle and can cause illness in cattle when they are stressed due to transportation, crowding, food deprivation and other stresses (Stevens et al., 2009). *Salmonella* of multiple serotypes can carry antibiotic resistance genes, and when these infect humans, treatment of illnesses is more difficult. Other pathogens affecting beef include *Campylobacter*, *Listeria monocytogenes*, *Mycobacterium avium* subsp. *paratuberculosis* (MAP) and *Clostridium difficile*.

The most rigorous testing of cattle, their environments and beef products is for *E. coli* O157:H7, non-O157 STEC and *Salmonella*. Furthermore, cattle and environments are often tested for antimicrobial-resistant (AMR) bacteria and *Campylobacter*. This chapter reviews methods for detecting *E. coli* O157:H7 and non-O157 STEC, *Salmonella*, *Listeria*

http://dx.doi.org/10.19103/AS.2016.0008.16

and *Campylobacter*, *Clostridium*, *Bacillus cereus* and MAP. Further, the chapter also reviews the development of standardised methods for regulatory testing of *E. coli*.

2 Methods used to detect *E. coli* O157:H7 and non-O157 STEC

Usually, methods for the detection of *E. coli* O157:H7 involve a multistep process. Initial samples, be they beef, faeces or a sponge swab, are enriched in a selective or semi-selective medium (modified tryptic soy broth (TSB) containing novobiocin, potassium tellurite and vancomycin) at selective conditions (42°C) for a period of 6–12 hours (Barkocy-Gallagher et al., 2005). The enriched sample is then screened for the presence of *E. coli* O157:H7 using rapid methods based on immunoassays or molecular detection (DNA or RNA amplification), often followed by culture isolation. Immunoassays generally include processing the enrichment to extract the desired antigen target while molecular assays involve steps of cell lysis and DNA or RNA purification. An immunoassay such as a lateral flow device can provide results in as little as 15 minutes, while enzyme-linked immunoassay (EIA) and molecular polymerase chain reaction (PCR) assays require 1–3 hours (Law et al., 2014). If an enrichment is to be processed for culture isolation of *E. coli* O157:H7, then immunomagnetic separation (IMS) is used to selectively concentrate *E. coli* of the O157 serogroup (Barkocy-Gallagher et al., 2005; Chapman and Ashton, 2003). Culture methods then rely on various growth media for simplified isolation and identification of *E. coli* O157:H7. The most rapid detection methods take approximately 12 hours while culture-based methods may take 48–72 hours.

Methods for detection of *E. coli* O157H7 are used throughout the beef chain, from monitoring the pre-harvest colonisation of cattle (Cernicchiaro et al., 2013; Ennis et al., 2012) and their environment (Johnson et al., 2014) to the peri-harvest (Moxley and Acuff, 2014) and post-harvest monitoring of processing plants (Arthur et al., 2004, 2008), carcasses, and final beef cuts and products (Mathusa et al., 2010). The primary methods used for the detection of *E. coli* O157:H7 in each area are largely based on the type of sample and the intended use of the result. For instance, in the pre-harvest arena, sample types are faeces from live animals or feedlot surface materials that have exceptionally high levels of background bacteria competing for growth during the enrichment phase of detection (Vimont et al., 2007), and the results of the testing are used for academic reasons over public health reasons. That is, nearly all cattle may shed *E. coli* O157:H7 at some point in their life, but an animal that is found positive is not removed from the beef herd or supply chain. Identifying cattle that are colonised or shed *E. coli* O157:H7 aids in building information on the production practices that may influence its presence (Arthur et al., 2009). On the other hand, final raw beef products are tested for *E. coli* O157:H7 to ensure that beef processing interventions are properly controlling the pathogen. Final beef products, such as boneless beef trimmings, ground beef and whole muscle cuts of beef have very little competing microflora; therefore, a brief enrichment period will result in a broth suitable for a rapid screening assay (Bosilevac et al., 2010). Since fresh beef is perishable, molecular and immunoassay identification of *E coli* O157, including H7 are used to ensure that meats are free of the pathogen and safe for consumption. Positive test results from beef products result in potentially dangerous products diverted to thermal inactivation treatments and reduce the number of recalls.

Examples of rapid methods used for *E. coli* O157:H7 on the farm are few in comparison with culture-based methods. To monitor *E. coli* O157:H7 in this environment, only the most robust rapid methods are useful. In addition to having high interfering background, samples collected on farms (pastures, feedlots) contain a number of complex molecules that are inhibitory to many PCR analyses (Luedtke et al., 2014). To properly use rapid DNA detection methods on samples from these environments, DNA template preparation requires affinity purification while reaction master mix solutions need to be designated for environmental samples (Luedtke et al., 2014). This usually encompasses addition of an increased concentration of a more highly purified hot-start DNA polymerase, as well as the addition of amplification cofactors such as chelators, reducing agents and detergents. Properly performed, *E. coli* O157:H7 detection assays not only can determine if the organism is present but also can provide quantitative results (Sharma, 2002; Ibekwe and Grieve, 2003). This aids in identifying cow–calf pairs on pasture and feedlot cattle that harbour and shed *E. coli* O157:H7 and allows the identification of super-shedder cattle, those shedding greater than 1×10^4 CFU *E. coli* O157:H7 per gram of faeces (Arthur et al., 2010). Detection from faecal and soil samples has been found to be improved by the addition of phosphate buffers to the TSB enrichment broth. It is thought that buffering of the enrichment broth provides a more suitable environment for growth (Barkocy-Gallagher et al., 2005).

Before, during and after processing steps is when the bulk of common *E. coli* O157:H7 detection testing takes place. Beef dressing practices are monitored by collecting sponge swab samples from hides and carcasses before splitting. The sponge samples are enriched and tested with most molecular or immunoassay testing platforms. The USDA FSIS conducted a national survey of such samples to generate data for modelling of pathogen transfer from hide to carcass during processing. Split beef carcasses were subjected to a series of antimicrobial steps such as spraying of lactic acid and hot water then chilled until fabricated into retail whole-muscle cuts of beef (Koohmaraie et al., 2005; Wheeler et al., 2014). During this fabrication process, s referred to as combos. The vast majority of beef processors mall bits and pieces of beef were sorted into one tonne (2000 pound) bins, referred to as combos. The vast majority of beef processors test each and every combo of trim for *E. coli* O157:H7. Single combo testing is performed by collecting a representative sample of the beef within the combo. In most cases, pieces of beef trim recognised as containing external carcass surface are targeted for sample collection by surface excision, and then composited into bags of 60 pieces (N60 sample) and weighing approximately 350 g (Murphy and Seward, 2004; Arthur et al., 2014). In other cases, multiple core samples are collected by drilling through the combo or an automated shaving device is driven into the combo to collect the beef trim surface sample (Koohmaraie et al., 2007).

Trim testing is rapid, with *E. coli* O157:H7-specific enrichments of 6 to 9 or to 12 hours typically used. The rapid detection of *E. coli* O157 in this sort of testing relies on PCR (Arthur et al., 2005). In the United States approximately one-third of beef trim testing is performed using an IMS step to selectively concentrate the organism for detection by real-time PCR. Another approximate third of this testing is performed using multiplexed PCR targets resolved on agarose gels, while the remainder of testing is generally a mixture of various methods supplied by numerous commercial suppliers. *E. coli* O157:H7 is targeted in PCR assays for genes known to be specific to the organism, such as intimin subtype gamma and Shiga toxin, as well as insertion sequences (IS elements) that follow the evolutionary development of this serotype (Deisingh and Thompson, 2004; Rivas et al., 2015).

Most rapid testing for *E. coli* O157:H7 is composed of two parts. An initial screen is used to rapidly identify beef products that may be contaminated, and then the enrichment broths from the suspect samples are confirmed by a subsequent test. For instance, one commercially available assay initially screens for gamma intimin that is found in O157:H7, and if present, the sample is tested for the presence of Shiga toxin genes because a number of *E. coli* of low pathogenic concern can also possess the gamma intimin gene (Arthur et al., 2005). Thus, the presence of a Shiga toxin gene then signals that the product may be hazardous. Other systems screen for multiple targets associated with *E. coli* O157:H7 then confirm using immunoassays for the O157 serogroup and vice versa.

Non-O157 STEC, especially those of serogroups O26, O45, O103, O111, O121 and O145 are considered adulterants and limited testing from farm to beef products occurs (Stromberg et al., 2015, 2016). Testing for non-O157 STEC follows the same trends as *E. coli* O157:H7 with farm testing being more of a research tool and based on culture, while final beef product testing uses rapid immunoassays and PCR for process monitoring and public health reasons. Whereas all beef processors test and monitor their products for *E. coli* O157:H7, only limited non-O157 STEC testing is currently taking place. The reason for this is two-fold. First, the reliability of detection tests for non-O157 STEC are limited, and secondly, processors are relying on their *E. coli* O157:H7 tests, which are very specific and sensitive to confirm that their processes are in control (Bosilevac, 2013). That is, a processing plant that is controlling *E. coli* O157:H7 is controlling the non-O157 STEC because the organisms all react similarly to the interventions in place. Wheeler et al. (2015) compared three commercial methods (BioControl Assurance Genetic Detection System MPX top 7 STEC, DuPont BAX System STEC Suite and the Pall GeneDisc System for STEC and EHEC) to the USDA Microbiology Laboratory Guidebook method for detection of non-O157 STEC in beef trim. Generally, results of the commercial methods were similar, use of IMS helped with confirmation procedures and modified Rainbow Agar (mRBA) used in the FSIS MLG performed better than other confirmation plates (MacConkey agar with or without rhamnose) used in the International Organization for Standardization protocols. The authors also found that the enrichment medium used for the GeneDisc assay, buffered peptone water (BPW), maintained the pH at ~7.0; however, the other enrichments that used other media (BAX MP and mTSB) became acidic.

The need for non-O157 tests that are sensitive and specific are paramount before routine testing is likely to be adopted. Currently, non-O157 STEC are detected using PCR to identify enrichment broths containing both genes encoding intimin (*eae*) and Shiga toxin (*stx*1, *stx*2). A broth found positive for both markers is then tested for the presence of the O-group genes of the six most common STEC that cause human infections. The problem in this scheme is that the multiplexed PCR assays cannot distinguish whether the genes detected are from one, two, three or more different organisms (Bosilevac, 2013). Nearly 25% of all ground beef and 10% of all beef trim can be identified as containing Shiga toxin genes (Bosilevac and Koohmaraie, 2011). Likewise about the same proportions can be determined to contain an intimin gene, or an O-group-specific gene (Bosilevac and Koohmaraie, 2012). The presence of all three non-O157 STEC targets is more likely a false positive than truly representing the presence of a pathogen.

Whole-genome sequencing (WGS) is beginning to reveal additional markers in this group of *E. coli* that may eventually lead to tests with improved specificity. One example is the gene *ecf-1* (Livezey et al., 2015). This target localises to the large virulence plasmid carried by STEC that also carry *eae* (Boerlin et al., 1998). It is the primary screen in a new assay that was released commercially. The presence of the *ecf-1* target is a good indicator for clinically significant non-O157 STEC, but follow up screens for the presence of *stx* and O-group specific genes are required to identify STECs of regulatory concern (Livezey et al., 2015). WGS has also identified a series of STEC-specific single-nucleotide polymorphisms (SNP) present in each individual serogroup (Norman et al., 2012). Using these SNPs, a commercially available assay offers a test that identifies over 70 targets simultaneously to assess whether the STEC-specific SNPs and other genes are present. This method is not approved as a screening assay, but is approved as a culture-free confirmation of a suspect broth. Examining feedlot cattle hide samples and their matched pre-intervention carcasses, Stromberg et al. (2015) used this commercial method to determine the prevalence of the top 6 STEC and O157 in these samples and compared results with a culture-based method. Fewer positive samples were detected by the culture method as compared to the commercial assay in faecal and hide samples, likely due to the culture methodologies used. In that same study, seven carcass sponge swabs were identified by SNP analysis to contain *E. coli* O157:H7 or other STEC. Of these seven samples, three were culturally confirmed.

Delannoy et al. (2012, 2013a,b, 2016a,b) identified novel markers, including clustered regularly interspaced short palindromic repeat (CRISPR) loci for more specific identification of the top 7 EHEC strains. Subpopulations of STEC can be segregated according to their large accessory genome composed of virulence genes and pathogenicity islands such as the locus of enterocyte effacement (LEE) and O-island (OI-57 and OI-122) pathogenicity islands (Michelacci et al., 2016). Through the use of novel markers *espK* and *espV* in conjunction with *stx* and *eae*, potentially positive beef enrichments were narrowed down by 49% (Delannoy et al., 2016b). Further inclusion of the *ureD* gene and the z2098 open reading frame were found to be the best combinations to identify *E. coli* O157:H7, the top 6 non-O157 STEC, as well as other *eae*-positive STEC that may be non-appreciated pathogens (Delannoy et al., 2013a). Targeting the CRISPR regions of STEC using simplex real-time PCR assays, Delannoy et al. (2012) found specificity estimates for many top 6 non-O157 STEC to be nearly 100%.

3 Culture isolation of *E. coli* O157:H7 and non-O157 STEC

E. coli O157:H7 has physiologic features that provide for relatively easy identification of its unique phenotype. This serotype of *E. coli* largely lacks the ability to ferment D-sorbitol and does not produce β-glucuronidase, which can hydrolyse MUG (4-methyl-umbelliferyl-D-glucuronide), a synthetic molecule. Therefore, when growing on sorbitol MacConkey (SMAC) agar supplemented with MUG, *E. coli* O157:H7 strains are easily identified. The selectivity for *E. coli* O157:H7 is increased by adding cefixime, potassium tellurite and vancomycin to SMAC agar to inhibit other microflora.

Other commercially available agar formulations such as CHROMagar™ O157 and Rainbow® Agar O157 provide additional chromogenic media to aid in the identification of *E. coli* O157:H7. CHROMagar STEC takes advantage of STEC tolerance for selective agents; however, since tellurite is used in this agar and in other selective agars, strains that are not resistant to tellurite may not grow (Hirvonen et al., 2012). Washed sheep blood agar has been shown to not only aid in identifying *E. coli* O157:H7 but also the non-O157 STEC, as many STEC strains possess the *ehxA* gene that results in a unique and identifiable enterohaemolytic phenotype on this agar (Sugiyama et al., 2001).

As alluded to, non-O157 STEC lack the unique physiology and phenotypes of *E. coli* O157:H7, and thus, culture identification and isolation is much more laborious. Some media formulations have been described that attempt to provide for chromogenic selection of the non-O157 STEC, but each non-O157 serogroup and serotype may have a distinct appearance. Agars described by Posse et al. (2007, 2008) and Kalchayanand et al. (2013) take advantage of differences in the carbohydrate-fermenting ability of STEC strains. Specifically, dulcitol, D-raffinose, D-saccharose, D-sorbitol and D-sorbose have the potential to serve as discriminative markers of STEC serotypes O26:H11, O45:H2, O103:H2, O111:NM, O121:H19, O145:NM and O157:H7 (Kalchayanand et al., 2013). The FSIS has reported that by using Rainbow® Agar O157, many of the non-O157 STEC can be distinguished or provide a suspect phenotype when the STEC serogroup under investigation is known (Fratamico et al., 2011). However, the levels of novobiocin and potassium tellurite recommended for use in the agar for isolation of *E. coli* O157:H7 inhibit growth of some non-O157 STEC; therefore, Tillman et al. (2012) modified Rainbow® Agar O157 to allow growth of more non-O157 STEC and referred to the agar as mRBA.

A tool to help discriminate non-O157 STEC growth on chromogenic media has been described (Windham 2013). It is a colony imaging technique based on hyperspectral imaging, an optical technique that combines conventional imaging and vibrational spectroscopy to acquire spectral fingerprints of the bacterial colonies. Multivariate classification models were developed from the spectral libraries to differentiate among colonies of non-O157 STEC. A label-free light scattering sensor known as BARDOT (Bacterial Rapid Detection using Optical scattering Technology) was evaluated for identifying STEC serogroups on selective agar plates (Tang et al., 2014). By shining a laser at 635 nm on the centre of the colony, the instrument generates scatter patterns in a two-dimensional plane, and the image is dependent on the colony macromolecular composition and structure, including shape, chromogen and O-antigen composition. The scatter patterns are analysed using an image classifier, and STEC belonging to various serogroups could be distinguished and identified.

4 Methods used to detect *Salmonella*

Although *Salmonella* is not considered an adulterant in beef, its presence is monitored and contaminated products are often diverted to lethality steps in the interest of public health. The presence of *Salmonella* in cattle from the pasture to the processor is an active area of study as there is vastly more *Salmonella* present in beef than *E. coli* (Bosilevac et al., 2009; Brichta-Harhay et al., 2008, 2011). Methods used for *Salmonella* detection are similar to those for *E. coli* O157:H7 and STEC. Culture, immunoassays and PCR are all used in similar situations but with modifications specific to *Salmonella*. The number of validated methods for *Salmonella* detection far outnumber those for *E. coli* O157:H7 detection. This is because *Salmonella* is also a crucial contaminant in poultry, eggs, seafood and produce.

Both rapid PCR and culture-based methods are used for detection of *Salmonella* in the pre-harvest environment. *Salmonella* can be identified in faeces, soil and the cattle-raising environment. *Salmonella* is a hardy organism and readily grows to detectable levels even in the presence of competing microflora. Immunoassay detection methods for *Salmonella* are generally reserved for food and beef product testing because *Salmonella* is a highly diverse group of organisms that cannot be simply detected with one or two antibodies to surface proteins. Most immunoassays for *Salmonella* use pools of numerous antibodies raised to the flagella of the organism (Cheung and Kam, 2012). For this reason, immunoassays for *Salmonella* can at times generate false-positive results due to the presence of other cross-reacting bacteria, and false negatives with non-motile *Salmonella*. Since foods may have lower levels of other bacteria compared to some other types of samples, the likelihood of cross-reactions is less. Nearly every molecular DNA and RNA assay for *Salmonella* targets either unique regions of the ribosome or specific *Salmonella* virulence factors such as the *invA* gene involved in invasion (Rahn et al., 1992; Chiu and Ou, 1996). Assays that target the detection of ribosomal RNA are very sensitive as each *Salmonella* can contain thousands of ribosomal targets for detection, while the *invA* gene target is present as a single copy in the organisms (Livezey et al., 2013).

Enrichment growth of *Salmonella* uses a common two-step process of non-selective growth in a medium, such as TSB or buffered peptone water (BPW) for 12–24 hours, followed by a dilution and further selective enrichment in Rappaport Vasiliadis soy (RVS) broth, tetrathionate (TT) broth or selenite cystine (SC) broth incubated at elevated temperatures of 42°C for 18–24 hours (Vassiliadis, 1983; Bailey et al., 1988). The selective enrichments are then streaked onto selective agars such as XLD (xylose lysine deoxycholate) agar, BG (Brilliant Green) agar or BGS (BG with the addition of sulphadiazine), or HE (Hektoen Enteric) agar (Vassiliadis, 1983; Bailey et al., 1988; Miller et al., 1994). In all cases, *Salmonella* grow with unique suspect phenotypes of colour or H_2S production on the agars. Recently in the area of beef product testing, there has been a desire to use a universal enrichment for the screening of *Salmonella* and *E. coli* O157:H7 (and non-O157 STEC). This universal enrichment broth is mTSB, as the detection of *Salmonella* has been validated from this broth. Thus, one sample of beef can be enriched for the detection of multiple pathogens (USDA FSIS, 2014).

5 Methods used to detect *Listeria* and *Campylobacter*

Although the primary food-borne pathogens of concern in the beef chain are *E. coli* O157:H7, non-O157 STEC and *Salmonella*, other pathogens may be present and pose contamination concerns to the cattle production environment, as well as to beef products. Of these, *L. monocytogenes* and *Campylobacter* species *jejuni* and *coli* have numerous rapid methods available for detection, as these pathogens are of considerable concern in other foods, as well. The other organisms considered in this section such as *Bacillus* and *Clostridium* are often considered spoilage organisms but these spore formers can also produce dangerous toxins. Multiple toxin detection assays are available to identify the presence of toxins using antibody-mediated capture methods. Others organisms such as *Mycobacterium avium* may be related to chronic human diseases that require more study. As such the latter group often lack commercial detection tests and rely upon described molecular detection techniques, as well as classical microbiological culture isolation methods.

5.1 Methods for detecting *L. monocytogenes*

L. monocytogenes is generally regarded as a food safety issue in ready-to-eat (RTE) foods and its presence in fresh beef that is to be cooked has been considered relatively inconsequential. However, *L. monocytogenes* can be traced through the beef chain, and often RTE processors request that *L. monocytogenes* analysis accompany raw materials before accepting them for further processing. Conventional methods used for the detection and isolation of *L. monocytogenes* in foods consist of primary and secondary enrichments in University of Vermont modification medium (UVM), Demi-Fraser broth and Fraser broth. These series of enrichments require 12–36 hours, and each stage can be used in a molecular screening assay. Molecular detection is preferred to culture isolation because culture confirmation requires biochemical identification and serotyping, which may add 7–8 days for test completion (Schlech et al., 1983; USDA FSIS, 2013; Gasanov et al., 2005). Alternate universal enrichment for detection has been demonstrated using TSB (Guerini et al., 2007) The TSB enrichments allow simultaneous detection of *L. monocytogenes* with other pathogens.

Rapid methods for *L. monocytogenes* generally target specific virulence genes unique to *L. monocytogenes* such as the *hyl*, *inlB* or *iap* genes (Norton, 2002) or the 16S ribosomal subunit. Often for environmental screening and sanitation monitoring, testing focuses on the detection of *Listeria* species, as an indicator of conditions favourable to *L. monocytogenes*. In this case, molecular targets common to all *Listeria* species are used, such as the *prs* gene. Some serovars of *L. monocytogenes* are more frequently identified in human infections and outbreaks. These serovars can be distinguished from one another using multiplexed PCR assays (Doumith et al., 2004).

Culture confirmation of *Listeria* and *L. monocytogenes* is improved though the use of agar media that reveal *Listeria* esculin hydrolysis observed by a blackening of the colony and surrounding medium. Oxford agar is one common agar for the isolation of *Listeria* that contains esculin and ferric ammonium citrate. The addition of ferric ions to the medium acts as an indicator of esculin-hydrolysing bacteria and results from the formation of 6,7-dihydroxycoumarin reacting with the ferric ions (Fraser and Sperber, 1988). Other agar media such as CHROMAgar *Listeria* show this *Listeria* phenotype as blue coloured colonies, and include egg yolk emulsion to reveal the lecithinase activity of *L. monocytogenes* that distinguish it from other *Listeria* species and background bacteria.

5.2 Detecting *Campylobacter*

Campylobacter species are the leading causes of bacterial food-borne illness in the United States. This organism is part of the natural gut microflora of cattle and has been studied in cattle and their production environments (Besser et al., 2005), through processing (Steele, 1998) and has been found in retail beef products (Noormohamed and Fakhr, 2013). The primary focus on *Campylobacter*, however, is as a contaminant of poultry. As such, extensive methods and test kits are available for poultry and are typically modified for use on samples collected from cattle and beef.

Campylobacter are microaerophilic and are detected by direct plating and enrichment procedures for isolation that identify *C. jejuni* and *C. coli*, the two most frequently occurring *Campylobacter* species of clinical significance in human infections. Samples from cattle and beef include faeces, hide swabs, carcass swabs, raw beef, organ meats and environmental swabs. Isolates are confirmed as *C. jejuni* or *C. coli* by microscopy revealing corkscrew morphology and darting motility or confirmed by immunological testing using latex agglutination tests. Rapid molecular methods can substitute for the lengthy culture and confirmation tests by directly testing the enrichments by PCR.

6 Methods used to detect *Clostridium*, *Bacillus cereus* and MAP

6.1 Methods for detecting *Clostridium perfringens*

Clostridium perfringens is a spore-forming, anaerobic bacterium that is widespread in soil, water and foods. It can be found in the intestinal tract of humans and animals. Viable, spore-forming strains that cause food-borne illness belong to *C. perfringens* Type A, which produces an enterotoxin. The toxin of *C. perfringens* can be rapidly detected with commercial ELISA or test strip immunoassay kits. The enterotoxin gene *cpe* of *C. perfringens* can be detected in PCR assays if the organism has reached a high enough level for molecular detection, usually about $1x10^5$ CFU/mL, following enrichment in Tryptone Glucose Yeast Extract (TGY) medium or Duncan – Strong medium incubated at 37°C overnight. Many PCR assays can be used to type the *C. perfringens* present to Type A or Type B (Li et al., 2007).

Culture confirmation of *C. perfringens* is based on classical isolation methods using membrane *Clostridium perfringens* (m-CP) agar or tryptose sulphite cycloserine (TSC) agar. Each of these agars presents problems, however, as the mCP method is labour intensive and results in colonies that cannot be used for biochemical testing. International Organization for Standardization (ISO) methods favour TSC agar methods, but, although TSC detects all sulphite-reducing clostridia, excessive blackening of the agar interferes with colony counts at low dilutions. To remedy this, there are now selective chromogenic media available, and one such medium is CP ChromoSelect.

6.2 Methods for detecting *Bacillus cereus*

Bacillus cereus is a spore-forming, aerobic bacterium widespread in soil and foods. *B. cereus* is not a common food-borne pathogen in the United States, but when it causes food-borne disease, the illness is one of two types, a more common self-limiting diarrhoea and an emetic (vomiting) illness. Of the two, the diarrhoeal syndrome is more commonly associated with meats that may include beef. The diarrhoeal syndrome is associated with the production of five different *B. cereus* toxins: haemolysin enterotoxin, non-haemolytic enterotoxin, cytotoxin K, enterotoxin T and enterotoxin FM (Agata et al., 1995; Asano et al., 1997; Beecher and MacMillan, 1991; Lund and Granum, 1996; Lund et al., 2000). Meat and beef have been shown to be an adequate environment for *B. cereus* spores to germinate and multiply if proper refrigeration is not maintained (Soares et al., 2012).

Rapid molecular detection of *B. cereus* can be performed by PCR or qPCR directed towards genes on the virulence plasmid that carries many of the toxin genes (Ehling-Schulz et al., 2006; Oliwa-Stasiak et al., 2010). The toxins of *B. cereus* can be detected using commercially available latex test kits from a wide range of foods and samples. Extraction of the toxin from the samples may require modifications for particular sample types. The main requirement is to achieve a non-turbid, fat-free extract. For optimum sensitivity, a low dilution factor is desirable. *B. cereus* toxin can also be recovered from culture filtrates. Classically, *B. cereus* is identified through culture isolation on *B. cereus* nutrient agar on which the colonies have a dull or frosted appearance. The organism is a mannitol non-fermentor, produces lecithinase and is resistant to polymixin. Therefore, common selective agars contain mannitol, egg yolk emulsion and polymixin for isolation of the pathogen. On the selective agars, colonies are typically pink in colour and surrounded by a precipitation zone.

6.3 Detecting *Clostridium difficile*

Clostridium difficile is an obligate-anaerobic, gram-positive, spore-forming bacillus that is ubiquitous in nature. The organism also has been identified as the aetiologic agent causing *C. difficile* infections (CDI), with manifestations ranging from mild diarrhoea to severe pseudomembranous colitis and even death (Sunenshine and McDonald, 2007). Severe CDI cases have been reported throughout North America and Europe (Loo et al., 2005; Warny et al., 2005). The increase in severe CDI is attributed to the emergence of a strain called North America pulsed-field gel electrophoresis type 1 (NAP1, also known as PCR ribotype 027), which is believed to be more virulent due to the elevated production of toxins A and B (TcdA, TcdB; Warny et al., 2005). In addition to these two toxins, this strain of *C. difficile* also produces a third toxin, the cytolethal distending toxin (CDT). Clostridial species are associated with spoilage of vacuum-packaged refrigerated fresh and roasted beef (Kalchayanand et al., 1989, 1993), and food-borne links to human outbreaks of CDI have not been confirmed (Harvey et al., 2011; Rodriguez-Palacios et al., 2007). Nevertheless, studies have identified *C. difficile* throughout the beef continuum from cattle through processing and in final retail beef products (Harvey et al., 2011; Rodriguez-Palacios et al., 2007; Kalchayanand et al., 2013b). Therefore, the detection of this potential pathogen is often examined to better elucidate the role of beef in its transmission. Detection from all sources of samples is performed by culture based on the isolation of spores through the process of generating an ethanol-treated homogenate (Rodriguez-Palacios et al., 2007; Kalchayanand et al., 1989) that is streaked on *C. difficile* selective medium, CDMN (Aspinall and Hutchinson, 1992), as well as enriched at 37°C for 10–15 days in CDMN broth containing 1 g/litre sodium taurocholate before streaking for isolation onto CDMN agar. The CDMN plates must be incubated under anaerobic conditions for up to 5 days at 37°C before identifying presumptive colonies.

Presumptive *C. difficile* strains are identified as colonies that are swarming, rough, non-haemolytic and fluorescent yellowish green under long-wavelength UV light (365 nm). Presumptive *C. difficile* strains produce a unique horse manure odour (Cato et al., 1986), because of *p*-cresol production on media containing *p*-hydroxyphenylacetic acid (Phillips and Roger, 1981) and lack lecithinase that can be identified by streaking presumptive colonies on CDMN supplemented with 80 ml of egg yolk emulsion per litre. Confirmation of *C. difficile* is completed using molecular means. PCR is used to identify the specific *tpi* gene (Lemee et al., 2004), as well as the *tcdA* and *tcdB* toxin genes (Rupnik, 1997) and the *cdtA* and *cdtB* toxin genes (Stubbs et al., 2000; Rupnik et al., 1997). The identification of the highly virulent NAP1 strain is accomplished through PCR ribotyping and comparison with reference strains (Bidet et al., 1999).

6.4 Methods for detecting MAP

MAP is the causative agent of Johne's disease in cattle (Valentin-Weigand and Goethe, 1999). Approximately 22% of dairy herds and 8% of beef herds in the United States are infected (Linnabary et al., 2001). MAP infection in humans may be linked to inflammatory bowel disease (Collins et al., 2000) and has been implicated as a possible causative agent of Crohn's disease, due to similar aetiologies with Johne's disease in ruminants (Mishina et al., 1996; Chiodini and Rossiter, 1996). Due to lack of direct evidence, most testing for this pathogen in cattle and beef is to clarify its potential role as a food-borne pathogen.

The presence of MAP can be determined through molecular screening of samples directly and after enrichment. Direct screening is performed by pelleting cells through centrifugation and extracting DNA from one-half of the pellet to amplify MAP-specific IS900 repeat genome sequences (Ellingson et al., 2003). The other half of the pellet is decontaminated by suspending in 0.7% hexadecylpyridinium chloride solution for 3 hours at room temperature. Following treatment with hexadecylpyridinium chloride, the pellet is recovered by centrifugation and suspended in an antibiotic cocktail containing 50 μg/ml amphotericin B, 100 μg/ml naladixic acid and 100 μg/ml vancomycin (ANV) and incubating at 37°C for 22–24 hours, followed by transferring a 200-μl aliquot onto Herrold's egg yolk agar slants with ANV and Mycobactin J (HEYA). The slants are then incubated at 37°C for 8–12 weeks until growth appears, at which time the surfaces of the slant can be scraped and the microbial growth transferred to sterile water for bacterial DNA extraction. Extracted DNA from MAP enrichments is subjected to PCR amplification using nested primer sets targeting MAP-specific IS-MAP02 repeat genome sequences (Stabel and Bannantine, 2005) that confirms the identity of the MAP present.

7 Developing standardised methods for pathogen detection: the case of the FSIS Microbiology Laboratory Guidebook method for testing for non-O157 STEC

E. coli O157:H7 was declared an adulterant in raw ground beef in 1994 by the USDA FSIS, following a large outbreak associated with undercooked hamburgers (Bell et al., 1994), and a raw ground beef testing programme was initiated. In 1999, *E. coli* O157:H7 was declared an adulterant in all non-intact beef products, and in 2004, FSIS began testing beef trim for *E. coli* O157:H7. Data from the Foodborne Diseases Active Surveillance Network (FoodNet) and from other sources indicate that six STEC serogroups (O26, O45, O103, O111, O121 and O145) cause the majority of non-O157 STEC infections, and illness caused by these STEC serogroups can be as severe as illness caused by *E. coli* O157:H7 (Gould et al., 2013). A report on outbreaks caused by non-O157 STEC from 1990 to 2010 showed that 46 outbreaks had been reported from 46 states; STEC O111 and O26 were the most common cause of single aetiology outbreaks and 17 outbreaks were food-borne (Luna-Gierke et al., 2014). Several of the outbreaks were linked to direct contact with cattle or consumption of unpasteurized milk, whereas single sporadic cases were linked to ground beef.

Because of the increasing importance of non-O157 STEC, in 2007, the FSIS, FDA and CDC hosted a public meeting to discuss the public health significance of these pathogens. Based on information regarding these pathogens and the possibility that they would be declared as adulterants in beef, in 2008, the USDA FSIS set up a collaboration with its sister agency, the USDA Agricultural Research Service to develop methods for detection of the top six STEC serogroups that could be employed for regulatory testing. In 2010, the first version of FSIS method for non-O157 STEC was released as MLG 5B.00. In 2011, there was a Federal Register Notice (76-FR58157) announcing that the top six non-O157 STEC would be declared as adulterants in raw, non-intact beef products (Anonymous, 2011). A Draft Risk Profile for Pathogenic Non-O157 Shiga toxin-producing *Escherichia coli* (non-O157 STEC) was published (http://citeseerx.ist.psu.edu/viewdoc/download;jsessionid=36FBCD85CCB6976E90F060BAF6C6E536?doi=10.1.1.231.581&rep=rep1&type=pdf) to clarify issues related to the importance of the top six STEC based on information in the scientific literature. In addition, a guidance document for test kit manufacturers and laboratories on how to evaluate the performance of new test kit methods was released (http://www.fsis.usda.gov/shared/PDF/Validation_Studies_Pathogen_Detection_Methods.pdf). In 2012, the USDA FSIS issued a Federal Register Notice (77 FR31975) announcing that a verification testing programme for the top six STEC would begin on 4 June 2012.

Fratamico et al. (2011) published a report describing a method that could be employed for testing for the top six non-O157 STEC, which basically consisted of testing enrichments by multiplex PCR assays targeting the *stx*1, *stx*2 and *eae* genes, and samples positive for both *stx* and *eae* were tested for the top six O-group-specific genes. If positive for one of the top six O-serogroups, the STEC strain was then isolated from the enrichment following IMS and plating onto selective agar. Colonies were then confirmed using the same multiplex PCR assays targeting *stx*1, *stx*2 and O-group and by biochemical testing. After the release of the first version of the MLG for non-O157 STEC, several modifications were made to the method, including the development of new primer sets for *stx*1/*stx*2 and *eae* that detected most of the subtypes of these genes, changes in the DNA extraction method, and use of a higher throughput PCR platform (Wasilenko et al., 2012).

As discussed above, Tillman et al. (2012) modified Rainbow Agar O157 by decreasing the recommended concentrations of 10 mg/L novobiocin and 0.8 mg/L potassium tellurite to 5 mg/L novobiocin, 0.15 mg/L potassium tellurite and adding 0.05 mg/L cefixime trihydrate to prevent growth inhibition of STEC strains belonging to the top six serogroups while preventing growth of some background organisms. In addition, an acid treatment step after IMS was incorporated to reduce background flora and increase the recovery of target STEC. To more easily identify colonies growing on mRBA or other selective agars, Medina et al. (2012) developed latex beads linked to antibodies produced against the top six STEC for latex agglutination assays. Also, to simplify performing the PCR screening assays, a commercial system for screening for the top six STEC that employs tableted PCR reagents was developed and incorporated into the MLG version 5B-04 after it was demonstrated that its performance was equivalent to that of the MLG method (Fratamico et al., 2014). Finally, use of the novobiocin was eliminated from the enrichment medium, and mTSB was adopted for simultaneous enrichment of *E. coli* O157:H7, the top six non-O157 STEC and *Salmonella* in the same beef sample.

8 Where to look for further information

There are several books and comprehensive review articles that cover pathogens of importance in beef and their detection technologies. Several of these are the following:

Advances in Microbial Food Safety. 2013. In J. Sofos (ed.), vol. 1, Woodhead Publishing in Food Science, Technology, and Nutrition, Cambridge, UK.

Advances in Microbial Food Safety. 2015. In J. Sofos (ed.), vol. 2, Woodhead Publishing in Food Science, Technology, and Nutrition, Cambridge, UK.

Dwivedi, H. P. and Jaykus, L.-A. 2011. Detection of pathogens in foods: the current state-of-the-art and future directions. *Crit. Rev. Microbiol.* 37:40–63.

Franz, E., Delaquis, P., Morabito, S., Beutin, L., Gobius, K., Rasko, D. A., Bono, J, French, N., Osek, J., Lindstedt, B. A., Muniesa, M., Manning, S., LeJeune, J., Callaway, T., Beatson, S., Eppinger, M., Dallman, T., Forbes, K. J., Aarts, H., Pearl, D. L., Gannon, V. P., Laing, C. R., and Strachan, N. J. 2014. Exploiting the explosion of information associated with whole genome sequencing to tackle Shiga toxin-producing *Escherichia coli* (STEC) in global food production systems. *Int. J. Food Microbiol.* 187:57–72.

Improving the Safety of Fresh Meat. 2005. In J. Sofos (ed.), Woodhead Publishing in Food Science, Technology, and Nutrition, Cambridge, UK.

Law, J. W.-F., Ab Mutalib, N.-S., Chan, K.-G. and Lee, L.-H. 2014. Rapid methods for the detection of foodborne bacterial pathogens: principles, applications, advantages and limitations. *Front. Microbiol.* 5:770. doi: 10.3389/fmicb.2014.00770.

Lee, K.-M., Runyon, M., Herrman, T. J., Phillips, R., and Hsieh, J. 2015. Review of *Salmonella* detection and identification methods: aspects of rapid emergency response and food safety. *Food Control.* 47:264–76.

Rapid detection, characterization, and enumeration of foodborne pathogens. 2011. J. Hoorfar (ed.). ASM Press, Washington, D.C.

Wang, Y. and Salazar, J. K. 2016. Culture-independent rapid detection methods for bacterial pathogens and toxins in food matrices. *Comp. Rev. Food Sci. Technol.* 15:183–205.

Wiedmann, M., Wang, S., Post, L., and Nightingale, K. 2014. Assessment of criteria and approaches for rapid detection methods to be used in the food industry. *J. Food Prot.* 77:670–90.

Websites:

USDA FSIS website: http://www.fsis.usda.gov/wps/portal/fsis/home

The USDA FSIS Microbiology Laboratory Guidebook: http://www.fsis.usda.gov/wps/portal/fsis/topics/science/laboratories-and-procedures/guidebooks-and-methods/microbiology-laboratory-guidebook/microbiology-laboratory-guidebook

U.S. Food and Drug Administration (FDA), Bad Bug Book (2nd edition): http://www.fda.gov/Food/FoodborneIllnessContaminants/CausesOfIllnessBadBugBook/

U.S. FDA, Microbiological Methods and Bacteriological Analytical Manual (BAM): http://www.fda.gov/Food/FoodScienceResearch/LaboratoryMethods/ucm114664.htm

Professional societies and organizations and journals:

The International Association for Food Protection (IAFP; https://www.foodprotection.org/) and the scientific journal of IAFP, The Journal of Food Protection.

The Institute of Food Technologists (IFT; https://www.ift.org/) and the scientific journal of IFT, The Journal of Food Science.

Beef Industry Food Safety Council (BIFSCo; https://www.bifsco.org/)

National Cattlemen's Beef Association (NCBA; https://www.beef.org/; http://www.beefusa.org/

9 References

Agata, N., Otha, M., Arakawa, Y. and Mori, M. 1995. The *bceT* gene of *Bacillus cereus* encodes an enterotoxic protein. *Microbiol. Read.* 41:983–8.

Anonymous. 2011. *Proposed Rules: Shiga Toxin-Producing Escherichia coli in Certain Raw Beef Products*, vol. 76. Federal Register, pp. 58157–65.

Arthur, T. M., Bono, J. L. and Kalchayanand, N. 2014. Characterization of *Escherichia coli* O157:H7 strains from contaminated raw beef trim during 'high event periods'. *Appl. Environ. Microbiol.* 80:506–14.

Arthur, T. M., Bosilevac, J. M, Brichta-Harhay, D. M., Kalchayanand, N., King, D. A., Shackelford, S. D., Wheeler, T. L. and Koohmaraie, M. 2008. Source tracking of *Escherichia coli* O157:H7 and *Salmonella* contamination in the lairage environment at commercial U.S. beef processing plants and identification of an effective intervention. *J. Food Prot.* 71:1752–60.

Arthur, T. M., Bosilevac, J. M., Nou, X. and Koohmaraie, M. 2005. Evaluation of culture- and PCR-based detection methods for *Escherichia coli* O157:H7 in inoculated ground beef. *J. Food Prot.* 68:1566–74.

Arthur, T. M., Bosilevac, J. M., Nou, X., Shackelford, S. D., Wheeler, T. L., Kent, M. P., Jaroni, D., Pauling, B., Allen, D. M. and Koohmaraie, M. 2004. *Escherichia coli* O157 prevalence and enumeration of aerobic bacteria, Enterobacteriaceae, and *Escherichia coli* O157 at various steps in commercial beef processing plants. *J. Food Prot.* 67:658–65.

Arthur, T. M., Brichta-Harhay, D. M., Bosilevac, J. M., Kalchayanand, N., Shackelford, S. D., Wheeler, T. L. and Koohmaraie, M. 2010. Super shedding of *Escherichia coli* O157:H7 by cattle and the impact on beef carcass contamination. *Meat Sci.* 86:32–7.

Arthur, T. M., Keen, J. E., Bosilevac, J. M., Brichta-Harhay, D. M., Kalchayanand, N., Shackelford, S. D., Wheeler, T. L., Nou, X. and Koohmaraie, M. 2009. Longitudinal study of *Escherichia coli* O157:H7 in a beef cattle feedlot and role of high-level shedders in hide contamination. *Appl. Environ. Microbiol.* 75:6515–23.

Asano, S. I., Nukumizu, Y., Bando, H., Iizuca, T. and Yamamoto, T. 1997. *Clostridium difficile* Cloning of novel enterotoxin genes from *Bacillus cereus* and *Bacillus thuringiensis*. *Appl. Environ. Microbiol.* 63:1054–7.

Aspinall, S. T. and Hutchinson, D. N. 1992. New selective medium for isolating from feces. *J. Clin. Pathol.* 45:812–14.

Bailey, J. S., Chiu, J. Y., Cox, N. A. and Johnston, R. W. 1988. Improved selective procedure for detection of salmonellae from poultry and sausage products. *J. Food Prot.* 51:391–6.

Barkocy-Gallagher, G. A., Edwards, K. K., Nou, X., Bosilevac, J. M., Arthur, T. M., Shackelford, S. D. and Koohmaraie, M. 2005. Methods for recovering *Escherichia coli* O157:H7 from cattle fecal, hide, and carcass samples: sensitivity and improvements. *J. Food Prot.* 68:2264–8.

Beecher, D. J. and MacMillan, J. D. 1991. Characterization of the components of hemolysin BL from *Bacillus cereus*. *Infect. Immun.* 59:1778–84.

Bell, B. P., Goldoft, M., Griffin, P. M., Davis, M. A., Gordon, D. C., Tarr, P. I., Bartleson, C. A., Lewis, J. H., Barrett, T. J., Wells, J. G., Baron, R. and Kobayashi, J. 1994. A multistate outbreak of *Escherichia coli* O157:H7-associated bloody diarrhea and hemolytic uremic syndrome from hamburgers. The Washington Experience. *JAMA.* 272:1349–53.

Besser, T. E., LeJeune, J. T., Rice, D. H., Berg, J., Stilborn, R. P., Kaya, K., Bae, W. and Hancock, D. D. 2005. Increasing Prevalence of *Campylobacter jejuni* in feedlot cattle through the feeding period. *Appl. Environ. Microbiol.* 71:5752–8.

Bidet, P., Barbut, F., Lalande, V., Burghoffer, B. and Petit, J. 1999. Development of a new PCR-ribotyping method for *Clostridium difficile* based on ribosomal RNA gene sequencing. *FEMS Microbiol. Lett.* 175:261–6.

Boerlin, P., Chen, S., Colbourne, J. K., Johnson, R., De Grandis, S. and Gyles, C. 1998. Evolution of enterohemorrhagic *Escherichia coli* hemolysin plasmids and the locus for enterocyte effacement in Shiga toxin-producing *E. coli*. *Infect. Immun.* 66:2553–61.

Bosilevac, J. M. 2013. When STEC are our target, Where do you aim? A review of the current state of STEC testing. *Food Safety Mag.* April–May:30–35.

Bosilevac, J. M., Arthur, T. M., Bono JL, Brichta-Harhay, .DM., Kalchayanand, N., King, D. A., Shackelford, S. D., Wheeler, T. L. and Koohmaraie, M. 2009. Prevalence and enumeration of *Escherichia coli* O157:H7 and *Salmonella* in U.S. abattoirs that process fewer than 1000 head of cattle per day. *J. Food Prot.* 72:1272–8.

Bosilevac, J. M., Kalchayanand, N., Schmidt, J. W., Shackelford, S. D., Wheeler, T. L. and Koohmaraie, M. 2010. Inoculation of beef with low concentrations of *Escherichia coli* O157:H7 and examination of factors that interfere with its detection by culture isolation and rapid methods. *J. Food Prot.* 73:2180–8.

Bosilevac, J. M. and Koohmaraie, M. 2011. Prevalence and characterization of non-O157 Shiga toxin-producing *Escherichia coli* isolates from commercial ground beef in the United States. *Appl. Environ. Microbiol.* 77:2103–12.

Bosilevac, J. M. and Koohmaraie, M. 2012. Predicting the presence of non-O157 Shiga toxin-producing *Escherichia coli* in ground beef using molecular tests for Shiga toxins, intimin, and O serogroups. *Appl. Environ. Microbiol.* 78:7152–5.

Brichta-Harhay, D. M., Arthur, T. M., Bosilevac, J. M., Kalchayanand, N., Shackelford, S. D., Wheeler, T. L., and Koohmaraie, M. 2011. Diversity of multidrug-resistant salmonella enterica strains associated with cattle at harvest in the United States. *Appl. Environ. Microbiol.* 77:1783-96.

Brichta-Harhay, D. M., Guerini, M. N., Arthur, T. M., Bosilevac, J. M., Kalchayanand, N., Shackelford, S. D., Wheeler, T. L. and Koohmaraie, M. 2008. *Salmonella* and *E. coli* O157:H7 contamination on hides and carcasses of cull cattle presented for slaughter in the United States – an evaluation of prevalence and load using immunomagnetic separation and direct plating methods. *Appl. Environ. Microbiol.* 74:6289–97.

Cato, E. P., George, W. L. and Finegold, S. M. 1986. Genus *Clostridium*. In P. H. A. Sneath, N. S. Mair and M. E. Sharpe (eds), *Bergey's Manual of Systematic Bacteriology*, vol. 2, pp. 1165–6. Williams and Wilkins, Baltimore.

Cernicchiaro, N., Cull, C. A., Paddock, Z. D., Shi, X., Bai, J., Nagaraja, T. G. and Renter, D. G. 2013. Prevalence of Shiga toxin-producing *Escherichia coli* and associated virulence genes in feces of commercial feedlot cattle. *Foodborne Pathog. Dis.* 10:835–41.

Chapman, P. A. and Ashton, R. 2003. An evaluation of rapid methods for detecting *Escherichia coli* O157 on beef carcasses. *Int. J. Food Microbiol.* 87:279–85.

Cheung, P.-Y. and Kam, K. M. 2012. *Salmonella* in food surveillance: PCR, immunoassays, and other rapid detection and quantification methods. *Food Res. Int.* 45:802–8.

Chiodini, R. J. and Rossiter, C. A. 1996. Paratuberculosis: a potential zoonosis? *Vet. Clin. North Am. Food Anim. Pract.* 12:457–67.

Chiu, C. H. and Ou, J. T. 1996. Rapid identification of *Salmonella* serovars in feces by specific detection of virulence genes, *invA* and *spvC*, by an enrichment broth culture-multiplex PCR combination assay. *J. Clin. Microbiol.* 34:2619–22.

Collins, M. T., Lisby, G., Moser, C., Chicks, D., Christensen, S., Reichelderfer, M., Hoiby, N., Harms, B. A., Thomsen, O. O., Skibsted, U. and Binder, B. 2000. Results of multiple diagnostic tests for *Mycobacterium avium* subsp. *paratuberculosis* in patients with inflammatory bowel disease and in controls. *J. Clin. Microbiol.* 38:4373–81.

Delannoy, S., Beutin, L. and Fach, P. 2012. Use of clustered regularly interspaced short palindromic repeat sequence polymorphism for specific detection of enterohemorrhagic *Escherichia coli* strains of serotypes O26:H11, O45:H2, O103:H2, O111:H8, O121:H19, O145:H28, and O157:H7 by real-time PCR. *J. Clin. Microbiol.* 50:4035–40.

Delannoy, S., Beutin, L. and Fach, P. 2013a. Discrimination of enterohemorrhagic *Escherichia coli* (EHEC) from non-EHEC strains based on detection of various combinations of type III effector genes. *J. Clin. Microbiol.* 51:3257–62.

Delannoy, S., Beutin, L. and Fach, P. 2013b. Towards a molecular definition of enterohemorrhagic *Escherichia coli* (EHEC): detection of genes located on O island 57 as markers to distinguish EHEC from closely related enteropathogenic *E. coli* strains. *J. Clin. Microbiol.* 51:1083–88.

Delannoy, S., Beutin, L., and Fach, P. 2016a. Improved traceability of Shiga toxin-producing *Escherichia coli* using CRISPRs for detection and typing. *Environ. Sci. Pollut. Res. Int.* 23:8163–74.

Delannoy, S., Chaves, B. D., Ison, S. A., Webb, H. E., Beutin, L., Delaval, J., Billet, I. and Fach, P. 2016b. Revisiting the STEC testing approach: using *espK* and *espV* to make enterohemorrhagic *Escherichia coli* more reliable in beef. *Front. Microbiol.* 7:1. Doi: 10.3389/fmicb.2016.00001.

Deisingh, A. K. and Thompson, M. 2004. Strategies for the detection of Escherichia coli O157:H7 in foods. *J. Appl. Microbiol.* 96:419–29.

Doumith, M., Buchrieser, C., Glaser, P., Jacquet, C. and Martin, P. 2004. Differentiation of the major *Listeria monocytogenes* serovars by multiplex PCR. *J. Clin. Microbiol.* 42:3819–22.

Ehling-Schulz, M., Fricker, M., Grallert, H., Rieck, P., Wagner, M. and Scherer, S. 2006. Cereulide synthetase gene cluster from emetic *Bacillus cereus:* structure and location on a mega virulence plasmid related to Bacillus anthracis toxin plasmid pXO1. *BMC Microbiol.* 6:20.

Ellingson, J. L., Cheville, J. C., Brees, D., Miller, J. M. and Cheville, N. F. 2003. Absence of *Mycobacterium avium* subspecies *paratuberculosis* components from Crohn's disease intestinal biopsy tissues. *Clin. Med. Res.* 1:217–26.

Ennis, C., McDowell, D. and Bolton, D. J. 2012. The prevalence, distribution and characterization of Shiga toxin-producing *Escherichia coli* (STEC) serotypes and virulotypes from a cluster of bovine farms. *J. Appl. Microbiol.* 113:1238–48.

Fraser, J. and Sperber, W. 1988. Rapid detection of *Listeria* in food and environmental samples by esculin hydrolysis. *J. Food Prot.* 51:762–5.

Fratamico, P. M., Bagi, L. K., Cray, W. C. Jr., Narang, N., Yan, X., Medina, M. and Liu, Y. 2011. Detection by multiplex real-time polymerase chain reaction assays and isolation of Shiga toxin-producing *Escherichia coli* serogroups O26, O45, O103, O111, O121, and O145 in ground beef. *Foodborne Pathog. Dis.* 8:601–7.

Fratamico, P. M., Wasilenko, J. L., Garman, B., DeMarco, D. R., Varkey, S., Jensen, M., Rhoden, K. and Tice, G. 2014. Evaluation of a multiplex real-time PCR method for detecting Shiga toxin-producing *Escherichia coli* in beef and comparison to the FSIS Microbiology Laboratory Guidebook method. *J. Food Prot.* 77:180–8.

Gasanov, U., Hughes, D. and Hansbro, P. M. 2005. Methods for the isolation and identification of *Listeria* spp. and *Listeria monocytogenes*: a review. *FEMS Microbiol. Rev.* 29:851–75.

Gould, L. H., Mody, R. K., Ong, K. L., Clogher, P., Cronquist, A. B., Garman, K. N., Lathrop, S., Medus, C., Spina, N. L., Webb, T. H., White, P. L., Wynmore, K., Gierke, R. E., Mahon, B. E., Griffin, P. M., for the Emerging Infectious Diseases Program FoodNet Working Group. 2013. Increased recognition of non-O157 Shiga toxin-producing Escherichia coli infections in the United States during 2000–2010: epidemiologic features and comparison with *E. coli* O157 infections. *Foodborne Pathog. Dis.* 10:453–60.

Guerini, M. N., Bosilevac, J. M. and Koohmaraie, M. 2007. Rapid enrichment strategy for isolation of *Listeria* from bovine hide, carcass, and meat samples. *J. Food Prot.* 70:53–7.

Harvey, R. B., Norman, K. N., Andrews, K., Norby, B., Hume, M. E., Scanlan, C. M., Hardin, M. D. and Scott, H. M. 2011. *Clostridium difficile* in retail meat and processing plants in Texas. *J. Vet. Diagn. Invest.* 23:807–11.

Hirvonen, J. J., Siitonen, A. and Kaukoranta, S. S. 2012. Usability and performance of CHROMagar STEC medium in detection of Shiga toxin-producing *Escherichia coli* strains. *J. Clin. Microbiol.* 50:3586–90.

Ibekwe, A. M. and Grieve, C. M. 2003. Detection and quantification of *Escherichia coli* O157:H7 in environmental samples by real-time PCR. *J. Appl. Microbiol.* 94:421–31.

Johnson, R. P., Holtslander, B., Mazzocco, A., Roche, S., Thomas, J. L., Pollari, F. and Pintar, K. D. 2014. Detection and prevalence of verotoxin-producing *Escherichia coli* O157 and non-O157 serotypes in a Canadian watershed. *Appl. Environ. Microbiol.* 80:2166–75.

Kalchayanand, N., Arthur, T. M,, Bosilevac, J. M,, Brichta-Harhay, D. M., Shackelford, S. D., Wells, J. E., Wheeler, T. L. and Koohmaraie, M. 2013b. Isolation and characterization of *Clostridium difficile* associated with beef cattle and commercially produced ground beef. *J. Food Prot.* 76:256–64.

Kalchayanand, N., Ray, B., Field, R. A. and Johnson, M. C. 1989. Spoilage of vacuum-packaged refrigerated beef by *Clostridium*. *J. Food Prot.* 52:424–6.

Kalchayanand, N., Ray, B. and Field, R. A. 1993. Characteristics of psychrotrophic *Clostridium laramie* causing spoilage of vacuum- packaged refrigerated fresh and roasted beef. *J. Food Prot.* 56:13–17.

Koohmaraie, M., Arthur, T. M., Bosilevac, J. M., Brichta-Harhay, D. M., Kalchayanand, N., Shackelford, S. D. and Wheeler, T. L. 2007. Interventions to reduce/eliminate *Escherichia coli* O157:H7 in ground beef. *Meat Sci.* 77:90–6.

Koohmaraie, M., Arthur, T. M., Bosilevac, J. M., Guerini, M. N., Shackelford, S. D. and Wheeler, T. L. 2005. Post-harvest interventions to reduce/eliminate pathogens in beef. *Meat Sci.* 71:79–91.

Law, J. W., Ab Mutalib, N. S., Chan, K. G. and Lee, L. H. 2014. Rapid methods for the detection of foodborne bacterial pathogens: principles, applications, advantages and limitations. *Front. Microbiol.* 5:770. doi: 10.3389/fmicb.2014.00770.

Lemee, L., Dhalluin, A., Testelin, S., Mattrat, M. A., Maillard, K., Lemeland, J. F. and Pons, J. L. 2004. Multiplex PCR targeting *tpi* (triose- phosphate isomerase), *tcdA* (toxin A), and *tcdB* (toxin B) genes for toxigenic culture of *Clostridium difficile*. *J. Clin. Microbiol.* 42: 5710–14.

Li, J., Sayeed, S. and McClane, B. A. 2007. Prevalence of enterotoxigenic *Clostridium perfringens* isolates in Pittsburgh (Pennsylvania) area soils and home kitchens. *Appl. Environ. Microbiol.* 73:7218–24.

Linnabary, R. D., Collins, M. T., Sweeney, R. W., Wells, S. J., Meerdink, G. L., Stabel, J. R. and Washington, M. K. 2001. Johne's disease in cattle. Council for Agricultural Science and Technology Issue Paper. 17 May 2001. http://www.cast-science.org/publications/?johnes_disease_in_cattle&show=product&productID=2866. Accessed 6 July 2016.

Livezey, K. W., Groschel, B. and Becker, M. M. 2015. Use of the *ecf1* gene to detect Shiga toxin-producing *Escherichia coli* in beef samples. *J. Food Prot.* 78:675–84.

Livezey, K., Kaplan, S., Wisniewski, M. and Becker, M. M. 2013. A new generation of food-borne pathogen detection based on ribosomal RNA. *Annu. Rev. Food Sci. Technol.* 4:313–25.

Loo, V. G., Poirier, L., Miller, M. A., Oughton, M., Libman, M. D., Michaud, S., Bourgault, A. M., Nguyen, T., Frenette, C., Kelly, M., Vibien, A., Brassard, P., Fenn, S., Dewar, K., Hudson, T. J., Horn, R., Rene, P., Monczak, Y. and Dascal, A. 2005. A predominantly clonal multi-institutional outbreak of *Clostridium difficile*–associated diarrhea with high morbidity and mortality. *N. Engl. J. Med.* 353:2442–9.

Luedtke, B. E., Bono, J. L. and Bosilevac, J. M. 2014. Evaluation of real time PCR assays for the detection and enumeration of enterohemorrhagic *Escherichia coli* directly from cattle feces. *J. Microbiol. Methods.* 105:72–9.

Luna-Gierke, R. E., Griffin, P. M., Gould, L. H., Herman, K., Bopp, C. A., Strockbine, N. and Mody, R. K. 2014. Outbreaks of non-O157 Shiga toxin-producing *Escherichia coli* infection: USA. *Epidemiol. Infect.* 142:2270–80.

Lund, T., De Buyser, M. L. and Granum, P. E. 2000. A new cytotoxin from *Bacillus cereus* that may cause necrotic enteritis. *Mol. Microbiol.* 38:254–61.

Lund, T. and Granum, P. E. 1996. Characterisation of a non-haemolytic enterotoxin complex from *Bacillus cereus* isolated after a foodborne outbreak. *FEMS Microbiol. Lett.* 141:151–6.

Mathusa, E. C., Chen, Y., Enache, E. and Hontz, L. 2010. Non-O157 Shiga toxin-producing *Escherichia coli* in foods. *J. Food Prot.* 73:1721–36.

Medina, M., Shelver, W., Fratamico, P. M., Fortis, L., Narang, N., Cray Jr., W. C., Esteban, E. and DebRoy, C. 2012. Latex agglutination assays for detection and confirmation of non-O157 Shiga toxin-producing *E. coli* serogroups O26, O45, O103, O111, O121, O145. *J. Food Prot.* 75:819–26.

Michelacci, V., Orsini, M., Knijn, A. Delannoy, S., Fach, P., Caprioli, A. and Morabito, S. 2016. Development of a high resolution virulence allelic profiling (HReVAP) approach based on the accessory genome of *Escherichia coli* to characterize Shiga toxin-producing *E. coli* (STEC). *Front Microbiol.*, 23 February 2016, 7:202.

Miller, R. G., Tate, C. R. and Mallinson, E. T. 1994. Improved XLT4 agar: small addition of peptone to promote stronger production of hydrogen-sulfide by salmonellae. *J. Food Prot.* 57:854–8.

Miller, R. R., Montoya, V., Gardy, J. L., Patrick, D. M. and Tang, P. 2013. Metagenomics for pathogen detection in public health. *Genome Med.* 5:81. doi: 10.1186/gm485.

Mishina, D., Katsel, P., Brown, S. T., Gilberts, E. C. and Greenstein, R. J. 1996. On the etiology of Crohn disease. *Proc. Natl. Acad. Sci. USA* 93:9816–20.

Moxley, R. A. and Acuff, G. R. 2014. Peri- and postharvest factors in the control of Shiga toxin-producing *Escherichia coli* in beef. *Microbiol. Spectr.* 2(6).

Murphy, R. Y. and Seward, R. A. 2004. Process control and sampling for *Escherichia coli* O157:H7 in beef trimmings. *J. Food Prot.* 67:1755–9.

Noormohamed, A. and Fakhr, M. K. 2013. A higher prevalence rate of *Campylobacter* in retail beef livers compared to other beef and pork meat cuts. *Int. J. Environ. Res. Public Health.* 10:2058–68.

Norman, K. N., Strockbine, N. A. and Bono, J. L. 2012. Association of nucleotide polymorphisms within the O-antigen gene cluster of *Escherichia coli* O26, O45, O103, O111, O121, and O145 with serogroups and genetic subtypes. *Appl. Environ. Microbiol.* 78:6689–703.

Norton, D. M. 2002. Polymerase chain reaction-based methods for detection of *Listeria monocytogenes*: toward real-time screening for food and environmental samples. *J. AOAC Int.* 85:505–15.

Oliwa-Stasiak, K., Molnar, C. I., Arshak, K., Bartoszcze, M. and Adley, C. C. 2010. Development of a PCR assay for identification of the Bacillus cereus group species. *J. Appl. Microbiol.* 108:266–73.

Phillips, K. D. and Roger, P. A. 1981. Rapid detection and presumptive identification of *Clostridium difficile* by p-cresol production on selective medium. *J. Clin. Pathol.* 34:642–4.

Possé, B., De Zutter, L., Heyndrickx, M. and Herman, L. 2008. Novel differential and confirmation plating media for Shiga toxin-producing *Escherichia coli* serotypes O26, O103, O111, O145 and sorbitol-positive and -negative O157. *FEMS Microbiol. Lett.* 282:124–31.

Possé, B., De Zutter, L., Heyndrickx, M. and Herman, L. 2007. Metabolic and genetic profiling of clinical O157 and non-O157 Shiga toxin-producing *Escherichia coli*. *Res. Microbiol.* 158:591–9.

Presad, A. K. and Lejeune, J. T. 2014. Animal reservoirs of Shiga toxin-producing *Escherichia coli*. *Microbiol Spectr.* 2(4):EHEC-0027-2014. doi: 10.1128/microbiolspec.EHEC-0027-2014.

Rahn, K., De Grandis, S. A., Clarke, R. C., McEwen, S. A., Galan, J. E., Ginocchio, C., Curtiss, R. and Gyles, C. L. 1992. Amplification of an *invA* gene sequence of *Salmonella typhimurium* by polymerase chain reaction as a specific method of detection of *Salmonella*. *Mol. Cell. Probes* 6:271–9.

Rivas, L., Mellor, G. E., Gobius, K. and Fegan, N. 2015. *Detection and Typing Strategies for Pathogenic Escherichia coli*, SpringerBriefs in Food, Health, and Nutrition, Springer, New York.

Rodriguez-Palacios, A., Stämpfli, H. R., Duffield, T. and Weese, J. S. 2007. *Clostridium difficile* in retail ground meat, Canada. *Emerg. Infect. Dis.* 13:485–7.

Rupnik, M., Braun, V., Soehn, F., Janc, M., Hofstetter, M., Laufenberg-Feldmann, R. and von Eichel-Streiber, C. 1997. Characterization of polymorphisms in the toxins A and B genes of *Clostridium difficile*. *FEMS Microbiol. Lett.* 148:197–202.

Schlech, W. F., Lavigne, P. M., Bortolussi, R. A., Allen, A. C., Haldane, E. V., Wort, A. J., Hightower, A. W., Johnson, S. E., King, S. H., Nicholls, E. S. and Broome, C. V. 1983. Epidemic listeriosis – evidence for transmission by food. *N. Engl. J. Med.* 308:203–6.

Sharma, V. K. 2002. Detection and quantitation of enterohemorrhagic *Escherichia coli* O157, O111, and O26 in beef and bovine feces by real-time polymerase chain reaction. *J. Food Prot.* 65:1371–80.

Soares, C. M., Kabuki, D. Y. and Kuaye, A. Y. 2012. Growth of enterotoxin producing *Bacillus cereus* in meat substrate at 10C and 30C. *Braz. J. Microbiol.* 43:1401–5.

Stabel, J. R. and Bannantine, J. P. 2005. Development of a nested PCR method targeting a unique multicopy element, ISMAP02, for detection of *Mycobacterium avium* subsp. *paratuberculosis* in fecal samples. *J. Clin. Microbiol.* 43:4744–50.

Steele, M., McNab, B., Fruhner, L., DeGrandis, S., Woodward, D. and Odumeru, J. A. 1998. Epidemiological typing of *Campylobacter* isolates from meat processing plants by pulsed-field gel electrophoresis, fatty acid profile typing, serotyping, and biotyping. *Appl. Environ. Microbiol.* 64:2346–9.

Stevens, M. P.,Humphrey, T. J. and Maskell, D. J. 2009. Molecular insights into farm animal and zoonotic *Salmonella* infections. *Philos. Trans. R. Soc. Lond. B. Biol. Sci.* 364:2709–23.

Stromberg, Z. R., Baumann, N. W., Lewis, G. L., Sevart, N. J., Cernicchiaro, N., Renter, D. G., Marx, D., Phebus, R. K. and Moxley, R. A. 2015. Prevalence of enterohemorrhagic *Escherichia coli* O26, O45, O103, O111, O121, O145, and O157 on hides and preintervention carcass surfaces of feedlot cattle at harvest. *Foodborne Pathog. Dis.* 12:631–8.

Stromberg, Z. R., Lewis, G. L., Aly, S. S., Lehenbauer, T. W., Bosilevac, J. M., Cernicchiaro, N. and Moxley, R. A. 2016. Prevalence and level of enterohemorrhagic *Escherichia coli* in culled dairy cows at harvest. *J. Food Prot.* 79:421-31.

Stubbs, S., Rupnik, M., Gilbert, M., Brazier, J. S., Duerden, B. and Popolf, M. 2000. Production of actin-specific ADP-ribosyltranferase (binary toxin) by strains of *Clostridium difficile*. *FEMS Microbiol. Lett.* 286:307–12.

Sugiyama, K., Inoue, K. and Sakazaki, R. 2001. Mitomycin-supplemented washed blood agar for the isolation of Shiga toxin-producing *Escherichia coli* other than O157:H7. *Lett. Appl. Microbiol.* 33:193–5.

Sunenshine, R. H. and McDonald, L. C. 2007. *Clostridium difficile*– associated disease: new challenges from an established pathogen. *Cleveland Clin. J. Med.* 73:187–97.

Tang, Y., Kim, H., Singh, A. K., Aroonnual, A., Bae, E., Rajwa, B., Fratamico, P. M. and Bhunia, A. K. 2014. Light scattering sensor for direct identification of colonies of *Escherichia coli* serogroups O26, O45, O103, O111, O121, O145 and O157. *PLoS ONE* 9:e105272. doi: 10.1371/journal.pone.0105272.

Tillman, G. E., Wasilenko, J. L., Simmons, M., Lauze, T. A., Minnicozzi, J., Oakley, B., Narang, N., Fratamico, P. M. and Cray Jr., W. C. 2012. Isolation of Shiga toxin-producing *Escherichia coli* serogroups O26, O45, O103, O111, O121, and O145 from ground beef using modified Rainbow Agar and post-immunomagnetic separation acid treatment. *J. Food Prot.* 75:1548–54.

USDA FSIS. 2014. Isolation and Identification of *Salmonella* From Meat, Poultry, Pasteurized Egg, and Catfish Products and Carcass and Environmental Sponges. MLG chap 4.08.

USDA FSIS. 2013. Isolation and Identification of Listeria monocytogenes from Red Meat, Poultry and Egg Products, and Environmental Samples. MLG chap 8.09.

Valentin-Weigand, P. and Goethe, R. 1999. Pathogenesis of *Mycobacterium avium* subspecies *paratuberculosis* infections in ruminants: still more questions than answers. *Microbes Infect.* 1:1121–7.

Vassiliadis, P. 1983. The Rappaport-Vassiliadis (RV) enrichment medium for the isolation of *Salmonellas*: an overview. *J. Appl. Bacteriol.* 54:69–76.

Vimont, A., Vernozy-Rozand, C., Montet, M. P., Bavai, C., Fremaux, B. and Delignette-Muller, M. L. 2007. Growth of Shiga toxin-producing *Escherichia coli* (STEC) and bovine feces background microflora in various enrichment protocols. *Vet. Microbiol.* 123:274–81.

Warny, M., Pepin, J., Fang, A., Killgore, G.,Thompson, A., Brazier, J., Frost, E. and McDonald, L. C. 2005. Toxin production by an emerging strain of *Clostridium difficile* associated with outbreaks of severe disease in North America and Europe. *Lancet* 366:1079–84.

Wasilenko, J. L., Fratamico, P. M., Narang, N., Tillman, G. E., Ladely, S. and Cray Jr., W. C. 2012. Influence of primer sequences and DNA extraction method on detection of non-O157 Shiga toxin-producing *Escherichia coli* in ground beef by real-time PCR targeting the *eae*, *stx*, and serogroup-specific genes. *J. Food Prot.* 75:1939–50

Wheeler, S. R., Heard, P., Dufour, C., Thevenot-Sergentet, D., Loukiadis, E., Flowers, R. S. and McMahon, W. 2015. Detection of non-O157 Shiga toxin-producing *Escherichia coli* in 375 grams of beef trim enrichments across multiple commercial PCR detection platforms. *J. Food Prot.* 78:196–202.

Wheeler, T. L., Kalchayanand, N. and Bosilevac, J. M. 2014. Pre- and post-harvest interventions to reduce pathogen contamination in the U.S. beef industry. *Meat Sci.* 98:372–82.

Windham, W. R., Yoon, S. C., Ladely, S. R., Haley, J. A., Heitschmidt, J. W., Lawrence, K. C., Park, B., Narrang, N. and Cray, W. C. 2013. Detection by hyperspectral imaging of Shiga toxin-producing *Escherichia coli* serogroups O26, O45, O103, O111, O121, and O145 on Rainbow agar. *J. Food Prot.* 76:1129–36.

Chapter 4

Food safety management on farms producing beef

Peter Paulsen, Frans J. M. Smulders and Friederike Hilbert, University of Veterinary Medicine, Austria

1 Introduction

Experience has shown that the production of safe and wholesome food requires measures along the whole chain – from 'farm to fork' or 'from gate to plate'. The necessary measures and structures can be broadly separated into a set of 'prerequisites', often termed 'Good Farming Practice', and measures targeted against specific hazards to human or animal health ('interventions').

This chapter will describe the key elements of 'Good Farming Practice' comprising (1) a traditional 'hygiene' approach to control potential routes of contamination and carry-over, (2) the prerequisites needed to ensure that exposure of animals to hazards is minimized and (3) practices to ensure that animals are 'healthy' and not exposed to unnecessary stress and strain in the production cycle. This will include also management of the farm–environment interface (e.g. emissions from farms) and animal welfare issues.

Hazard-specific control measures in the sense of 'interventions' will be presented for only one selected biological hazard; details on all relevant biological and chemical hazards are discussed elsewhere in this volume.

http://dx.doi.org/10.19103/AS.2016.0008.07

2 Good farming practices and biosecurity for beef cattle farms

There is no universal definition of 'Good Farming Practices' (GFP), but it is obviously one facet of general 'Good Agricultural Practice' (GAP). According to FAO (2003), GAP 'applies available knowledge to addressing environmental, economic and social sustainability for on-farm production and post-production processes resulting in safe and healthy food and non-food agricultural products'. This means that in practice, legislation and standards addressing GFP often will address not only food safety, food security and animal welfare, but also environmental protection, rural development, common agricultural market and trade issues. The European Union gives a good example how complex the ramifications in legislation on GFP can be in large economic and political associations of nations (Bergschmidt et al., 2003).

The World Organisation of Animal Health (OIE) has developed a guide on GFP, with a focus on animal health (OIE Animal Production Food Safety Working Group, 2006). The guide addressed infectious and parasitic agents ('biological hazards'), chemical contaminants and residues ('chemical hazards') and foreign objects ('physical hazards') as well. The tools, namely codes of hygienic practice and Hazard Analysis and Critical Control Points (HACCP) or HACCP-based procedures, should be applied in order to implement preventive actions in eight areas of primary production (Table 1). The guide was not intended for direct use by farmers, but rather as an instrument for the competent authorities to assist farmers to assume their responsibility in the food chain, for example, by codes of practice. This guide from 2006 evolved into a joint FAO-OIE recommendation of 2009, which is targeted at biological, chemical and physical hazards that can enter the food-producing animal or products of animal origin. They addressed this issue in six chapters, namely (1) General farm management, (2) Animal health management, (3) Veterinary medicines and biologicals, (4) Animal feeding and watering, (5) Environment and infrastructure and, (6) Animal and product handling. As regards hazards and corresponding control points, the guide does not elaborate on specific hazardous agents, but more on events/situations likely to occur on farms, for example, under 'biohazards': 'livestock not well adapted to conditions' or under 'physical hazards': 'ingestion of dangerous/harmful objects'; 'broken needles or other penetrating objects'. For the listed hazards (or, more specifically, hazardous situations), the 'points' where the hazard occurs and can be managed, and finally, a set of recommended actions are given. For implementing the recommended actions or measures in specific production systems, it must be evaluated if the measures predefined in the guide are applicable either 'as is' or with some rewording. Such 'validated' measures are regarded as of 'critical priority' or as 'highly advisable', respectively. This implementation procedure should allow consistent application of GFP within countries and also facilitate trade between countries. The guide should be considered in conjunction with others from the FAO on food, feed, veterinary drugs and animal handling and slaughter. In practice, a number of GFP-related practices are already part of standing legislation or are included in codes of marketing organizations, but it is above the scope of this chapter to address all national legislation and codes or guides on this matter.

There are several textbooks providing Good Hygiene Practice (GHP) elements for specific pre-harvest settings. As regards cleaning and disinfection and treatment of animal waste, the reader may refer to Motarjemi and Lelieveld (2014).

Table 1 Summary of the OIE Guide to Good Farming Practice (OIE, 2006), key areas and measures to be implemented. Implementation will have to consider national legislation, and some actions will require involvement veterinarians and/or official authorities

Area	Type of hazard	
Buildings and other facilities	b c p	Control of surroundings: • farming activities distant from industry likely to be a source of pollution • farm buildings separated from residential accommodation, waste storage • restrict access for visitors • farms should be separated from each other Control of environment on the farm: • livestock buildings should be adequate in size and well ventilated; surfaces must not release toxic substances • separation of clean from soiled, of storage and working areas from animal production areas • newly arrived animals and sick animals separated from the others • facilities easy to clean and disinfect • collection of waste and effluents • barriers against pests and wild or stray animals • restricted access for non-authorized persons • control measures should still work in case of occurrence of natural disasters
Health conditions for introduction of animals into the farm	b	• accept only animals on known health status (serology, vaccination etc.), and keep them separated for a period of time so as to allow acclimatization, to observe health status and carry out medical measures, if necessary (quarantine) • refuse animals showing suspicious clinical signs on delivery • animals should originate from farms working under GFP • animals must be clearly identified (tag, health document) • 'history' of animals must be documented (origin, transport etc.)
Animal feeding	b c p	Pasture and grassland • Ensure that land is suitable for pasture, esp. as regards toxic plants, residues of pesticides, pollutants from industry, parasites (e.g. liver fluke) or bacteria/bacterial spores (e.g. *Bacillus anthracis*) Commercial feed • Feed must be correctly labelled • Visual examination for signs of spoilage, esp. mould growth • Correct storage (e.g. protected from pests) • Traceability: feed batch numbers must be recorded and a feed sample be kept in case feed-related problems are suspected and tests are necessary Manufacture of animal feed on the farm • requirements listed above for 'commercial feed' apply to finished feed and raw materials, procedures and devices used for mixing and storage • composition (nutrients) should be checked regularly

(*Continued*)

Table 1 (*Continued*)

Area	Type of hazard	
		General issues • feed must be adequate for the animal species • troughs must not be overfilled and unused feed must be removed. Troughs and feeders must be cleaned regularly
Animal watering	b c	• domestic or wild/stray animals must not contaminate safe water reserves or watering points • livestock must not have access to polluted water • water reserves must be protected from pollution (e.g. by pesticides, herbicides, slurry, manure) • water distribution systems must be maintained and cleaned regularly • water quality should be tested regularly (microbiological and physico-chemical characteristics)
Veterinary drugs	c (p)	Drugs must be used on the basis of a sound diagnosis, ensuring that • dosage, mode of administration are correct • withdrawal times, if applicable, are respected • appropriate records are kept
Farm management		Training, conduct and health status of staff • workers must receive training to be able to do their job • training in biosecurity principles and practices • suitable working clothes and sanitary measures • monitoring to detect healthy carriers of biohazards transmissible to animals Maintenance, cleaning and disinfection of equipment, premises and immediate surroundings • cleaning and disinfection procedures based on manufacturer's instructions • tests to ascertain if procedures are effective Measures to control pests and stray animals and to prevent unauthorized access • pest control plan • no access of stray animals or pets to livestock buildings • no access for unauthorized persons • minimize contact of livestock with wild animals Stock management for feed and drugs Management of waste materials, effluents and expired products Storage of chemical products Production monitoring of animals • animals on the farm are permanently identified and records kept up-to-date • daily surveillance for health condition of production animals • production performance monitoring to detect abnormalities which could indicate the onset of disease

		Health monitoring of animals and disease prevention programmes • development and implementation of disease control programmes • standardized operational procedures for detection, management and treatment of diseases • take advantage of information gathered during ante- and post-mortem inspection at the abattoir • sampling in the course of official surveillance programmes Animal movements Isolation of sick animals and their products • sick animals should be separated from healthy ones • hygiene regulations regarding contacts between humans and sick animals or animals undergoing treatment • products from sick animals not used for food or animal feed Storage and disposal of dead animals • disposal in a way that pathogens cannot spread
Preparation of animals for slaughter	b	General measures • animals must be fit for slaughter • animals must be correctly identified (tagged) • loading, unloading areas and transport devices should be kept clean to avoid soiling of animals • animals should be handled humanely and not be stressed • in the days prior to slaughter, the content of hay or straw should be raised • feed be withdrawn 24h before transport • water access should be given until transport Extensively grazed livestock • should be kept protected from mud at the end of the fattening phase (e.g. new pasture, watering points well distributed, or in a fenced shedded area) Livestock housed on slatted flooring • stocking density should be correct • flooring be cleaned regularly • cattle kept on straw bedding the weeks before slaughter Livestock housed on litter • stocking density should be correct • clean litter must be supplied • ventilation and drainage of effluents must be effective Health measures • sick animals or medicated animals in the withdrawal period must not be sent to slaughter
Common measures for record keeping and traceability		Traceability system for animals, feed and animal products Record keeping

b, c, p refer to biological, chemical and physical hazards, respectively

Table 2 Principles to implement biosecurity for grazing beef cattle (Animal Health Australia, 2012)

Principle	What to achieve	Recommended practices (i.e. 'how to achieve') (selected)
Livestock	Manage the introduction and movement of livestock in a way that minimizes the risk of introducing or spreading infectious disease	• Check animals for health status before purchasing • Purchase animals from suppliers with a quality assurance system • Segregate and observe newly introduced animals
People, equipment and vehicles	People, equipment and vehicles entering the property are controlled to minimize the potential for property contamination	• Define permitted access areas for farm contractors • Farm contractors and visitors to use protective clothing and clean hands and boots when entering/leaving the permitted access areas • Keep a visitor register
Feed and water	Quality of stockfeed and water is fit for purpose	• Water management to reduce risks of transmitting disease agents or weeds • Ruminants have no access to restricted animal material (i.e. material from vertebrates except gelatin, tallow, milk products, oils)
Feral animals/ wildlife/weed control	Minimize the potential for wildlife and domestic or feral animals to introduce diseases to livestock	• Actions to control pests and wildlife should be coordinated with neighbouring farms • Minimize access by feral animals, domestic animals and wildlife to waste
Animal health management	Regular monitoring of livestock health to prevent control diseases	• Staff on the farm should be vaccinated against risk diseases (e.g. tetanus), • Inspection of the health of the livestock should be regular and health history of the herd be recorded
Staff instruction	All staff must understand the importance of biosecurity and implement the agreed practices	Staff knows • its role in implementation of biosecurity and knows how to recognize and handle sick and injured animals • how to contact competent authorities in case of disease outbreaks
Carcass, effluent and waste management	Disposal of dead animals and waste is managed to minimize the spread of disease	• Disposal areas not accessible for wild or domestic animals • No spread of contaminants by water

In practice, some aspects of GHP are dealt with in separate guides. One major aspect is biosecurity, which comprises preventive measures to reduce the risk of transmission of infectious diseases in agricultural products and livestock (Koblentz, 2010). In this sense, Laurence presented a comprehensive review on biosecurity measures on beef cattle farms in Australia (Laurence, 2014). The two pillars, bioexclusion and biocontainment, refer to the measures to keep pathogens and potential carriers of pathogens out and to the measures to prevent the spread of pathogens within a farm or to neighbouring farms. Similar to OIE, the author presents a list of requirements and the measures how to achieve this. For example, the requirement 'control visitors' can be achieved by the following ways: 'provide visitors with clean boots and clothes or coveralls'; 'install footbaths', 'Do not let visitors step into feed troughs or wash in water troughs'; 'limit contact by visitors with the herd'. Likewise, 'minimize vehicle traffic' can be achieved by 'keeping a separate vehicle for use on the farm' or 'prevent off-farm vehicles from driving in areas where animals travel'. Principally, the same messages are conferred to farmers in the 'National Farm Biosecurity Reference Manual – Grazing Livestock Production' (Animal Health Australia, 2012), see Table 2.

Given the differences in farming systems and farm sizes throughout the world, it is obvious that GHP at national and regional level will need some adaptions in terms of cost–benefit analyses. Perhaps the most important issue is how to communicate GFP principles to all involved parties. Expectedly, the presentation of GFP measures to the competent authority (e.g. the FAO-OIE guide from 2009) will be more comprehensive than that to specialized farmers (e.g. the grazing livestock manual, Animal Health Australia, 2012), whereas to the less knowledgeable rural population in developing countries, only the essential messages will be delivered, preferably relying on a mix of illustrations and text rather than on text only (e.g. the 'guide of best farming practices' for Haiti; OXFAM, 2014). Didactic approaches, including interaction of trainees and trainers, should be considered (see the 'Pre-Harvest Food Safety in 4-H Animal Science' training programme of the University of California, Smith et al., 2014).

3 Animal handling and animal welfare

Perhaps the most crucial facet of GFP is that of animal welfare and animal rights. Animal welfare principles received wide interest in Western societies ca. 100 years ago, and the concept of 'animal rights' was founded only ca. 40 years ago. The basis for animal rights is that there is no justification that animals with the ability to suffer should not deserve the same consideration as humans (Singer, 1975). Basic rights for sentient animals were formulated in the 1960s, formalized by the UK Farm Animal Welfare Council in 1979 and further refined to the commonly known 'five freedoms', namely, the freedom from hunger and thirst; freedom from discomfort (e.g. by providing shelter and protection); freedom from pain, injury and disease; freedom to express normal behaviour and freedom from fear and distress. The main shift from traditional animal welfare to the five freedoms was to consider not only 'freedom from ...' but also 'freedom to ...'. These requirements are – reworded and amended – incorporated in guides of various intergovernmental bodies like the OIE (2015). The European Food Safety Authority (EFSA) has addressed animal welfare in several scientific opinions based on the five freedoms approach and using risk assessment approaches for various commodities and farm animal species (see 'further reading'). Apart from these scientific opinions, the textbook of Smulders and Algers (2009) gives a comprehensive picture of the innumerable facets of assessment of animal welfare risks and assurance of animal welfare. Since stressed or injured animals will be more susceptible to infections, or may yield meat with unfavourable characteristics (e.g. dark-firm-dry beef, DFD), it is common sense to handle animals in a way that minimizes stress and trauma. This includes, for instance, avoiding sudden and jerking movements near the head of the cattle, avoiding sudden noise and respecting the flight zone. There are a number of literature and national guides on this subject [e.g. the 'Cattle care & handling guidelines' or the 'Master cattle transporter guide' of the US Beef Quality Assurance program (BQA, 2015, undated)], but the presumably most impressive source of practical information is the website of Dr. Temple Grandin (www.grandin.com).

4 Clean cattle policy

GFP requires that cattle should be clean and not be delivered soiled, muddy and wet to slaughter. Bacterial numbers are higher on visually 'dirty' than on 'clean' hides (Hauge et al., 2015).

It could be expected that higher loads of (enteric) bacteria on cattle hides will result in higher numbers of bacteria on beef carcass surfaces, and a number of national programmes are targeted on preventing contamination of hides and grading of cattle sent to slaughter according to the degree of visual contamination. Based on this assessment, additional measures are taken (washing animals, clipping hairs, separate slaughter). In practice,

Table 3 What to consider when implementing 'clean cattle policy' pre-harvest (SAC, 1997)

Diet	Low moisture diet, esp. in the finishing phase, correct composition of diet, no abrupt changes in diet
Housing	Slatted flooring: correct stocking density, no build-up of dung near troughs, along walls, underneath divisions Straw-bedded pens: sufficient amount of dry, clean bedding (straw) Ventilation: to ensure that bedding and cattle are dry. It is suggested to clip the backs of finishing cattle to prevent sweating
Clipping	Clipping of belly, back and tail; handling facilities required for belly clipping
Washing	No power hoses (high pressure)
Transport	No water restriction prior to transport; cattle must be dry; correct stocking density; no mixing of animals from different holdings; good ventilation; no contamination from upper to lower floors in multi-tier transporters; straw bedding instead of sawdust or wood shavings
Winter-finishing cattle	Transfer from slatted flooring to pen with bedding (2–3 weeks before slaughter); if necessary, shave before transfer; change diet gradually to dry feeds
Cattle finished off-grass	Use anthelminthics to prevent scouring from parasitic infections, respecting the withdrawal period; if young, lush grass is foraged, offer straw *ab libitum*; move feeders regularly to avoid poaching and muddiness

Table 4 Categories for assessment of cattle cleanliness (SAC, 1997; FSA, 2007)

Category	Description	Measures
1	Dry, no dung/dirt, very minor amounts of loosely adherent straw/bedding	None required
2	Dry/damp, light contamination (dung, dirt), small amounts of loosely adherent straw/bedding	None required
3	Dry/damp, significant contamination with dirt/ faeces, and/or significant amounts of adherent straw/bedding	Precautions at slaughter
4	Dry/damp, heavy contamination with dirt/dung, and/or significant amounts of adherent bedding	Precautions at slaughter
5	Very wet, very heavily contaminated with dirt/ dung, and/or much bedding adherent to the coat	Animal not admitted to slaughter

assessment procedures have 4–5 categories, but there is evidence that less categories would suffice, when the relation of bacterial numbers on hides to visual contamination of hides is studied (Blagojevic et al., 2012; Hauge et al., 2015; McEvoy et al., 2000). Also, there is not always a clear relation of bacterial numbers on the skin to that on the dressed carcass, partly due to sampling procedures (in particular, which regions of the body are examined), partly due to the slaughter process as such or already implemented interventions (Hauge et al., 2015). This illustrates the general issue that not all hygiene practices in GFP need to have a direct impact on bacterial numbers. Still, it can be concluded that cattle arriving visually 'clean' at slaughter originates from holdings with a certain standard of GFP. National agencies provide guides on 'clean cattle' for farmers (e.g. FSI, 2007; SAC, 1997; TEAGASC, 2005), and illustrated grading schemes, together with suggested or required corrective steps. As an example, the main elements laid down in an UK guide are presented in Table 3. In the United Kingdom, five categories of cleanliness have been defined (Table 4). However, in practice, there will be only three ways to manage cattle at slaughter: (1) slaughter using the routine dressing procedures; (2) slaughter under extra-defined hygienic dressing controls and (3) application of extra pre-slaughter measures, for example, clipping or washing (e.g. SAC, 1997). As regards clipping at the slaughterhouse, clipping at lairage exerts some stress on animals (reflected in increases of ultimate muscle pH) and is a safety issue for operatives (McCleery et al., 2008).

In a nutshell, 'clean cattle' is one of the prerequisites if beef safety is to be improved by implementing interventions targeted against specific enteric biological hazards.

5 From GFP to pre-harvest food safety management: the case of *E. coli* O157

The need to augment GFP with targeted interventions became obvious with the emergence of *Escherichia coli* O157 in the beef chain in the United States (Sofos, 2002;

Soon et al., 2011), and it was driven by beef processors and competent authorities as well. Some of the peculiarities of *E. coli* O157 in cattle and other ruminants are that infection will not cause signs of disease in infected animals, shedding can be intermittent, some animals will shed the bacterium in high numbers (i.e. ≥4 log cfu/g faeces; 'super-shedder'; Arthur et al., 2010) and prevalence within pens can vary considerably. This invalidates traditional methods of disease control (i.e. identification, separation and culling/treatment of infected animals; Matthews et al., 2013). The 'history' of pre-harvest management of *E. coli* O157 in cattle in the United States and the points and methods of intervention have been described by Sofos (2002). Since the problem of contamination of hides and shedding of pathogenic bacteria occurs pre-harvest, it is common sense to prevent shedding and contamination at the earliest point of the food chain instead of solely relying on post-harvest measures. The Food Safety and Inspection Services (FSIS) have issued a guidance document on this issue (FSIS, 2010), clearly defining which areas or practices are critical for control of *E. coli* O157 pre-harvest. In addition to 'principle-based husbandry' (broadly equalling GFP), interventions can be grouped into: water and feed management; water and food additives; management practices (in particular, separation of calves from adults, since calves are more likely to excrete the pathogen) and transportation; live animal treatments. As regards water and feed additives, five types of intervention were considered: (1) supplementing diet with seaweed extracts; (2) administration of ß-agonists; (3) antibiotic feed additives (e.g. ionophores), (4) probiotics (*Lactobacillus*-based) and (5), *E. coli* producing bacteriocins (colicin). Live animal treatments comprise administration of bacteriophages onto the hides/skin of cattle, competitive exclusion by feeding non-O157 *E. coli*, washing of cattle hides with biodegradable agents, and finally, vaccination. The document was revised in 2014 (FSIS, 2014), and the array of measures was regrouped into (1) exposure reduction, (2) exclusion strategies and (3) direct anti-pathogen strategies.

Not all studied interventions are suitable in practice, and some are even conflicting, for example, fasting, which increases shedding, but reduces the amount of faeces. Vaccinations are presumably the most effective single intervention to control *E. coli* O157 on cattle farms, if at all a single intervention instead of a multiple-hurdle approach is considered. In principle, two types of vaccines are available, one targeted against the iron-transport system of Gram-negative bacteria, as *E. coli* and *Salmonella* (siderophore receptor and porins, SRP), and the others inducing immunity against type III secreted proteins of *E. coli* O157:H7. Most recently, the latter type of vaccine has been augmented by the inclusion of other antigenic bacterial structures. This may overcome the limited effect against non-O157 *E. coli*, which are most prevalent in European countries (e.g. 72% of Austrian sheep farms were contaminated predominantly with non-O157 shiga toxin-producing *E. coli* in 2013; BMG/AGES, 2014). However, not all of these vaccines protect against colonization and/or shedding in clinical and/or field trials, or they are effective only against one serotype (independent of shiga toxin production).

A commercial vaccine against type III secretory proteins vaccines is approved in Canada, and has already been tested in other countries. It requires three injections per animal. When administered in three doses in 3-week intervals, the vaccine proved successful against experimental infections, reducing prevalence of faecal shedding by more than 50% (Potter et al., 2004). Most studies reported a significant effect in terms of reduction of prevalence of shedding (Moxley et al., 2009; Smith et al., 2009a,b), whereas Van Donkersgoed et al. (2005) found no significant effect of vaccination on in-pen-prevalence of faecal shedding. Both the different quality and the administered quantity of the vaccine may have accounted for this discrepancy. If administered, all animals of a pen need to be vaccinated to effectuate reduced shedding and hide contamination (Smith et al., 2009a). As mentioned above, a major issue is to what extent such a vaccine will be effective against different strains of *E. coli* O157 or non-O157 verotoxinogenic *E. coli* ('cross-protection'), in other words, whether the vaccine will be effective against non-O157 pathogenic *E. coli* prevalent in Europe and other regions. Nevertheless, the commercially available vaccine against type III secreted proteins has been tested in studies and has reduced shedding and colonization in the cattle of shiga toxin-producing *E. coli* O157, but has also cross-reacted with antigens of other serotypes likewise and thus has cross-protective potential against non-O157 (McNeilly et al., 2015). But further field trials are needed to prove its effectiveness.

A way to overcome this limitation is to include other cell compounds/virulence factors, as H7-flagellin or intimin. Recent research indicates that such vaccines would be very efficient at reasonable production costs (McNeilly et al., 2010, 2015).

Subcutaneous administration of an anti-SRP vaccine in two doses evoked significantly higher anti-SRP antibodies in calves, and, after experimental infection with *E. coli* O157:H7, prevalence of cattle shedding *E. coli* O157:H7 was lower than in a non-vaccinated control group. However, numbers of pathogens per g faeces were only moderately reduced (1.6 vs. 1.9 log cfu/g) (Thornton et al., 2009). Interestingly, in naturally infected cattle, a single dose was able to significantly reduce prevalence of shedding by 50% as well as the duration of shedding/high-shedding (Fox et al., 2009a,b). Efficacy of immunization was improved when calves born to vaccinated cows received a vaccination, but differences in prevalence of shedding *E. coli* O157 were not significant (Wileman et al., 2011). It must be noted that not all studies administered the vaccine thrice, as required by the manufacturer. However, a two-dose treatment combined with changes in the finishing diet was effective (Cull et al., 2012). A commercial formulation of this vaccine is conditionally approved (to be used by licensed veterinarians only).

Other types of vaccines have not been successfully introduced in the market but have proved effective at the experimental level. Most of these have not been characterized in detail for the antigenic structures they target or have targets common in many *E. coli*. Vaccines targeted against non-O157 shiga toxin-producing *E. coli* may also affect commensal *E. coli*, but these have been shown to be necessary for the development of a functioning immune system of the intestinal tract (especially the balance and the intestinal barrier between the gut microbiome and the cellular intestinal immune system) (Mirpuri et al., 2010). Vaccines that target loci directly involved in shiga toxin production or gene transfer like specific (pro)phage structures should be considered in the future.

The effect of vaccines is usually assessed by the ability to induce the formation of antibodies or by suppression of the targeted pathogens (frequency of animals excreting the pathogen and/or number of bacteria/g faeces). In case of symptomless carriage and shedding, as is the case for *E. coli* O157, the effect of vaccination is not really evident for farmers, but the additional work and costs are. From a public health viewpoint, reduction of the prevalence of shedding and super-shedding cattle by 50% is predicted to effectuate an 85% reduction in human cases of *E. coli* O157 infection (Matthews et al., 2013), but Tonsor and Schroeder (2015) could demonstrate that for farmers, this effect is not obvious or directly experienced and an economic incentive is missing to implement vaccination. In essence, this holds true for any specific measure targeted towards identification and management of super-shedders, although this would be very efficient in disease control (Munns et al., 2015).

6 Quality assurance programmes for beef production

Those involved in beef production will primarily seek advice from national organizations providing comprehensive quality assurance programmes (including certification) and training packages. The aim of such organizations is to implement best practices based on scientific knowledge and common sense husbandry techniques in cattle and beef production. To this end, information for farmers, information related to transport, etc., must be presented in a clear and concise way. In the United States, the National Cattlemen's Beef Association, linked to the 'Beef Quality Assurance' program (BQA), and the 'Beef Cattle Institute' at Kansas State University, provides such a service. Likewise, a branch of the Australian 'Livestock Production Assurance – Quality Assurance' program (LPA-QA) deals with cattle. The 'LPA Quality on-farm manual' is one example of their guides (LPA, 2014).

7 Summary

Pre-harvest food safety in beef production requires a multidisciplinary approach. HACCP-based systems can be implemented on beef cattle farms. Good hygiene practices (e.g. traceability and record keeping, bioexclusion and biocontainment, animal welfare and sustainability) form the basis that allows risk-assessment-based prioritization of hazards and the implementation of evidence-based interventions targeted against specific pathogens.

8 Future trends in research

Symptomless carriage and shedding of pathogenic bacteria is a major threat to beef safety. Despite around two decades of research on pre-harvest control of pathogenic *E. coli* in beef cattle, this issue is still not completely resolved at farm level. Although a sequence of interventions along the meat chain provide a high degree of food safety, these multiple interventions are mainly at post-harvest level and not universally portable, for example, to developing countries. Whereas toxinogenic *E. coli* in the digestive tract have been studied extensively, much less is known about other symptomlessly shed pathogens, as *Campylobacter* in the digestive tract of ruminants. It can be assumed that pre-harvest interventions will rely less on application of drugs, but more on interventions based on feed additives (e.g. probiotics), competitive exclusion and vaccination. More research on the microbiome in the digestive tract would help develop, implement and evaluate targeted control measures.

9 Where to look for further information

Further information can be taken from international bodies, as the FAO (www.fao.org) and the OIE websites (www.oie.int). Within countries, national official organizations as well as stakeholder/professional organizations provide information. Examples are: the National Cattlemen´s Beef Association (http://www.beefusa.org/) for the U.S., and Livestock

Production Assurance within Meat&Livestock Australia (http://www.mla.com.au/Meat-safety-and-traceability/Livestock-Production-Assurance). As regards welfare of production animals, national organizations usually provide very concise and practical training material and manuals. The website of Temple Grandin of handling of production animals is a comprehensive source of practical information (www.grandin.com). In the European Union, EFSA has issued several scientific opinions on animal welfare of production animals, and, more recently (2012), a methodological guidance to assess risks for animal welfare, by taking into account different husbandry systems, management procedures and animal welfare issues (www.efsa.europa.eu).

10 References

Animal Health Australia (2012), *National Farm Biosecurity Reference Manual. Grazing Livestock Production*. ISBN 978-1-921958-05-2.

Arthur, T. M., Brichta-Harhay, D. M., Bosilevac, J. M., Kalchayanand, N., Shackelford, S. D., Wheeler, T. L. and Koohmaraie, M. (2010), 'Super shedding of *Escherichia coli* O157:H7 by cattle and the impact on beef carcass contamination', *Meat Sci.*, 86(1), 32–7.

BQA (Beef Quality Assurance) (2015), *Cattle Care and Handling Guidelines*. http://www.bqa.org/Media/BQA/Docs/cchg2015_final.pdf (accessed 1 April 2016).

BQA (Beef Quality Assurance) (undated), *Master Cattle Transporter Guide*. http://www.bqa.org/Media/BQA/Docs/master_cattle_transporter_guide-digital.pdf (accessed 1 April 2016).

Bergschmidt, A., Nitsch, H. and Osterburg, B. (2003), *Good Farming Practice – Definitions, Implementation, Experiences*. Federal Agricultural Research Centre (FAL), Braunschweig, Germany.

Blagojevic, B., Antic, D., Ducic, M. and Buncic, S. (2012), 'Visual cleanliness scores of cattle at slaughter and microbial loads on the hides and the carcases', *Vet. Rec.*, 170, 563.

BMG/AGES (2014), *Bericht über Zoonosen und ihre Erreger in Österreich im Jahr 2013 (Report on Zoonoses in Austria, 2013)*, Federal Ministry of Health Publ., Vienna.

Cull, C. A., Paddock, Z. D., Nagaraja, T. G., Bello, N. M., Babcock, A. H. and Renter, D. G. (2012), 'Efficacy of a vaccine and a direct-fed microbial against fecal shedding of *Escherichia coli* O157:H7 in a randomized pen-level field trial of commercial feedlot cattle', *Vaccine*, 30(43), 6210–15.

FAO (2003), http://www.fao.org/docrep/meeting/006/y8704e.htm (accessed 27 January 2016).

FAO-OIE (2009), *Guide to Good Farming Practices for Animal Production Food Safety*. ISBN 978-92-5-006145-0.

Fox, J. T., Thomson, D. U., Drouillard, J. S., Thornton, A. B., Burkhardt, D. T., Emery, D. A. and Nagaraja, T. G. (2009), 'Efficacy of *Escherichia coli* O157:H7 siderophore receptor/porin proteins-based vaccine in feedlot cattle naturally shedding *E. coli* O157', *Foodborne Pathog. Dis.*, 6(7), 893–9.

FSA (Food Standards Agency) (2007), *Clean Beef Cattle for Slaughter. A Guide for Producers*, Publ. FSA, UK.

FSIS (Food Safety and Inspection Service) (2010), *Pre-Harvest Management Controls And Intervention Options For Reducing Escherichia Coli O157:H7 Shedding In Cattle*, Publ. FSIS, USA.

FSIS (Food Safety and Inspection Service) (2014), *Pre-Harvest Management Controls and Intervention Options for Reducing Shiga Toxin-Producing Escherichia coli Shedding in Cattle: An Overview of Current Research*, Publ. FSIS, USA.

Hauge, S. J., Nesbakken, T., Moen, B., Røtterud, O., Dommersnes, S., Nesteng, O., Østensvik, Ø. and Alvseike, O. (2015), 'The significance of clean and dirty animals for bacterial dynamics along the beef chain', *Int. J. Food Micro.*, 214, 70–6.

Koblentz, G. D. (2010), 'From biodefence to biosecurity: the Obama administration's strategy for countering biological threats', *International Affairs*, 88, 1.

Laurence, M. (2014), 'Biosecurity and beef cattle health, husbandry and welfare', in D. Cottle and L. Kahn (eds), *Beef Cattle Production and Trade*, CSIRO Publishing, Collingwood, Australia.

LPA (Livestock Production Assurance) (2014), *LPA Quality On-farm Manual*. http://www.mla.com.au/Meat-safety-and-traceability/Livestock-Production-Assurance/LPA-Quality-Assurance (accessed 1 April 2016).

Matthews, L., Reeve, R., Gally, D. L. J., Low, C., Woolhouse, M. E. J., McAteer, S. P., Locking, M. E., Chase-Topping, M. E., Haydon, D. T., Allison, L. J., Hanson, M. F., Gunn, G. J. and Reid, S. W. R. (2013), 'Predicting the public health benefit of vaccinating cattle against Escherichia coli O157', *Proc. Natl. Acad. Sci.*, 110(40), 16265–70.

McCleery, D. R., Stirling, J. M. E., McIvor, K. and Patterson, M. F. (2008), 'Effect of ante- and postmortem hide clipping on the microbiological quality and safety and ultimate pH value of beef carcasses in an EC-approved abattoir', *J. Appl. Microbiol.*, 104(5), 1471–9.

McEvoy, J. M., Doherty, A. M., Finnerty, M., Sheridan, J. J., McGuire, L., Blair, I. S., McDowell, D. A. and Harrington, D. (2000), 'The relationship between hide cleanliness and bacterial numbers on beef carcasses at a commercial abattoir', *Lett. Applied Microbiol.*, 30(5), 390–5.

McNeilly, T. N., Mitchell, M. C., Rosser, T., McAteer, S., Low, J. C., Smith, D. G., et al. (2010), 'Immunization of cattle with a combination of purified intimin-531, EspA and Tir significantly reduces shedding of *Escherichia coli* O157:H7 following oral challenge', *Vaccine*, 28, 1422–8.

McNeilly, T. N., Mitchell, M. C., Corbishley, A., Nath, M., Simmonds, H., McAteer, S. P., Mahajan, A., Low, J. C., David, G., Smith, E., Huntley, J. F. and Gally, D. L. (2015), 'Optimizing the protection of cattle against *Escherichia coli* O157:H7 colonization through immunization with different combinations of H7 Flagellin, Tir, Intimin-531 or EspA', *PLoS ONE*, 10(5): e0128391. doi:10.1371/journal.pone.0128391.

Mirpuri, J., Brazil, J. C., Berardinelli, A. J., Nasr, T. R., Cooper, K., Schnoor, M., Lin, P. W., Parkos, C. A. and Louis, N. A. (2010), 'Commensal Escherichia coli reduces epithelial apoptosis through IFN-alphaA-mediated induction of guanylate binding protein-1 in human and murine models of developing intestine', *J. Immunol.*, 184(12), 7186–95.

Motarjemi, J. and Lelieveld, H. (2014), *Food Safety Management: A Practical Guide for the Food Industry*, Academic Press – Elsevier.

Moxley, R. A., Smith, D. R., Luebbe, M., Erickson, G. E., Klopfenstein, T. J. and Rogan, D. (2009), '*Escherichia coli* O157:H7 vaccine dose-effect in feedlot cattle', *Foodborne Pathog. Dis.*, 6, 879–84.

Munns, K. D., Selinger, L. B., Stanford, K., Guan, L., Callaway, T. R. and McAllister, T. A. (2015), 'Perspectives on super-shedding of *Escherichia coli* O157:H7 by cattle', *Foodborne Pathog. Dis.*, 12(2), 89–103.

OIE (2015), *Terrestrial Animal Code. Chapter 7.1: Introduction to the Recommendations for Animal Welfare*, OIE, Paris.

OIE Animal Production Food Safety Working Group (2006), 'Guide to good farming practices for animal production food safety', *Rev. Sci. Tech. Off. Int. Epiz.*, 25(2), 823–36.

OXFAM (2014), *Guide of Best Farming Practices*. http://eird.org/pr14/cd/documentos/espanol/CaribeHerramientasydocumentos/Agricultura/oxfamguideagriculture2014eng.pdf (accessed 10 January 2016).

Potter, A. A., Klashinsky, S., Li, Y., Frey, E., Townsend, H., Rogan, D., Erickson, G., Hinkley, S., Klopfenstein, T. J., Moxley, R. A., Smith, D. R. and Finlay, B. B. (2004), 'Decreased shedding of *Escherichia coli* O157:H7 by cattle following vaccination with type III secreted proteins', *Vaccine*, 22, 362–9.

SAC (1997), *Technical Note T468: Producing Clean Slaughter Cattle*, SAC, Edinburgh, Scotland.

Singer, P. (1975), *Animal Liberation: A New Ethics for Our Treatment of Animals*, Harper-Collins, New York, USA.

Smith, D. R., Moxley, R. A., Klopfenstein, T. J. and Erickson, E. G. (2009a), 'A randomized longitudinal trial to test the effect of regional vaccination within a cattle feedyard on *E. coli* O157:H7 rectal colonization, fecal shedding and hide contamination', *Foodb. Pathog. Dis.*, 6(7), 885–92.

Smith, D. R., Moxley, R. A., Peterson, R. E., Klopfenstein, T. J., Erickson, G. E., Bretschneider, G., Berberov, E. M. and Clowser, S. (2009b), 'A two-dose regimen of a vaccine against type III secreted proteins reduced *Escherichia coli* O157:H7 colonization of the terminal rectum in beef cattle in commercial feedlots', *Foodb. Path. Dis.*, 6, 155–61.

Smith, M. H., Meehan, C. L., Techahun, J., Castaneda, K., Harrigan, J. and Lembrikova, J. (2014), *Pre-Harvest Food Safety in 4-H Animal Science*, University of California Publication numbers 8508–8511. http://anrcatalog.ucanr.edu/Details.aspx?itemNo=8491 (accessed 10 January 2016).

Smulders, F. J. M. and Algers, B. (eds) (2009), *Food Safety Assurance and Veterinary Public Health. Vol.5. Welfare of Production Animals: Assessment and Management of Risks*, Wageningen Academic Publishers, Wageningen, The Netherlands.

Sofos, J. (2002), '*Approaches to pre-harvest food safety assurance*', in F. J. M. Smulders and J. D. Collins (eds), *Food Safety Assurance and Veterinary Public Health. Vol.1. Food Safety Assurance in the Pre-Harvest Phase*, Wageningen Academic Publishers, Wageningen, The Netherlands.

Soon, J. M., Chadd, S. A. and Baines, R. N. (2011), '*Escherichia coli* O157:H7 in beef cattle: on farm contamination and pre-slaughter control methods', *Anim Health Res. Rev.*, 12(2), 197–211.

TEAGASC (2005), *Producing Clean Cattle: A Guide for Farmers*. TEAGASC, Carlow, Ireland.

Thornton, A. B., Thomson, D. U., Loneragan, G. H., Fox, J. T., Burkhardt, J. D. T., Emery, D. A. and Nagaraja, T. G. (2009), 'Effects of a siderophore receptor and porin proteins-based vaccination on fecal shedding of *Escherichia coli* O157:H7 in experimentally inoculated cattle', *J. Food Prot.*, 2, 866–9.

Tonsor, G. T. and Schroeder, T. C. (2015), 'Market impacts of E. coli vaccination in U.S. Fedlot cattle', *Agricultural and Food Economics*, 3, 7. doi: 10.1186/s40100-014-0021-2.

Van Donkersgoed, J., Hancock, D., Rogan, D. and Potter, A. A. (2005), '*Escherichia coli* O157:H7 vaccine field trial in 9 feedlots in Alberta and Saskatchewan', *Can. Vet. J.*, 46, 724–8.

Wileman, B. W., Thomson, D. U., Olson, K. C., Jaeger, J. R., Pacheco, L. A., Bolte, J., Burkhardt, D. T., Emery, D. A. and Straub, D. (2011), '*Escherichia coli* O157:H7 shedding in vaccinated beef calves born to cows vaccinated prepartum with *Escherichia coli* O157:H7 SRP vaccine', *J. Food. Prot.*, 74(10), 1599–604.

Chapter 5

Ensuring the safety of feed for beef cattle

Grant Dewell, Iowa State University, USA

1 Introduction

Feed products are potential sources for food safety risks (could be biological or chemical in nature) in beef cattle. Unlike many other food animal operations, beef cattle operations utilize an extensive source of raw materials that may not have assurances of safety and quality compared with that feed purchased from a commercial feed mill. Additionally, beef cattle are able to utilize a wide range of by-product or alternative feeds that are not utilized in other livestock production systems. Cattle are able to eat and digest this wide array of feed because of the ruminant digestive system. These ruminants rely on bacteria in the rumen to digest and ferment feed ingredients into products that can be absorbed or further digested in the intestinal tract. Beef producers can implement some basic steps to make sure the feed they provide to cattle does not pose a food safety risk for people. Fortunately, the digestive tract acts as a barrier to prevent many potential risks that may be consumed over the life of the animal from contaminating edible product.

The first step is to know what potential risks may be present in the animal's feed. This chapter will provide an overview of some potential contaminants but is not an exhaustive list of all potential risks. Governmental agencies can help provide assistance on identifying these potential contaminants and what maximum tolerance levels are. Once the potential contaminants are known, measures can be taken to prevent them from becoming an unacceptable risk. The next step would be to obtain feed from safe and reliable sources. Without knowing what the risks are to begin with, it is impossible to know if the purchased feed is safe. Once raised or purchased, feed ingredients should be stored to prevent contamination and monitored regularly to assure continued safety.

There are three major areas where food safety risks can be introduced via feed material: purchase of contaminated feed products, contamination during storage and contamination

http://dx.doi.org/10.19103/AS.2016.0008.08

of feed due to mixing or delivery errors. Good management practices such as prerequisite programmes and standard operating procedures (SOPs) are fundamental in limiting food safety risks and should be sufficient to protect food safety. However, many beef operations purchase raw feed ingredients without a standard prerequisite programme and human error during mixing and delivery of feed is not uncommon.

2 Food safety risks in purchased or raised feeds

Human food safety risks that could be carried in feed ingredients could be either biological or chemical. Physical contaminants generally would not be transferred from feed material to edible portions of livestock. The primary biological agent of concern in cattle feed would be *Salmonella* species. *Salmonella* infections in cattle are characterized by bloody diarrhoea, dehydration, fever and septicaemia (Haskell, 2008). *Salmonella* is estimated to be responsible for 35% of hospitalizations from foodborne illnesses (Scallan et al., 2011).

The prevalence of *Salmonella* in lymph nodes of slaughtered finished cattle has been reported to range from 59% to 88% (Haneklaus et al., 2012; Gragg et al., 2013; Brown et al., 2015). The original source of *Salmonella* in these studies has not been definitely determined but faecal oral contamination, biting flies, use of run-off water for heat abatement and contaminated feed are possibilities. Magwedere et al. (2015) reported that 7.2% of feed samples (raw material or processed feeds) were positive for *Salmonella* in South Africa. Further, cereal grains have been shown to be contaminated with *Salmonella* (Davies and Wales, 2013). Management practices at the farm of origin were implicated as a possible reason for increased *Salmonella* contamination. Grain farms that also had an associated livestock component were implicated to have a higher risk. Practices such as sharing equipment between the grain and livestock operations, using manure or poultry litter for fertilizer were identified. Additionally, wildlife (rodents or birds) or farm pets may carry pathogens between the livestock operation and grain storage areas.

One feed of particular concern for harbouring *Salmonella* is poultry litter. Although this feedstuff is somewhat controversial, it does have nutritional value that ruminants like cattle can utilize. Poultry litter is a high protein feed, which can provide an environment for bacterial growth such as *Salmonella*. In order to decrease the bacterial load in raw poultry litter, it must be processed before being fed. Although pelleting with heat, treating with acid or ensiling with corn are possible treatment methods, the most common method is deep stacking, which requires stacking litter in mounds 2–3 m high for at least 3 weeks. Microbial fermentation will raise the temperature in excess of 60°C, which will kill potential pathogens. Stacks of litter should be monitored to make sure that adequate heat levels are obtained throughout the stack before feeding to cattle. Additionally, it is recommended to not feed poultry litter in the 21 days before slaughter, as residues of pharmaceuticals used in poultry production may be present in the litter (Daniel and Olson, 2005)

Other bacteria such as *E. coli* O157, *Campylobacter*, *Clostridium botulinum* and so on could potentially contaminate raw feed ingredients, but evidence for this being a source for potential human food safety issue has not been established.

Prions capable of causing bovine spongiform encephalopathy (BSE) can also be a contaminant in feed material (Harman and Silva, 2009). However, policies prohibit feeding ruminant meat and bonemeal (MBM) to cattle. After the initial ruling of ruminant MBM, there was the potential that cattle feed could be contaminated at a commercial feed mill that also formulated feeds for other species. However, in 2008 the Food and Drug Administration (FDA) modified regulations preventing all specified risk materials (nervous tissue, tonsils, etc.) from being included in MBM for monogastric species decreasing the risk that cross-contamination can occur. Initially, there was also risk of exposure to the BSE prion for cattle that consumed poultry litter as spilled chicken feed is a component of poultry litter. However, with the new regulations poultry litter has been deemed to be zero risk for BSE. Needless to say, beef producers need to be cognizant of the risk and ascertain that purchased processed feeds do not contain ruminant MBM, food waste or pet food waste. Many slaughter establishments require an affidavit that cattle have not received ruminant MBM during the feeding period.

3 Chemical food safety risks

Of all the potential chemical food safety risks, mycotoxins would be the most likely to occur. These mycotoxins are produced by naturally occurring fungi that can grow on grains during certain conditions. Most mycotoxins such as deoxynivalenol, zearalenone and ergot alkaloids, although problematic for cattle, do not pose a human food safety risk (EFSA, 2004). Aflatoxins, however, can pose a human food safety risk because of their ability to cause acute liver necrosis or hepatic carcinomas, and are considered mutagenic (Cornell, 2015). The established safe levels for human consumption (4 mcg/kg in European Union (EU)) is over 20 mcg/kg. Since aflatoxins can be excreted in milk, dairy animals are at risk of transferring aflatoxins to humans. The FDA has established that raw ingredients for dairy cattle cannot have excess of 20 mcg/kg. However, for finishing beef cattle the level is 300 mcg/kg (50 mcg/kg in EU) for raw ingredients such as corn or peanut products. In certain special circumstances the FDA may allow feed mills or producers to blend raw feed ingredients that exceed established control levels. Otherwise, feed would have to be detoxified by thermal inactivation, solvent extraction, fermentation or adsorption (Cornell, 2015).

Other chemical contaminants are also associated with uptake of the chemical from the environment; the most concerning of these chemical contaminants is pesticide residues. Pesticides can result in adverse health risks for humans, and feedstuffs could be contaminated with pesticides either through application to the growing plant prior to harvest or potentially by uptake from the soil. Pesticides such as chlorpyrifos have been detected in cattle feeds at levels above the maximum residue level (Walorczyk, 2007; Fernandez et al., 2010). Other agricultural chemicals such as herbicides or fungicides could also be potential contaminates. In some areas environmental contamination with heavy metals, dioxin or PCBs could be potential risk. In 2008 elevated levels of dioxins and PCBs were found in pigs and subsequently cattle associated with using contaminated fuel in the drying system (Tlustos et al., 2014). Also, studies have shown that plants can uptake mercury from contaminated soil (Henriques et al., 2013). Additionally, although not studied, baling of crop residues such as corn stalks or soya bean stubble for animal feed does commonly incorporate dirt into bale that could be mixed into the animals feed. Other studies have reported that chemicals such as perfluorooctane sulfonic acid (PFOS) could be transferred from contaminated feed to milk (van Asselt et al., 2013). Although it would be impractical and cost prohibitive to test all feeds for all potential contaminates, it is important to know where the feed comes from and obtain assurances that it was handled appropriately before feeding livestock.

Finally, lead contamination is a concern. There are reported cases in the literature of lead intoxication of cattle from forages grown next to a clay pigeon range (Payne et al., 2013; Bischoff et al., 2014). There have been other similar cases that have not been published. In feed exposures such as this there may be many cattle that are asymptomatic but still have lead levels that could be a risk for humans. Since lead can be deposited in bone (Osweiler, 1996) and other tissue, it has the potential to be a food safety risk, even when animal blood levels have returned to normal.

By-product or co-product feeds are of special concern as they are not always a consistent product and can come from multiple sources. Ruminants such as cattle are able to utilize many by-product feeds that other animals cannot and as long as they remain an economical alternative they will be utilized (Rannkins, 2002). Standard by-products such as peanut or soya bean hulls, hominy or bakery waste would have the same risks associated with raw product (aflatoxins, pesticides, etc.). However, co-product feeds such as distiller's grains may have added risk. Antibiotics are normally added during the fermentation process to control bacterial growth (Sankarlal et al., 2015). Additionally, some corn ethanol facilities may process waste products from other industries that could contaminate distiller's grains with other compounds (Basaraba et al., 1999). Although *E. coli* O157 is not carried in corn co-product feeds, there is evidence that feeding co-product feed may increase the shedding of *E. coli* O157 by cattle (Callaway et al., 2009).

Feed material should be inspected before storing on the operation. Assurance of quality and safety should be obtained from purchased feeds. Prerequisite programmes are fundamental to food safety programmes in all sectors of the industry and can minimize risk associated with purchasing feedstuffs. Written specifications should include a description of the product, requirements for transportation and storage, and any required testing (Sperber et al., 1998). Approving suppliers of feed ingredients to be in compliance with good production practices and assuring quality and any regulatory issues will minimize the transfer of potential food safety risks through the cattle operation.

4 Storage, mixing and delivery of feed

Feed material needs to be stored in a manner that prevents contamination or spoilage of the feedstuff. As mentioned previously, storage of feed in the vicinity of livestock operations was a risk for increased *Salmonella* contamination from birds and rodents travelling back and forth from animal and feed storage areas. Starlings have been identified as potential carriers of *Salmonella* (Carlson et al., 2015) and could contaminate either stored feed or feed bunks. Proper pest control would be vital to protecting feed resources from potential contamination. Since many feeding operations have multiple lots of cattle, preventing transfer of biological transfer from infected lots of cattle to uninfected lots is an important part of a biosecurity/biocontainment programme (Brandt et al., 2008). Besides pest control, another important part of a biocontainment programme is traffic control. Allowing a piece of equipment used for manure handling to be used for feed handling without cleaning and disinfecting is a potential risk (Fig. 1 and 2). Storage conditions should also prevent the feed from spoiling or allowing mould growth. *Aspergillus* mould on corn could begin producing aflatoxins if storage conditions allow for continued growth of the mould (Cornell, 2015).

Figure 1 Loader bucket with manure on it scooping corn.

One of the biggest food safety risks for pharmaceutical residues is associated with the proper mixing and delivery of feed at the feed mill or cattle operation. For feed mills, chemical hazards from cross-contamination of medicated feed or incorrect mixing of a medicated feed are most frequently identified in hazard analysis and critical control point (HACCP) plans (Herrman et al., 2007). Critical control points designed to prevent introduction of a food safety risk during feed mixing are important. Control measures such as proper weighing steps, flushing practices in-between different batches and weigh back of medicated ingredients can assure that feed was mixed properly and identify when a potential problem has arisen. Additionally, SOPs should be developed so that there are check-offs to make sure that proper feed is delivered to right group of cattle. The last step in assuring that cattle do not have any chemical residues that would be a food safety risk is to make sure that appropriate withdrawals are followed.

Occasionally, accidents happen and feed can be contaminated. Cattle operations should develop SOPs to deal with rare or unknown contaminations when they occur. For

Figure 2 Modified distiller's grain with manure contamination from a wheel loader.

Figure 3 Fluid leak from feed mixing wagon in feed storage area.

example, a ruptured hydraulic hose could contaminate a load of feed as it is being mixed. Hydraulic fluid is potentially toxic to humans and should not be fed to cattle in any amount. Cattle operations should plan ahead with how to handle such accidental contaminations as managers may not always be available and all employees need to know the proper steps to handle such accidents. Furthermore, the cattle operations should have general maintenance programmes to make sure equipment is in good condition and any leaks are identified and fixed (Fig. 3).

5 Summary

The key to ensuring food safety quality for beef cattle in regard to feed ingredients is to follow established regulations and implement procedures to meet those regulations. Implementing basic practices such as HACCP, prerequisite programmes, SOPs and good production practices will be fundamental to meeting any human food safety issues related to animal feed. Someone with the cattle operation should be tasked to make sure that government regulations are followed to not only prevent legal issues but also prevent any food safety issues for consumers. Not knowing what the risks are is not an excuse for potentially exposing consumers to food safety risk no matter how small that risk might be. Developing programmes to assure safe food and documenting that safeguards are met will assure government regulators and consumers that your operation is producing a quality product with minimal risk for food safety issues.

6 Where to look for further information

The foundation of assuring the safety of beef regarding risks associated with animal feed is best addressed by preventive measures. Many segments of the food production industry have used HACCP principles to identify potential risks and then develop critical control points to mitigate the risk to an acceptable level. There are many resources associated with HACCP and related programmes (prerequisite programmes, SOPs, and good manufacturing practices and good agricultural practices) available to understand how to develop a complete food safety prevention programme.

Governmental agencies have multiple resources available regarding safety of feed ingredients. Maximum tolerance levels are established for food and for animal feed for many substances or biological agents. Additionally, agencies that have mandated HACCP or HACCP-like programmes on various segments of the food industry have guidelines for developing such programmes. Allied industry and stakeholders are also a good source for material or resources.

7 References

Basaraba, R. J., Oehme, F. W., Vorhies, M. W. and Stokka, G. L. 1999. Toxicosis in cattle from concurrent feeding of monensin and dried distiller's grains contaminated with macrolide antibiotics. *J. Vet. Diagn. Invest.* 11(1):79–86.

Bischoff, K., Higgins, W., Thompson, B. and Ebel, J. G. 2014. Lead excretion in milk of accidentally exposed dairy cattle. *Food Addit. Contam. Part A Chem. Anal. Control Expo. Risk Asses.* 31(5):839–44.

Brandt, A. W., Sanderson, M. W., DeGroot, B. D., Thomson, D. U. and Hollis, L. C. 2008. Biocontainment, biosecurity, and security practices in beef feedyards. *J. Am. Vet. Med. Assoc.* 232(2):262–9.

Brown, T. R., Edrington, T. S., Loneragan, G. H., Hanson, D. L., Malin, K., Ison, J. J. and Nisbet, D. J. 2015. Investigation into possible differences in salmonella prevalence in the peripheral lymph nodes of cattle derived from distinct production systems and of different breed types. *J. Food Prot.* 78(11):2081–4.

Callaway, T. R., Carr, M. A., Edrington, T. S., Anderson, R. C. and Nisbet, D. J. 2009. Diet, Escherichia coli O157:H7, and cattle: a review after 10 years. *Curr. Issues Mol. Biol.* 11(2):67–79.

Carlson, J. C., Hyatt, D. R., Bentler, K., Mangan, A. M., Russell, M., Piaggio, A. J. and Linz, G. M. 2015. Molecular characterization of Salmonella enterica isolates associated with starling-livestock interactions. *Vet. Microbiol.* 179(1–2):109–18.

Cornell University. 2015. Aflatoxins: Occurrence and health risks. 14:58:41.

Daniel and Olson. 2005. Feeding poultry litter to beef cattle. University of Missouri Extension. G 2077

Davies, R. H. and Wales, A. D. 2013. Salmonella contamination of cereal ingredients for animal feeds. *Vet. Microbiol.* 166(3–4):543–9.

EFSA (European Food Safety Authority). 2004. Opinion of the scientific panel on contaminants in the food chain on a request from the commission related to aflatoxin B1 as undesirable substance in animal feed. *EFSA J.* 39:1–27.

Fernandez-Alvarez, M., Lamas, J. P., Garcia-Chao, M., Garcia-Jares, C., Llompart, M., Lores, M. and Dagnac, T. 2010. Monitoring of pesticide residues in dairy cattle farms from NW Spain. *J. Environ. Monit.* 12(10):1864–75.

Gragg, S. E., Loneragan, G. H., Nightingale, K. K., Brichta-Harhay, D. M., Ruiz, H., Elder, J. R., Garcia, L. G., Miller, M. F., Echeverry, A., Ramírez Porras, R. G. and Brashears, M. M. 2013. Substantial within-animal diversity of Salmonella isolates from lymph nodes, feces, and hides of cattle at slaughter. *Appl. Environ. Microbiol.* 79(15):4744–50.

Haneklaus, A. N., Harris, K. B., Griffin, D. B., Edrington, T. S., Lucia, L. M. and Savell, J. W. 2012. Salmonella prevalence in bovine lymph nodes differs among feedyards. *J. Food Prot.* 75(6):1131–3.

Harman, J. L. and Silva, C. J. 2009. Bovine spongiform encephalopathy. *J. Am. Vet. Med. Assoc.* 234(1):59–72.

Haskell, S. R. 2008. *Salmonelllosis*. Blackwell's Five-Minute Veterinary Consult: Ruminant, Wiley-Blackwell, Ames, IA.

Henriques, B. I., Rodrigues, S. M., Coelho, C., Cruz, N., Duarte, A. C., Römkens, P. F. and Pereira, E. 2013. Risks associated with the transfer of toxic organo-metallic mercury from soils into the terrestrial feed chain. *Environ. Int.* 59:408–17.

Herrman, T. J., Langemeier, M. R. and Frederking, M. 2007. Development and implementation of hazard analysis and critical control point plans by several U.S. feed manufacturers. *J. Food Prot.* 70(12):2819–23.

Magwedere, K., Rauff, D., De Klerk, G., Keddy, K. H. and Dziva, F. 2015. Incidence of nontyphoidal salmonella in food-producing animals, animal feed, and the associated environment in South Africa, 2012–2014. *Clin. Infect. Dis.* 61 (Suppl. 4):S283–9.

Osweiler, G. D. 1996. *Toxicology*. Williams & Wilkins, Philadelphia, PA.

Payne, J. H., Holmes, J. P., Hogg, R. A., van der Burgt, G. M., Jewell, N. J. and Welchman Dde, B. 2013. Lead intoxication incidents associated with shot from clay pigeon shooting. *Vet. Rec.* 173(22):552.

Rankins Jr., D. L. 2002. The importance of by-products to the US beef industry. *Vet. Clin. North Am. Food Anim. Pract.* 18(2):207–11.

Sankarlal, V. M., Testroet, E. D., Beitz, D. C. and Clark, S. 2015. Short communication: No antimicrobial effects from one source of commercial dried distillers grains with solubles. *J. Dairy Sci.* 98(12):8554–9.

Scallan, E., Hoekstra, R. M., Angulo, F. J., Tauxe, R. V., Widdowson, M. A., Roy, S. L., Jones, J. L. and Griffin, P. M. 2011. Foodborne illness acquired in the United States–major pathogens. *Emerg. Infect. Dis.* 17(1):7–15.

Sperber, W. H., Stevenson, K. E., Bernard, D. T., Deibel, K. E., Moberg, L. J., Hontz, L. R. and Scott, V. N. 1998. The role of prerequisite programs in managing a HACCP system. *Dairy Food Environ. Sanit.* 18:418–23.

Tlustos, C., Anderson, W., Flynn, A. and Pratt, I. 2014. Additional exposure of the Irish adult population to dioxins and PCBs from the diet as a consequence of the 2008 Irish dioxin food contamination incident. *Food Addit. Contam. Part A Chem. Anal. Control Expo. Risk Assess.* 31(5):889–904.

van Asselt, E. D., Kowalczyk, J., van Eijkeren, J. C., Zeilmaker, M. J., Ehlers, S., Fürst, P., Lahrssen-Wiederholt, M. and van der Fels-Klerx, H. J. 2013. Transfer of perfluorooctane sulfonic acid (PFOS) from contaminated feed to dairy milk. *Food Chem.* 141(2):1489–95.

Walorczyk, S. I. 2007. Development of a multi-residue screening method for the determination of pesticides in cereals and dry animal feed using gas chromatography-triple quadrupole tandem mass spectrometry. *J. Chromatogr. A.* 1165(1–2):200–12.

Chapter 6

Detecting antibiotic residues in animal feed: the case of distiller's grains

Lynn Post, Food and Drug Administration, USA

1 Introduction

The potential risks of the use of veterinary drugs in farm animal production have long been a concern to consumers (Anon., 2016a; Beyene, 2016). There has also been concern about chemicals and other contaminants from indirect sources such as animal feed entering the food chain (Fink-Gremmels, 2012). This chapter reviews the particular challenges of detecting antibiotic residues in animal feed. The biggest potential risk posed by antibiotic residues from sources such as feed is in promoting the development of antibiotic resistance by zoonoses and other harmful microorganisms. The human and animal health threat posed by antimicrobial resistance (AMR) has been recognized at national, regional and global levels (Meek et al., 2015). Antibiotics are used in animal husbandry, bee-keeping, fish farming and other forms of aquaculture, ethanol production, horticulture, antifouling paints and food preservation. In May 2014, The World Health Organization (WHO) emphasized the global extent of AMR and repeated the need for urgent action (Anon., 2014). In addition to the paucity of hard evidence for more accurate quantification of the social and economic burdens associated with AMR regionally and globally (Fletcher,

http://dx.doi.org/10.19103/AS.2016.0008.05

2015), the WHO reports that major gaps continue to exist in understanding in order to find solutions to the problem (Anon., 2014).

In 2006, the European Union (EU) banned the use of growth-promoting antibiotics that were relevant to human health by implementation of an EU regulation (Anon., 2003). As of January 2017, a new United States (US) regulation will not allow growth-promotion claims for medicated feed containing medically important drugs that are also used in human medicine (Anon., 2015). Therapeutic uses will continue to be allowed, but under the supervision of a veterinarian. The Food and Drug Administration (FDA) developed the new regulation to safeguard animal and human health with respect to AMR.

The purpose of this chapter is to review the methods for detection of antibiotic residues in one type of animal feed, distiller's grains. There is a growing concern about the use of antibiotics in the process of which distiller's grains are a co-product (De Alwis and Heller, 2010; Kaklamanos et al., 2013b). Antibiotic residues in distiller's grains may contribute towards AMR in animals and humans.

2 Distiller's grains as animal feed

Distiller's grains (hereinafter, generally referred to as DG, unless otherwise indicated) are a cereal by-product of distillation processes; traditionally in brewing, but increasingly in the manufacture of the biofuel ethanol. DG are sold as a high protein livestock feed. They are a unique matrix that presents different analytical challenges compared to regular animal feed. This has made the development of a multiclass and multiresidue method even more challenging than ordinary animal feed, including medicated feed (De Alwis and Heller, 2010).

In 2008, the quantity of distiller's grains as ethanol fuel co-products that were marketed for use in animal feed was estimated to be about 20 million metric tons in the US (Schroeder et al., 2010). In 2015, there were 214 bioethanol plants in the US. In 2015, DG had doubled to 40 million metric tons, consisting of 4% syrup, 10% dried DG (DDG), 10% modified wet DG (MWDG), 30% wet DG (WDG) and 46% dried DG with solubles (DDGS, Anon. 2016b). Approximately 0.75 and 0.8 kg DDGS are derived from 2.4 and 2.7 kg of maize and wheat, respectively, with each litre of ethanol produced from these cereals (Smyth, 2010). From cereal grain (maize, wheat) bioethanol production, the main DG are DDGS and DWG, which are used as feeds for ruminant and monogastric animals (Corrigan and Mass, 2009; Monceaux and Kuhner, 2009; Pilgrim, 2009).

Fuel alcohol production is a non-sterile process that requires contamination control and good hygienic operations are important (Narendranath, 2003). Poor yeast viability and vitality can significantly reduce ethanol yields (Walker, 2011). Chronic bacterial contamination constantly drains the sugar available for conversion to ethanol, and bacteria scavenge essential nutrients needed for optimum yeast growth (Narendranath, 2003). Acute infections are unpredictable and bacterial by-products, such as acetic and lactic acids, inhibit yeast growth (Makanjuola et al., 1992; Narendranath et al., 1997).

According to Bischoff et al. (2016), it is generally believed that lactic acid bacteria are the primary bacterial contaminants due to a variety of Gram-positive and Gram-negative bacteria that were isolated from fuel ethanol fermentations: to include species of *Lactobacillus*, *Lactococcus*, *Pediococcus*, *Enterococcus*, *Acetobacter*, *Glucobacter* and *Clostridium* (Narendranath, 2003; Skinner and Leathers, 2004; Lush and Heist, 2005). For every molecule of bacterial lactic acid produced in a fermenter, there is a corresponding loss of an ethanol molecule (Walker, 2011). Consequently, bioethanol distilleries employ a variety of measures to control microbial contamination, including antibiotics. While a variety of antimicrobial agents have been reported to control bacterial contamination during ethanol fermentation, the most common commercial products in the US are penicillin and virginiamycin (Narendranath, 2003).

In addition to bacterial contamination, poor yeast viability can reduce ethanol yields (Walker, 2011). Selection of specific starter cultures of *S. cerevisiae* with bactericidal activity to control contaminants in bioethanol distilleries is an alternative to adding antibiotics (De Olivia Neto et al., 2004). Stress-resistant yeasts for fuel alcohol fermentations are needed, especially strains that are able to survive in the presence of toxins in the substrate and co-products (Bettiga et al., 2008). It is now possible to produce over 20% ethanol in wheat fermentations (Ingledew, 2009; Thomas and Ingledew, 1992). Industrial *S. cerevisiae* strains grow best from 20 to 30°C and between pH 4.5 and 5.5 (Walker, 2011). Heat-tolerant yeasts are sought after for fuel alcohol plants in tropical countries, such as Brazil.

Several cell physiological approaches can be adopted to improve stress-tolerance of yeasts for bioethanol production (Walker, 2011). These are non-genetic conditions that include sterol pre-enrichment by mild aeration, mineral preconditioning of yeast with magnesium and zinc, ethanol adaptation, pre-heat-shocks to confer heat tolerance and salt preconditioning to confer osmotic tolerance (Logothetis et al., 2011). *S. cerevisiae* is not entirely a facultative anaerobe and is unable to grow well under completely anaerobic conditions, because oxygen is needed for membrane biosynthesis of fatty acids. Hence, some bioethanol fermentation processes may benefit from mild aeration (Alfenore et al., 2004; Amorim et al., 2004). Under ideal laboratory conditions, *S. cerevisiae* reproduces quickly at approximately every 90 minutes, but in industrial fermenters this takes considerably longer due to the stressful physical-chemical environment (Walker, 2011). In Brazilian semi-continuous fermentations, yeast growth and budding is inhibited, but is offset by very high yeast cell densities.

3 Multiresidue methods for analysing veterinary and other residues in food and feed

Multiresidue methods have been cited for the determination of a wide range of veterinary drugs in human food products (meat, milk and eggs) in various review articles (Careri et al., 2002; Balizs and Hewitt, 2003; Stolker et al., 2007; Blasco et al., 2007; Le Bizec et al., 2009; Marazuela and Bogialli, 2009; Guillarme et al., 2010; Malik et al., 2010; Peters et al., 2010; Farré et al., 2013; Marazuela and Bogialli, 2013). Sample preparation included extraction of the analytes or simple dilution steps, purification with solid phase extraction (SPE) and determination with liquid chromatography combined with tandem mass spectrometry (LC-MS/MS) or high-resolution mass spectrometry (HRMS). Time-of-flight (TOF) and Orbitrap mass spectrometry (Orbitrap MS) have been applied in various food matrices (Stolker et al., 2008; Kaufmann et al., 2008; Ortelli et al., 2009; Peters et al., 2009; Kaufmann et al., 2011a; Villar-Pulido et al., 2011; Kaufmann et al., 2011b; Romero-Gonzalez et al., 2011; Gomez-Perez et al., 2012). Full scan mass spectrometry (FSMS) techniques, such as Orbitrap-HRMS or TOF-MS, are the alternatives to LC-MS/MS techniques (van der Heeft et al., 2009; Kellmann et al., 2009; Kaklamanos et al., 2013b).

In contrast to food products, only a limited number of corresponding methods in animal feed have been published. For feed matrices, published analytical methods for antimicrobial residues were scarce before 2008 (De Alwis and Heller, 2010). Most of the analytical methods that existed for feeds focused on medicated feeds or premixes that were intended for checking compliance with the label specifications and claims (De Alwis and Heller, 2010). Unapproved uses of compounds in medicated feed were described by Samanidou and Evagglopulous (2008). The methods targeted either a single drug or a single class of drugs. A review by Borras et al. (2011) presented analytical methods reported over a ten-year period for the analysis of antimicrobial agents in animal feed.

The complex composition of feed matrices and the wide polarity range among the different chemical groups (physical–chemical properties) of the veterinary drugs have promoted the use of LC–MS/MS-based analytical techniques for the development of multiresidue methods (Boscher et al., 2010; Kantiani et al., 2010; Cronly et al., 2010; Mol et al., 2008). Recent studies of the use of liquid chromatography to screen animal feed include Aguilera et al. (2013) and Boix et al. (2014), whilst MS is discussed by Gómez-Pérez et al. (2015).

4 Multiresidue analysis in practice: ion-trap tandem mass spectrometry

4.1 Overview

Any analytical method needs to meet a range of criteria. For example, De Alwis and Heller (2010) wanted to avoid false-negative results in developing a method for the analysis of DG, but not overlook the presence of any target compound. A chemical method based on MS ensured that the data had a high degree of reliability. The further goal was to evaluate the method's quantitative performance to provide an estimate of concentration. The final goal was to establish confirmation criteria for target compounds with a wide range of chemical properties.

Thirteen commonly used antibiotics were chosen as targets for detection were included as target analytes (Fig. 1): ampicillin, penicillin G, tetracycline, oxytetracycline, chlortetracycline, bacitracin A, virginiamycin M1, chloramphenicol, erythromycin A, clarithromycin, tylosin A, monensin and streptomycin. These compounds belong to seven chemical classes consisting of β-lactams, tetracyclines, polypeptides, amphenicols, macrolides, polyethers and aminoglycosides.

Considering the range of chemical properties of the 13 compounds, developing a viable method that could simultaneously extract and determine these analytes was challenging. It was essential to identify the right balance of analytical conditions through careful consideration of analyte physical–chemical properties, such as solubility, pK, chemical and thermal stability, and polarity to maximize the analyte recoveries. Therefore, it was logical for De Alwis and Heller (2010) to caution that any multiple methods may involve a compromise in optimal conditions to accommodate all of the compounds.

An example of a physical–chemical property is extraction into neutral/organic solvents. Some antibiotics can be extracted with neutral aqueous buffers or polar organic solvents. Chloramphenicol, an amphenicol, is a highly polar, stable compound typically extracted under neutral conditions in aqueous buffers and/or polar organic solvents (Stolker and Brinkman, 2005; Botsoglu and Fletouris, 2001). Aminoglycosides, such as streptomycin, are hydrophilic, highly polar, and resistant to acids, bases and heat. They are extracted from bio-matrices with acidic or basic aqueous solutions or aqueous/organic mixtures to

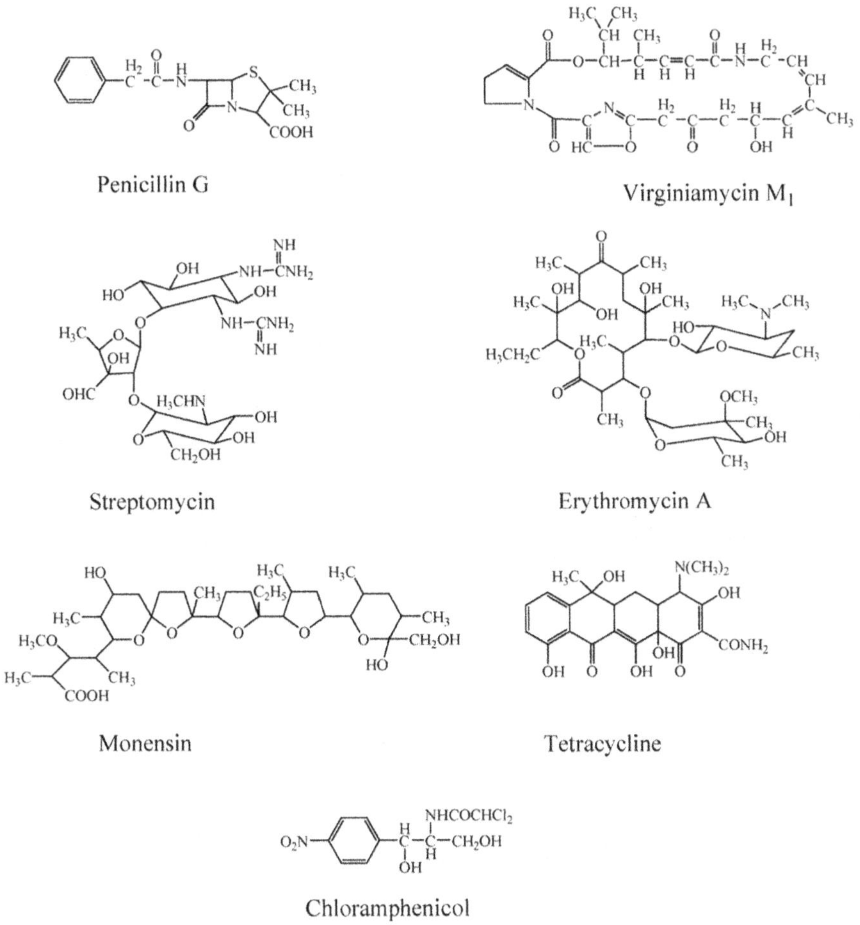

Figure 1 Structures of some representative antimicrobial compounds (reproduced from De Alwis and Heller, 2010).

release them from proteins (McGlinchey et al., 2008; Schenck and Callery, 1998; Botsoglu and Fletouris, 2001).

In general, neutral or slightly basic conditions are desirable to extract macrolides to avoid degradation in an acidic environment (Wang, 2009; Botsoglu and Fletouris, 2001). For example, erythromycin degrades to anhydro-erythromycin (anhydro-ERY). Monensin, a polyether, is both acid and base labile due to its hemi-ketal group, and neutral conditions have been commonly employed in its extraction (Cha et al., 2005).

The bacitracin commercial formulation is a mixture of closely related compounds comprised of bacitracin A, B1 and B2 with A constituting >50% of the mixture (Botsoglu and Fletouris, 2001). Bacitracin A was selected as the surrogate marker to analyse the mixture for the ion-trap tandem mass spectrometer method (ion-trap MS/MS method). Virginiamycin is a combination of two unrelated cyclic polypeptides (M1 and S1) that are non-polar and practically insoluble in water while soluble in organic solvents such as methanol (Botsoglu and Fletouris, 2001). The major percentage of the active formulation is the virginiamycin M1 form. Similar to bacitracin, De Alwis and Heller (2010) selected M1 as a surrogate for detection of the combination product (Table 1).

4.2 Method validation

The 13 residues were extracted with a solution of ethylenediaminetetraacetic acid (EDTA) and trichloroacetic acid followed by methanol (De Alwis and Heller, 2010). The combined extract was subjected to a two-track clean-up and concentration on either hydrophilic polymeric (12 residues) or weak cation (streptomycin) exchange SPE cartridges. The extracts were analysed by liquid chromatography/ion-trap tandem mass spectrometry

Table 1 Specifications of the ion-trap analytical method using fortified distiller's grains (reproduced from De Alwis and Heller, 2010)

	R^{2a}	LOQ[b]	SPE recovery	Extraction recovery ± SD[c] (%; *n* = 4)			Accuracy[d]	RSD[e,f]
Analyte	(*n* = 4)	(mg/kg)	(%; *n* = 3)	(1 mg/kg)	(2.5 mg/kg)	(7.5 mg/kg)	(%)	(%)
AMP	0.9953	0.25	97	97 ± 6	94 ± 5	99 ± 5	95	6.5
OTC	0.9961	0.25	97	84 ± 4	72 ± 3	80 ± 4	96	5.8
TC	0.9915	0.5	93	71 ± 6	68 ± 5	73 ± 4	89	11[f]
CTC	0.9961	0.1	93	71 ± 4	65 ± 1	68 ± 5	97	4.3
BACI*	0.9853	0.5	89	89 ± 7	93 ± 13	76 ± 5	98	16[g]
CAP*	0.9763	0.5	102	91 ± 12	103 ± 2	89 ± 4	106	17[g]
TYL	0.9975	0.1	95	99 ± 1	88 ± 8	96 ± 5	98	5.4
ERY[i]	0.9956	0.1	92	122 ± 7	108 ± 4	107 ± 10	99	6.2
CLARI	0.9942	0.1	91	108 ± 3	87 ± 2	96 ± 7	97	10
Pen G*	0.9686	1.0	95	64 ± 30	70 ± 21	63 ± 11	97	10[h]
VIRG	0.9947	0.1	104	84 ± 6	86 ± 10	87 ± 17	90	11
MON	0.9900	0.1	93	63 ± 11	51 ± 13	80 ± 6	111	30
STREP	0.9934	0.5	86	79 ± 4	77 ± 6	73 ± 1	100	16

AMP, ampicillin; Pen G, penicillin G; TC, tetracycline; OTC, oxytetracycline; CTC, chlortetracycline; BACI, bacitracin A; VIRG, virginiamycin M1; CAP, chloramphenicol; ERY, erythromycin A; CLARI, clarithromycin; TYL, tylosin A; MON monensin; STREP, streptomycin.

[a] R^2 of calibration line. These values were poor for the less sensitive compounds marked with asterisk.

[b] LOQ: Limit of quantitation.

[c] SD: standard deviation.

[d] Accuracy: % relative recovery was determined by enriching control samples at three concentrations; 0.4, 1.0 and 4.0 mg/kg and comparing the calculated and expected concentrations.

[e] RSD: relative standard deviation.

[f] Fortification levels: 0.4, 1.5 and 4.0 mg/kg, *n* = 24, unless otherwise noted.

[g] Fortification levels: 1.5 and 4.0 mg/kg, *n* = 16.

[h] Fortification level: 4.0 mg/kg, *n* = 8.

[i] Computed using anhydro-ERY.

(LC-MS/MS). The method was validated in a DDG matrix. Extraction recoveries for the analytes averaged 65–112% for monensin and anhydro-ERY, respectively (Table1). The relatively poor recovery of monensin was likely associated with the acid-labile hemi-ketal group, because heating was completely avoided.

Adequate quantification was achieved even though the ion-trap technology was less than ideal (Table 1). The ion-trap MS/MS method accuracy (% relative recovery) ranged from 89 to111%, with a relative standard deviation <17% for all analytes except for monensin (30%) providing acceptable precision (Table 1). Quantitation limits ranged from 0.1 to 1.0 mg/kg depending on the residue (Table 1). The calibration curve prepared from extracted matrix standards was linear ranging from 0.1 to 10 mg/kg with the coefficient of determination (R^2) exceeding 0.99 for ten analytes (Table 1). The reduced linearity coefficient of determination (R2) values observed for bacitracin, chloramphenicol and penicillin were attributed to their relatively poor mass spectrometric sensitivity. Compared to ion trap, the detection limit was at least five-fold lower for penicillin when extracts were analysed on a linear tandem mass spectrometer.

5 Multiresidue analysis in practice: orbitrap high-resolution mass spectrometry

5.1 Overview

Kaklamanos et al. (2013b) developed a method that fulfilled some of De Alwis and Heller (2010) criteria for improvement of their generic method. The Orbitrap-HRMS method demonstrated a multiclass/multiresidue capability for the simultaneous quantitation of wide range of antibiotic compounds (96 analytes), testing antibiotic presence in other matrices besides corn, and producing higher selectivity and sensitivity than the ion-trap MS/MS. The method was able to determine residues from a wide range of chemical classes: ionophore and non-ionophore, authorized coccidiostats, banned coccidiostats, macrolides, tetracyclines, quinolones, sulphonamides, nitroimidazoles, benzoimidazoles, non-steroidal anti-inflammatory drugs, tranquillizers and amphenicols in corn DDGS. β-Lactams, such as penicillin, were not analysed. The purpose of this analytical method was similar to the ion-trap MS/MS but sought improvements over the generic method.

The Orbitrap-HRMS method had two major differences from the ion-trap MS/MS method:

1 There was no intermediate SPE clean-up step in the Orbitrap-HRMS method. Sample preparation was minimal with an organic/aqueous (acetonitrile/methanol/water) extraction of the analytes and a dilution step, prior to the injection into the HRMS system.
2 The technological advantage of the Orbitrap-HRMS was a more selective and sensitive detection of the 96 analytes.

5.2 Method validation

Several parameters were evaluated to include selectivity, linearity, sensitivity, precision and accuracy using corn DDGS. Data in Table 2 is excerpted from Kaklamanos et al. (2013) for

9 of the 96 veterinary drugs. Eight compounds are part of the 13 antibiotics analysed in the ion-trap MS/MS method. One antibiotic, virginiamycin S1, was not tested in the ion-trap MS/MS method validation. Additionally, all 96 analytes were successfully identified in wheat and barley DDGS matrices similar to the validation of the spiked analytes in corn DDGS.

The R^2 for most analytes was higher than 0.9903 except for clopidol (0.9875), florfenicol (0.9862) and thiamphenicol (0.9805; Kaklamanos et al., 2013). The lack of fit calculation for the calibration curves was much lower than 20% for most of the veterinary drugs. To improve the precision of quantitative analysis and correct variability due to analyte loss, internal standards were used for 55 analytes.

Sensitivity was determined by comparing the calculated LOQ with the experimental LOQ (LOQexp). Orbitrap-HRMS technology behaves differently near the LOD, compared to LC-MS/MS, and extrapolation does not apply (Kaklamanos et al., 2013a). Uncertainty due to high variability of the LOD (no background noise), the intercept, the standard deviation of the intercept and the non-linearity effect on the intercept was avoided by experimentally verifying the LOQexp (see examples in Table 2). The calculated LOQ were confirmed by analysing blank samples fortified with the analyte. LOQexp values ranged between 25 and 250 μg/kg for all 96 compounds (Kaklamanos et al., 2013). Two amphenicols showed substantially poor ionization resulting in decreases in sensitivity. For florfenicol and thiamphenicol but not chloramphenicol, positive ionization was not applicable. The less abundant negative ionization mode was selected resulting in much higher LOQs.

Table 2 Validation parameters for 9 of the 96 veterinary drugs extracted from corn DDGS samples that were analysed using the Orbitrap high-resolution mass spectrometry method (adapted from Kaklamanos et al., 2013b)

Compounds	LOQ[a] (μg/kg)	LOQexp[b] (μg/kg)	RSDr[cf] (%)	RSDint[df] (%)	RR[ef] (%)
Chloramphenicol	94.1	100	6.4–6.2	7.6–6.2	106.6–106.3
Chlortetracycline	91.6	150	6.0–12.2	56.0–13.6	116.4–102.2
Clarithromycin	60.9	50	5.7–1.2	6.2–12.2	108.1–89.4
Erythromycin	29.1	25	10.1–9.2	13.9–19.3	60.2–59.4
Monensin	73.3	50	6.5–1.9	12.0–6.0	97.6–106.6
Tetracycline	50.1	100	5.8–4.2	19.2–13.0	52.9–50.2
Tylosin	87.6	100	4.0–3.2	8.7–15.4	121.0–93.3
Virginiamycin M1	65.8	50	5.0–2.2	6.5–15.2	110.5–83.8
Virginiamycin S1	78.2	100	5.1–2.9	5.1–5.9	115.6–86.0

[a] LOQ: limit of quantitation.
[b] LOQexp: experimental limit of quantitation.
[c] RSDr: relative standard deviation for repeatability.
[d] RSDint: relative standard deviation of intermediate precision.
[e] RR%: percentage recovery rate.
[f] Four concentration levels were evaluated at 0.5 × VL, 1 × VL, 1.5 × VL and 2 × VL; VL = validation level; $n = 27$ results for each analyte and concentration level. The validation level varied among the 96 analytes.

Precision was assessed by calculating the relative standard deviation (RSD) of the analytical results (Anon., 1994). The relative standard deviation for repeatability (RSDr) ranged in concentration from 0.54 to 12.17% (Kaklamanos et al., 2013; see examples in Table 2). The relative standard deviation of intermediate precision (RSDint, reproducibility) ranged between 1.38 and 20.75%. The results of RSDr and RSDint (RSD less than ~20%) for the 96 analytes indicated that the precision of the method was satisfactory.

Accuracy was expressed in terms of the percentage relative recovery rate (RR%, see examples in Table 2). The recoveries were adequate with 90 of the 96 compounds showing satisfactory recoveries ranging from 71 to 124%; four of the compounds (halofuginone, amprolium, erythromycin and minocycline) showed recoveries from 6 to 89%, and tetracycline around 50%. Owing to the strong binding between matrix components of the DDGS sample, such as tetracycline minocycline, lower recoveries were observed.

6 Testing samples of distiller's grains

6.1 Food and Drug Administration 2008 survey

In 2008, the Food and Drug Administration (FDA) wanted to test 60 DG samples from domestic ethanol processors for research purposes only, specifically to assist with ion-trap MS/MS method development (Luther, 2012a). Increasing public concern over the use of antibiotics in animal production prompted the United States FDA to test DG. The FDA's Center for Veterinary Medicine (CVM) acknowledged a potential risk of DG products contaminated with antibiotic residues to animals and the potential human food safety concerns, if antibiotic residues accumulate in the milk, meat and eggs. Therefore, the FDA developed the ion-trap MS/MS method to survey a limited number of DG. Samples were collected for informational purposes in order to determine the extent of antibiotic residues and to estimate their levels in DG produced by US ethanol production facilities.

From the 60 domestic samples that were requested, 49 domestic samples were sent to the CVM and were used to test the ruggedness (robustness) of the method undergoing the single lab method validation process (De Alwis and Heller, 2010; Luther, 2012a). The 49 domestic samples included 28 dried DG samples, 17 wet distiller's grains samples, and 4 DG solubles (stillage) samples. Preliminary results showed antibiotic residues were detected in 28 of the samples with the most prevalent being virginiamycin (17 samples), erythromycin (10 samples) and tylosin (5 samples). Although antibiotic residues were detected in 28 samples, quantifiable antibiotic residues were found in only 17 of the samples. Virginiamycin was estimated in ten samples at levels of 0.1–0.5 mg/kg, on a dry matter (DM) basis. Erythromycin was estimated in seven samples at levels of 0.1–1.5 mg/kg, on a DM basis. There were no other antibiotics found at levels above the LOQ.

6.2 Food and Drug Administration 2010 survey

In 2010, The FDA repeated the 2008 survey by analysing 46 samples (18 imported and 28 domestic DDGS) for residues of 13 antibiotics consisting of ampicillin, penicillin, chlortetracycline, oxytetracycline, tetracycline, clarithromycin, erythromycin, streptomycin, virginiamycin M1, bacitracin A, chloramphenicol, monensin (an ionophore and not a medically important human drug) and tylosin (De Alwis and Heller, 2010; Luther, 2012b).

The samples were analysed using the ion-trap MS/MS method developed by De Alwis and Heller (2010).

FDA analysed the 46 samples for residues of 13 antibiotics (ampicillin, penicillin G, chlortetracycline, oxytetracycline, tetracycline, clarithromycin, erythromycin, streptomycin, virginiamycin M1, bacitracin A, chloramphenicol and tylosin) and monensin using the ion-trap MS/MS method developed by De Alwis and Heller (2010). Although 5 of the 46 samples collected were initially reported as positive for antibiotics above LOQ, one sample was found to have a level of virginiamycin M1 that was below the LOQ verified by the laboratory. The sample was considered to be negative for the 13 antibiotics. The LOQ was 0.1 mg/kg for virginiamycin M1, 0.5 mg/kg for erythromycin, and 1.0 mg/kg for penicillin. Consequently, the FDA final analysis found that 4 of the 46 samples were quantifiable in the laboratory for antibiotics. The four positive samples contained a total of five antibiotic residues. Of the four positive samples, three were domestic samples and one was domestic-import sample. The positive domestic-import sample was from Canada and ~0.18 mg/kg of virginiamycin M1 was detected by the FDA laboratory on a DM basis.

The results of the three positive domestic samples were: Virginiamycin M1 was detected at ~0.16 mg/kg on a dry weight basis in one sample. Erythromycin was detected at ~0.58 mg/kg on a dry weight basis in another sample. The final positive domestic sample contained virginiamycin M1 at ~0.15 mg/kg on a dry weight basis and penicillin G at ~0.24 mg/kg on a dry weight basis. Although the amount of penicillin found (0.24 mg/kg) was lower than the LOQ for penicillin, the laboratory was able to accurately quantify this sample for penicillin below 1 mg/kg.

6.3 University of Minnesota 2011 survey

While the presence of antibiotic residues in DG is possible, it is likely that the high pH and temperatures encountered during the production process may render the residues biologically inactive (Compart et al., 2013). The objective of Compart et al. (2013) study was to quantify concentrations of various antibiotic residues in commercial DG and determine if those residues were biologically active. If biologically active amounts of antibiotic residues were present, they could theoretically have antimicrobial effects in food-animals consuming DG.

A total of 159 samples were collected and examined in Compart et al. (2013) study. Eighty of the samples were DDG and 79 were DWG. Samples were analysed for the presence of erythromycin, penicillin G, tetracycline, tylosin, and virginiamycin M1 using the ion-trap MS/MS method. Antibiotic residues were found in 18 of the DG samples in this study, with the majority (15) present in DDG samples and 3 in WDG samples. Erythromycin was found in 14 DDG samples and 2 WDG sample with a means and standard deviations of 0.35 mg/kg ± 0.26 and 0.35 mg/kg ± 0.01, respectively. One sample contained about 0.11 mg/kg of penicillin G, and one sample contained about 1.12 mg/kg of tetracycline.

The M1 subunit of virginiamycin was quantitated by the ion-trap MS/MS method, but only a biological assay results were included in the Compart et al. (2013) study. Instead of measuring only one subunit of virginiamycin with the ion-trap MS/MS method, biological activity to inhibit growth of *Kocuria rhizophila* was determined to be a more realistic measurement. Two samples contained virginiamycin at 0.6 and 0.5 mg/kg. Both positive samples were DDG from the same ethanol plant. Results of the microbiological assays revealed that of the 159 samples, one sample inhibited growth of *Escherichia coli*. This sample did not contain any detectable residues analysed in this study. No residue extracts inhibited *Listeria monocytogenes.*

Compart et al. (2013) detected the virginiamycin M1 subunit in eight samples analysed using ion-trap MS/MS. Using the bioassay procedure, only two samples had detectable concentrations of virginiamycin. Only one sample containing virginiamycin M1 was validated with the FDA-approved biological assay (0.33 vs 0.5 mg/kg, respectively), whereas the other sample found to contain the M1 fraction was not detected through the bioassay method. All of the unconfirmed samples testing positive for the virginiamycin M1 were also less than the LOQ of the bioassay.

6.4 European Union 2013 survey

Sixty-nine unknown DG samples from biofuel and alcoholic beverages were analysed using the Orbitrap-HRMS method for the presence of the 96 veterinary drugs (Kaklamanos et al., 2013b). Fifty samples out of 69 were found negative and 19 samples were positive. Among the 19 samples, there were positive findings for virginiamycin S1 (11 samples) ranging from 83 to 411 μg/kg, virginiamycin M1 (1 sample) at 204 μg/kg, erythromycin (3 samples) from 117 to 427 μg/kg, tetracycline (1 sample) at 171 μg/kg, doxycycline (13 samples) from 113 to 253 μg/kg, lasalocid (3 samples) from 61 to 156 μg/kg and narasin (11 samples) from 272 to 2313 μg/kg. It is noteworthy that many of the 19 DG samples contained more than one antibiotic. Kaklamanos et al. (2013b) did not designate the source of the 19 positive samples from either biofuel or alcoholic beverage production. Presumably, the positive samples were from the DG used during ethanol fuel production.

6.5 Comparing surveys

The presence of several different classes of antibiotics from biofuel and alcoholic beverage production in relatively small number of EU samples is remarkable compared to the findings from the three US surveys. Among the 19 samples from biofuel and alcoholic beverage production in the EU survey, there were a total of 42 positive findings for virginiamycin S1 and M1, erythromycin, tetracycline, doxycycline, lasalocid and narasin. Kaklamanos et al. (2013) did not designate the source of the residues, either from brewer's grain or ethanol fuel DG. It is more likely that they are from ethanol fuel DG. Most of the samples were below the trace level set at <0.5 mg/kg, but narasin was detected at concentrations well above 0.5 mg/kg in 10 samples.

While narasin is not a medically important antibiotic (used in animal and human medicine), other antibiotics in feed are medically important. Medically important antibiotics that are used to manufacture medicated feed may carry-over in trace amounts to other complete feeds including medicated feeds (Kaklamanos et al., 2013). Preventing cross-contamination and unsafe drug contamination can be controlled by procedures outlined in FDA, Compliance Policy Guide Section 680.600 (Anon., 1981). Nonetheless, trace amounts of antibiotics may be detectable in animal feed and may or may not contribute to AMR. Analogous to the findings of Kaklamanos et al. (2013) in the EU, trace levels of a wide range of medically important drugs in complete feed may unknowingly enter the US animal feed supply. Surveys of complete feeds for antibiotic residues are critical to rule in/out the influence of trace amounts of antibiotics from complete feeds. The further evaluation of AMR due to environmental antibiotics and other factors would be very beneficial to correlate with the outcome of the complete feed survey.

7 The survival of antibiotic residues in distiller's grains

In the Compart et al. (2013) study, DG samples had relatively low concentrations of antibiotic residues. Furthermore, 17 of the 20 antibiotic residue concentrations, using the ion-trap MS/MS method, were biologically inactive. Antibiotic residues inhibited bacterial growth in 3 of the 20 samples. Lack of inhibition in all other antibiotic-positive samples (17) was likely due to antibiotic degradation. Penicillin and ampicillin are sensitive to acids and bases during sample preparation and analysis (De Alwis and Heller, 2010). The probability of degradation varies with the nature of the side chain (Stolker and Brinkman, 2005; Botsoglu and Fletouris, 2001). Tetracyclines degrade through epimerization, dehydration and isomerization in strong acids and bases (Anderson et al., 2005; Stolker and Brinkman, 2005; Botsoglu and Fletouris, 2001). Penicillin, erythromycin and tylosin all degrade at a pH <4.0, during ethanol fermentation and render them to be inactive (Aksenova et al., 1984; Brisaert et al., 1996; Islam et al., 1999). Bacitracin, a peptide, transforms to epi-bacitracin in strong acid (Wan et al., 2006).

Erythromycin degrades to anhydro-ERY with a loss of a water molecule when exposed to acidic conditions (Wang, 2009; Yang and Carlson, 2004). Under conditions where pH <7, erythromycin exists mainly as anhydro-ERY (Botsoglu and Fletouris, 2001; Yang and Carlson, 2004; Yang et al., 2006). DG is an acidic matrix with a pH range of 3–6 depending on the individual co-product where some erythromycin is present as degradation product, anhydro-ERY (De Alwis and Heller, 2010). Any remaining erythromycin may also degrade with acidic extraction (pH ~3–5) and acidic chromatographic conditions (pH 2.4) using the ion-trap MS/MS method.

De Alwis and Heller (2010) conducted a stability study using the DG matrix that demonstrated a significant degradation of some antimicrobials under certain storage conditions (Fig. 2). Figure 2 summarizes the analyte degradation pattern after 5 weeks. Ampicillin was the most unstable in the DG matrix, and it nearly disappeared by the end of 5 weeks at both room temperature and refrigerator conditions. This was followed by tetracycline, chlortetracycline, bacitracin, tylosin and virginiamycin with 30–60% and 10–40% loss at room temperature and under refrigeration. Nine of the 13 analytes exhibited good stability at −15°C up to the ten-week period covered, except ampicillin showed slight degradation (Fig. 2). Across all storage conditions oxytetracycline, clarithromycin, chloramphenicol and monensin displayed the most stability. Penicillin G and streptomycin were not included in this study, and the stability of the fragile erythromycin was not assessed.

High temperatures (>200°C) can degrade penicillin, tetracycline and tylosin during ethanol fermentation, distillation and drying (Brisaert et al., 1996; Hynes et al., 1997; Kheirolomoom et al., 1999; Juranek and Duquette, 2007; Wang et al., 2008). On the other hand, a number of studies demonstrated the resilience of virginiamycin in ethanol production under laboratory conditions (Hamdy et al., 1996; Hynes et al., 1997; Islam et al., 1999). Virginiamycin was stable at pH 3.8–4.8 and 25–35°C; retaining greater than 87% of its initial activity levels during experimental beer production for up to 72 hours (Islam et al., 1999). However, heating for 30 minutes at 100°C reduced activity to 2.6% of the initial value and showed that virginiamycin activity in stillage was reduced to 13.2% of the beer (Hamdy et al., 1996). Dryers in ethanol plants may range from 93 to 232°C, suggesting that the heat of drying would destroy most of the virginiamycin, and other residues in DG (Bischoff et al., 2016; Shurson et al., 2003).

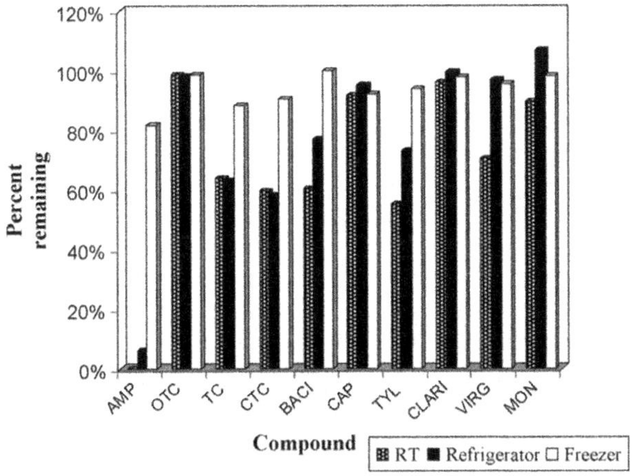

Figure 2 Loss of 10 analytes in DDG matrix on storage for five weeks (reproduced from De Alwis and Heller, 2010). AMP, ampicillin; TC, tetracycline; OTC, oxytetracycline; CTC, chlortetracycline; BACI, bacitracin A; VIRG, virginiamycin M1; CAP, chloramphenicol; CLARI, clarithromycin; TYL, tylosin A; and MON monensin.

A study to evaluate the fate of virginiamycin during the ethanol production process was conducted by Bischoff et al. (2016) at a pilot plant with an industrial scale fermentation vat. Three consecutive runs consisted of one with no antibiotic, one with 2 mg/kg and a run at 20 mg/kg were tested. The typical use rate for the commercial virginiamycin-based product was 0.1–3 mg/kg of the fermentation mixture with a maximum of 6 ppm. Samples of DDGS and process intermediates (stillage, syrup and wet cake) were collected from each run and analysed for virginiamycin M1 and S1 using a LC-MS method. The LC system was connected to a triple quadrupole QTRAP mass spectrometer that provided sensitivity, quantitation and confirmation. The method was validated by evaluating linearity ($R^2 > 0.98$), accuracy (recovery, 70–130%), precision (instrumental, 15%) and LOQ (0.1–0.3 mg/kg).

Virginiamycin M1 was detected in all process intermediates of the 20 mg/kg run (Bischoff et al., 2016). On a DM basis, virginiamycin M1 concentrations decreased approximately 97% from 41 mg/kg in the fermenter to 1.4 l mg/kg in the DDGS. Biological activity (bioassay) was detected in DDGS from both the 2 and 20 mg/kg runs at 0.69 and 8.9 mg/kg, respectively. No antibiotic activity (<0.6 mg/kg) was detected in the DDGS, fermenter and process intermediate samples from the 2 mg/kg run.

The results from Bischoff et al. (2016) demonstrated that low concentrations of biologically active antibiotic may persist in DG treated with virginiamycin to control bacterial growth during fermentations. The analytical and bioassay data from the 20 mg/kg run indicate that most of the virginiamycin was inactivated during processing. The bioassay data presented by Bischoff et al. (2016) implied that the concentration of virginiamycin residues in DDGS produced from fermentations with less than 2 mg/kg would be below the LOD. However, processing conditions for each step may vary, that is, time and temperature. Heat of distillation, evaporation and drying of DG may not be sufficient to fully inactivate virginiamycin, especially with concentrations of virginiamycin >3 mg/kg.

8 Assessing the risk of antimicrobial resistance

Whereas DG do not have measurable concentrations of bacteria immediately after production (Pedersen et al., 2004; Lehman and Rosentrater, 2007), samples of wheat WDG were found to contain up to 8.4 CFU/mL of lactobacilli after storage (Pedersen et al., 2004). WDG have been shown to have bacterial counts of 107 to 108 cells/g of DM within nine days of production (Lehman and Rosentrater, 2007). Jacob et al. (2008) showed that monensin- and tylosin-resistant *Enterococcus* species were present in one source of DG. Furthermore, da Costa et al. (2007) examined *Enterococcus* isolates in cereal DG and found resistance to tetracycline, erythromycin, rifampicin, gentamicin, chloramphenicol, nitrofurantoin and ciprofloxacin. This implies that bacteria gained AMR genes due to external environmental factors, such as exogenous bacteria in the environment (Compart et al., 2013).

Although many antibiotics are used both in animals and humans, resistance in humans is largely due to human use (Fletcher, 2015). The frequency with which resistant bacteria colonize the human gut, and resistance genes are transferred is yet to be determined (Kemper, 2008; Phillips et al., 2004). However, the risk of concentrating resistant bacteria in food-animal gastrointestinal tracts may increase, if active antibiotic residues are present in DG (Compart et al., 2013). If resistant bacteria were to develop in DG, those bacteria would have to make it through several barriers, such as acid and heat degradation, in order to become a public health threat (Bischoff et al., 2016; Hurd et al., 2004).

According to Jacob et al. (2008) and Edrington et al. (2010), a connection between use of DG in cattle or swine diets and resistant bacteria in the gastrointestinal tract of those species has been demonstrated. Resistant bacteria can be found in all food-animals and are often associated with medicated feed (Sunde et al., 1998). The risk from the low dosages used for animal growth promotion is relatively small but there might be disadvantages to human and animal health that are unknown (Phillips et al., 2004).

Hurd et al. (2004) found that the risk of illness due to macrolide-resistant campylobacteriosis, resulting from exposure of swine to tylosin, was <1 in 10 million for all meat commodities combined; 1 in 14 million for chicken, 1 in 53 million for beef, and 1 in 236 million for pork. This risk should decrease even further when considering the additional barriers, such as acid environment and heat, involved in ethanol production (Compart et al., 2013). On the basis of the Hurd risk assessment and considering the low concentration of antibiotic residues in DG, the risk of bacterial AMR developing and affecting the human population as the result of the use of antibiotics in ethanol production is extremely low (Compart et al., 2013).

9 Summary and future trends

The LOQ of virginiamycin M1 was about two-fold less for the Orbitrap-HRMS method compared to the ion-trap MS/MS method (Tables 1 and 2). While the ion-trap MS/MS method did not assess virginiamycin S1, the Orbitrap-HRMS method tested both M1 and S1 with good results. The ion-trap MS/MS method for the M1 faction of virginiamycin also resulted in a false negative in the Compart et al. (2013) study. FDA's, CVM Guidance for Industry #118 recommends that methodology demonstrate $\leq 10\%$ false-negative rate at or above the tolerance level (Anon., 2003). Avoidance of false negatives was one of the main goals of De Alwis and Heller (2010) in order to not overlook residues. Compart et al. (2013) suggested that a larger number of samples greater than the LOQ of the bioassay is needed to fully determine if that criterion is being satisfied by the ion-trap MS/MS procedure.

Although superior in other respects, the Orbitrap-HRMS method did not test penicillin or other β-lactams. Since β-lactams, especially penicillin, are likely to be present in DG, a multiresidue method that lowers the LOQ of penicillin below 0.5 mg/kg would be desirable.

In the past several decades, advances in analytical chemistry have pushed the limits of detection of chemicals in food and feed to minuscule amounts. More recently, the advances in molecular biology have allowed the rapid DNA fingerprinting of bacteria. Undoubtedly, our research tools for detecting residues and identifying bacteria in animal feed and environmental matrices will continue to improve. The characterization of environmental antibiotics and bacteria is essential to understanding the complexity of the interrelationships of AMR between animals and humans.

The antibiotics in DG will likely be inactivated by low pH and high temperature during the fermentation and subsequent drying process in ethanol fuel plants. Penicillin, erythromycin, tylosin, tetracyclines and bacitracin are all subject to degradation at these environmental extremes under experimental conditions. The results of the Bischoff et al. (2016) study suggested that research on the fate of other antibiotics, besides virginiamycin, during fuel ethanol production is warranted. A commercial source of penicillin, erythromycin, tylosin, tetracyclines and bacitracin are readily available to control bacterial contamination.

The concentration of antibiotic residues in DG, especially DDG, is extremely low, as shown by four US and EU surveys. More DG surveys may be useful to determine if there are a variety of compounds, as detectable residues, that are biologically active. If AMR bacteria survive through harsh conditions of extreme pH and high heat, they could pose a threat to public health. Do the harsh conditions influence the selection of more hardy AMR bacteria? Alternatively, antibiotic residues and bacteria may be destroyed by the harsh conditions during the fermentation process. *Studies to pursue other antibiotics, such as penicillin, erythromycin, tylosin, tetracyclines and bacitracin, used in ethanol fuel production may be beneficial to determine if their fate is similar to virginiamycin.*

Barceló and Bennett (2016) have raised broader questions concerning the AMR bacteria in the environment. Similarly, Fletcher (2015) discussed some environmental issues that contribute to AMR, including soil related factors, animal husbandry and waste management, potable and wastewater, and food safety, with examples drawn mainly from the Asian region. While their concerns are mainly directed towards the EU and Asia, the US has similar issues. In the EU according to Barceló and Bennett (2016), water contaminated with untreated effluents from livestock operations, or aquaculture is positive for enteric AMR bacteria. Manure and sewage wastewater that is recycled onto agricultural land can contain substantial amounts of various antibiotics with AMR bacteria (Barceló and Bennett, 2016; Fletcher, 2015).

The development and dissemination of bacterial resistance to antibiotics used in human medicine is an equally important public health issue (Fletcher, 2015). The spread and contamination of the environment through hospital wastewaters or incompletely treated urban wastewaters is considered a serious public health problem (Barceló and Bennett, 2016). Evidence in the current literature shows the widespread dependency on antibiotics and complex interactions between human health, animal husbandry, veterinary medicine, environment and food supply (Fig. 3; Fletcher, 2015). This dependency has contributed to

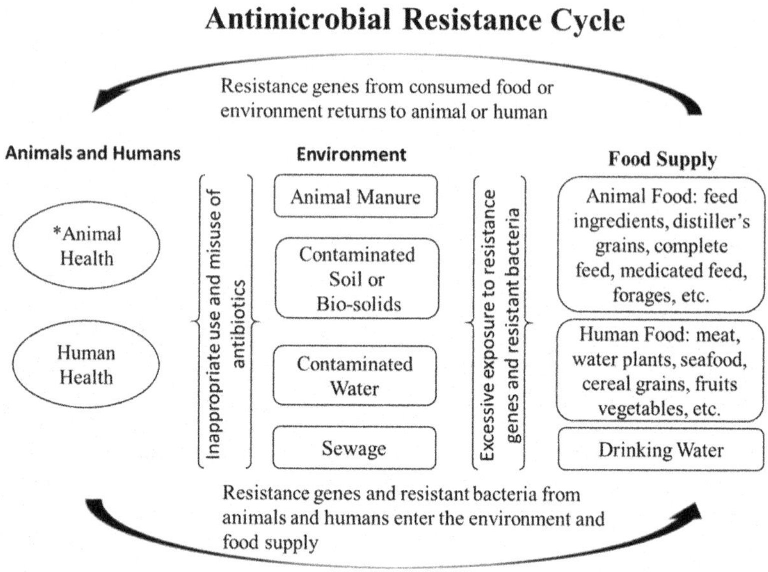

Figure 3 The antimicrobial-resistance cycle is complex consisting of environmental, animal and human health, food supply and bacterial interrelationships. Various interlinked factors contribute to the contamination of the environment and food supply with resistance genes and resistant bacteria. The transfer of resistant bacteria in both directions perpetuates the cycle of antimicrobial resistance. *Animal health represents all sources of antibiotic use (adapted from Fletcher, 2015).

the propagation and spread of resistant organisms with many gaps in knowledge about the links between AMR and the environment (Fig. 3; Fletcher, 2015).

Finally, optimal conditions for maintaining the yeast viability should be further explored in future research. For example, heat tolerant yeasts may be beneficial for ethanol fuel production in the US. The 'right' combination of stress-tolerant yeast strains and their ideal physiological adaptations may boost ethanol fuel yields. Hopefully, the net result would be reduction of antibiotic use to control bacterial contaminants.

10 Where to look for further information

10.1 *Science of The Total Environment*

The *Science of The Total Environment* contains articles that were selected to cover four areas (see Barceló et al., 2016):

(1) The origin of the problem. Presence in hospitals and wastewater treatment plants.

Alexander, J., Bollmann, A., Seitz, W. and Schwartz, Th. (2015), Microbiological characterization of aquatic microbiomes targeting taxonomical marker genes and antibiotic resistance genes of opportunistic bacteria, *Science of the Total Environment*, 512–13, 316–25.

Birosová, L., Mackulak, T., Bodík, I., Ryba, J., Skubák, J. and Grabic, R. (2014), Pilot study of seasonal occurrence and distribution of antibiotics and drug resistant bacteria in wastewater treatment plants in Slovakia, *Science of the Total Environment,* 490, 440–4.

Harnisz, M., Korzeniewska, E., Ciesielski, S. and Golás, I. (2015), Tet genes as indicators of changes in the water environment: Relationships between culture-dependent and culture-independent approaches, *Science of the Total Environment*, 505, 704–11.

Harris, S., Morris, C., Morris, D., Cormican, M. and Cummins E. (2014), Antimicrobial resistant Escherichia coli in the municipal wastewater system: Effect of hospital effluent and environmental fate, *Science of the Total Environment*, 468–9, 1078–85.

Kim, S., Zuwhan Yun, Ha, U., Lee, S., Park, H., Kwon, E., Cho, Y., Choung, S., Oh, J., Medriano, C. and Chandran K. (2014), Transfer of antibiotic resistance plasmids in pure and activated sludge cultures in the presence of environmentally representative micro-contaminant concentrations, *Science of the Total Environment*, 468–9, 813–20.

Tao, C., Hsu, B., Ji, W., Hsu, T., Kao, P., Hsu, C., Shen, S., Shen, T., Wan, T. and Huang, Y., (2014), Evaluation of five antibiotic resistance genes in wastewater treatment systems of swine farms by real-time PCR, *Science of the Total Environment*, 496, 116–21.

Varela, A., Ferro, G., Vredenburg, J., Yanik, M., Vieira, L., Rizzo, C. Lameiras, L. and Manaia C. (2013), Vancomycin resistant enterococci: From the hospital effluent to the urban wastewater treatment plant, *Science of the Total Environment,* 450–1, 155–61.

(2) Occurrence and risk in water, soil, sediments and river catchments

Bellanger, X., Guilloteau, H., Bonot, S. and Merlin, C. (2014), Demonstrating plasmid-based horizontal gene transfer in complex environmental matrices: A practical approach for a critical review, *Science of the Total Environment*, 493, 872–82.

Casey, C., Hernandez, S., Yabsley, M., Smith, K. and Sanchez, S. (2015), The carriage of antibiotic resistance by enteric bacteria from imported tokay geckos (Gekko gecko) destined for the pet trade, *Science of the Total Environment*, 505, 299–305.

Costa de Luca Rebello, R. and Hamond Regua-Mangia, A. (2014), Potential enterovirulence and antimicrobial resistance in Escherichia coli isolates from aquatic environments in Rio de Janeiro, Brazil, *Science of the Total Environment*,490, 19–27.

Harnisz, M., Korzeniewska, E., Ciesielski, S. and Golás, I. (2015), Tet genes as indicators of changes in the water environment: Relationships between culture-dependent and culture-independent approaches, *Science of the Total Environment*, 505, 704–11.

Huerta, B., Marti, E., Gros, M., López, P., Pompeo, M., Armengol, J., Barceló, D., Balcázar, J., Rodríguez-Mozaz, S. and Marcé, R. (2013), Exploring the links between antibiotic occurrence, antibiotic resistance, and bacterial communities in water supply reservoirs,*Science of the Total Environment*, 456–7, 161–70.

Johnson, A., Keller, V., Dumont, E. and Sumpter J. (2013), Assessing the concentrations and risks of toxicity from the antibiotics ciprofloxacin, sulfamethoxazole, trimethoprim and erythromycin in European rivers, *Science of the Total Environment*, 511, 747–55.

Joy, S., Li, X., Snow, D., Gilley, E., Woodbury, J. and Bartelt-Hunt, S. (2014), Fate of antimicrobials and antimicrobial resistance genes in simulated swine manure storage, *Science of the Total Environment*, 481, 69–74.

Maal-Bared, R., Bartlett, K., Bowie, W. and Hall E. (2013), Phenotypic antibiotic resistance of Escherichia coli and E. coli 0157 isolated from water, sediment and biofilms in an agricultural watershed in British Columbia, *Science of the Total Environment*, 443, 315–23.

Maal-Bared, R., Bartlett, K., Bowie, W. and Hall, E. (2013), Phenotypic antibiotic resistance of Escherichia coli and E. coli O157 isolated from water, sediment and biofilms in an agricultural watershed in British Columbia, *Science of the Total Environment*, 443, 315–23.

Peng, S., Wang, Y., Zhou, B. and Lin, X. (2015), Long-term application of fresh and composted manure increase tetracycline resistance in the arable soil of eastern China, *Science of the Total Environment*, 506–7, 279–86.

Pereira, A., Santos, A., Tacao, M., Alves, A., Henriques, I. and Correia, S. (2013), Genetic diversity and antimicrobial resistance of Escherichia coli from Tagus estuary (Portugal), *Science of the Total Environment*, 461–2, 65–71.

Pitondo-Silva, A., Martins, V., Tonelli Fernandes, A. and Stehling, E. (2014), High level of resistance to Aztreonam and Ticarcillin in Pseudomonas aeruginosa isolated from soil of different crops in Brazil, *Science of the Total Environment*, 473–4, 155–8.

Sapkota, A., Kinney, E., George, A., Hulet, R., Cruz-Cano, R., Schwab, K., Zhang, G. and Joseph, S. (2014), Lower prevalence of AMR Salmonella on large-scale U.S. conventional poultry farms that transitioned to organic practices *Science of the Total Environment*, 476–7, 387–92.

Sidrach-Cardona, R., Hijosa-Valsero, M., Marti, E., Balcázar, E. and Becares, J. (2014), Prevalence of AMR fecal bacteria in a river impacted by both an antibiotic production plant and urban treated discharges, *Science of the Total Environment*,488–9, 220–7.

Su, H., Pan, C., Ying, G., Zhao, J., Zhou, L., Li, Y., Tao, R., Zhang, R. and He, L. (2014), Contamination profiles of antibiotic resistance genes in the sediments at a catchment scale *Science of the Total Environment*, 490, 708–14.

Zhang, Y., Zhang, C., Parker, D., Snow, D. Zhou, Z. and Li, X. (2013), Occurrence of antimicrobials and antimicrobial resistance genes in beef cattle storage ponds and swine treatment lagoons, *Science of the Total Environment*, 463–4, 631–8.

(3) Human health risk through drinking water

Guo, X., Li, J., Yang, F., Yang, J. and Yin, D. (2014), Prevalence of sulfonamide and tetracycline resistance genes in drinking water treatment plants in the Yangtze River Delta, China, *Science of the Total Environment*, 493, 626–31.

Jiang, L., Hu, X., Xu, T., Zhang, H., Sheng, D. and Yin, D. (2013), Prevalence of antibiotic resistance genes and their relationship with antibiotics in the Huangpu River and the drinking water sources Shanghai, China *Science of the Total Environment*, 458–60, 267–72.

Narciso-da-Rocha, C., Vaz-Moreira, I. and Manaia C. (2014), Genotypic diversity and antibiotic resistance in Sphingomonadaceae isolated from hospital tap water, *Science of the Total Environment*, 466–7, 127–35.

Ribeiro, A., Bodilis, J., Alonso, L., Buquet, S., Feuilloley, M., Dupont, J-P. and Pawlak, B. (2014), Occurrence of multi-antibiotic resistant Pseudomonas spp. In drinking water produced from karstic hydrosystems, *Science of the Total Environment*, 490, 370–8.

Vaz-Moreira, I., Nunes, O. and Manaia C. (2012), Diversity and antibiotic resistance in Pseudomonas spp. From drinking water, *Science of the Total Environment*, 426, 366–74.

Vicenti, S., Quaranta, G., de Meo, C., Bruno, S., Ficarra, M., Carovillano, S., Ricciardi, W. and Laurenti, P. (2014), Non-fermentative gram-negative bacteria in hospital tap water and water used for haemodialysis and bronchoscope flushing: Prevalence and distribution of antibiotic resistant strains, *Science of the Total Environment*, 499, 47–54.

(4) Removal technologies

Berglund, B., Ali Khan, G., Weisner, S., Ehde, P., Fick, J. and Lindgren P-E. (2014), Efficient removal of antibiotics in surface-flow constructed wetlands with no observed impact on antibiotic resistance genes, *Science of the Total Environment*, 476–7, 29–37.
Cardinal, P., Anderson, J., Carlson, J., Low, J., Challis, J., Beattie, S. Bartel, C., Elliott, A., Montero, O., Lokesh, S., Favreau, A., Kozlova, T., Knapp, C., Hanson, M. and Wong, C. (2014), Macrophytes may not contribute significantly to removal of nutrients, pharmaceuticals, and antibiotic resistance in model surface constructed wetlands, *Science of the Total Environment*, 482–3, 294–304.
Karaolia, P., Michael, I., García-Fernández, I., Agüera, A., Malato, S., Fernández-Ibánez, P. and Fatta-Kassinos, D. (2014), Reduction of clarithromycin and sulfamethoxazole-resistant Enterococcus by pilot-scale solar-driven Fenton oxidation, *Science of the Total Environment*, 468–9, 19–27.
Morató, J., Codony, F., Sánchez, O., Martín Pérez, L., García, J. and Mas J. (2014), Key design factors affecting microbial community composition and pathogenic organism removal in horizontal subsurface flow constructed wetlands, *Science of the Total Environment*, 481, 81–9.
Rizzo, L., Manaia, C., Merlin, C., Schwartz, T., Dagot, C., Ploy M., Michael, I. and Fatta-Kassinos, D. (2013), Urban wastewater treatment plants as hotspots for antibiotic resistant bacteria and genes spread into the environment: A review, *Science of the Total Environment*, 447, 345–60.
Zhang, Y., Zhuang, Y., Geng, J., Ren, H. and Zhang, Y., Ding, L. and Xu, K. (2015), Inactivation of antibiotic resistance genes in municipal wastewater effluent by chlorination and sequential UV/chlorination disinfection, *Science of the Total Environment*, 512–13, 125–32.

10.2 The Association of American Feed Control Officials (AAFCO)

AAFCO is a voluntary membership association of local, state and federal agencies charged by law to regulate the sale and distribution of animal feeds and animal drug remedies. They publish an annual manual that lists current feed ingredients definitions, generally-regarded-as safe substances, approved Food Additive Petitions for animal food/feed; http://www.aafco.org/ (accessed May 2016).

10.3 Food and Drug Administration, Center for Veterinary Medicine

A safe animal feed supply helps ensure healthy animals and people. The website provides information for complying with federal regulations concerning animal feeds and informational updates to keep interested parties informed on current issues.

- monitors and establishes standards for feed contaminants,
- approves safe food additives, and
- manages the FDA's medicated feed and pet food programmes.

http://www.fda.gov/AnimalVeterinary/default.htm (accessed May 2016).

10.4 The American Feed Industry Association (AFIA)

The US feed industry is committed to producing animal food in an efficient and sustainable manner. AFIA shares this mission through four focus points:

- The optimization of energy and our natural resources;
- The enhancement of precision to our animal nutrition programmes through new, efficient and productive technologies;
- To provide consumers with a better understanding and appreciation of today's food production system; and
- To give support to local communities.

http://www.afia.org/ (accessed May 2016).

10.5 The European Medicines Agency, Veterinary Regulatory

The website provides information for companies and individuals involved in developing and marketing medicines for veterinary use in the EU. For further information on EU legislation and procedures for the regulation of veterinary medicines see volumes 4–9 of the rules governing medicinal products in the EU (accessed May 2016).

http://www.ema.europa.eu/ema/index.jsp?curl=pages/regulation/landing/veterinary_medicines_regulatory.jsp&mid= (accessed May 2016).

10.6 The International Feed Industry Federation (IFIF)

Represents and promotes the global feed industry as an essential participant in the food chain that provides sustainable, safe, nutritious and affordable food for a growing world population.

http://ifif.org/ (accessed May 2016).

10.7 The European Feed Manufacturers' Federation (FEFAC)

FEFAC was founded in 1959 by five national compound feed associations from France, Belgium, Germany, Italy and the Netherlands. FEFAC membership today consists of national associations from EU Member States as full members and of an increasing number of observer members from non-EU countries. FEFAC is the only independent spokesman of the European Compound Feed Industry at the level of the European Institutions. FEFAC holds observer status in Codex Alimentarius.

http://www.fefac.eu/ (accessed May 2016).

10.8 The Canadian Food Inspection Agency (CFIA)

CFIA aims to mitigate risks to Canada's animal resource base, animal feeds and animal products, which are integral to a safe and accessible food supply system as well as to public health.

http://www.inspection.gc.ca/animals/eng/1299155513713/1299155693492 (accessed May 2016).

10.9 National Grain and Feed Association (NGFA)

NGFA is the national trade association of grain elevators, feed and feed ingredient manufacturers, grain and oilseed processors, biofuel producers, exporters, livestock and poultry integrators, and firms providing products and services to the industry.

https://www.ngfa.org/(accessed May 2016).

10.10 Codex Alimentarius (Codex)

Codex or 'Food Code' was established by Food and Agriculture Organization of the United Nations and the World Health Organization (WHO) in 1963 to develop harmonized international food standards, which protect consumer health and promote fair practices in food trade.

Codex international food standards, guidelines and codes of practice contribute to the safety, quality and fairness of this international food trade. Consumers can trust the safety and quality of the food products they buy and importers can trust that the food they ordered will be in accordance with their specifications.

http://www.fao.org/fao-who-codexalimentarius/codex-home/en/(accessed May 2016).

10.11 Food and Agriculture Organization (FAO) of the United Nations

Antimicrobial resistance (AMR) is a major global public and animal health issue of increasing concern. The risk appears to be particularly high in countries where legislation, regulatory surveillance and monitoring systems regarding the use of antimicrobials and the prevention and control of antimicrobial resistance, are weak or inadequate. This is the reason why FAO plays a key role in supporting government, producers, traders and other stakeholders to adopt measures to minimize the use of antimicrobials and to prevent the development of antimicrobial resistance.

http://www.fao.org/antimicrobial-resistance/en/ (accessed May 2016).

10.12 World Health Organization (WHO)

The primary role of WHO is to direct and coordinate international health within the United Nations' system.

Main areas of work:

- Health systems
- Promoting health through the life course
- Non-communicable diseases
- Communicable diseases
- Corporate services
- Preparedness, surveillance and response.

Antimicrobial resistance: http://search.who.int/search?q=antimicrobial+resistance&ie=utf8&site=who&client=_en_r&proxystylesheet=_en_r&output=xml_no_dtd&oe=utf8&getfields=doctype (accessed May 2016).

11 References

Aguilera-Luiz, M., Romero-González, R., Plaza-Bolaños, P., Martinez Vidal, J. and Garrido Frenich, A. (2013), 'Wide-scope analysis of veterinary drug and pesticide residues in animal feed by liquid chromatography coupled to quadrupole-time-of-flight mass spectrometry', *Anal. Bional. Chem.*, 405(20), 6543–53.

Aksenova, I., Ter-Sarkisian, E., Soifer, R., Florova, G. and Iustratova, L. (1984), Effect of the pH of the medium and of temperature on tylosin stability, *Antibiotiki*, 29, 179–82.

Lenore S., Camel Eyre, X., Benbadis, L., Bideaux, C., Uribelarrea, J., Goma, G., Molina-Jouve, C. and Guillouet, S. (2004), Aeration strategy: a need for very high ethanol performance in *Saccharomyces cerevisiae* fed-batch process. *Appl. Microbiol. Biotechnol.*, 63, 537–42.

Amorim, H., Basso, L. and Lopes, M. (2004), Evolution of ethanol fermentation in Brazil, *Distilled Spirits: Tradition and Innovation*, Nottingham University Press, pp. 143–8.

Anderson, C., Rupp, H. and Wu, W. (2005), Complexities in tetracycline analysis-chemistry, matrix extraction, cleanup, and liquid chromatography, *J. Chromatogr. A*, 1075, 23–32.

Anon. (1981), Compliance Policy Guide (CPG) Sec. 680.600. Sequencing as a Means to Prevent Unsafe Drug Contamination in the Production, Storage, and Distribution of Feeds http://www.fda.gov/ICECI/ComplianceManuals/CompliancePolicyGuidanceManual/ucm074700.htm (accessed May 2016).

Anon. (1994), ISO 5725-3: 1994 (E), Accuracy (trueness and precision) of measurement methods and results – Part 3: Intermediate measures of the precision of a standard measurement method, ISO, Standards, ICS 03.120 30, p. 1.

Anon. (2003), CVM Guidance for Industry #118: Mass Spectrometry for Confirmation of the Identity of Animal Drug Residues, http://www.fda.gov/downloads/animalveterinary/guidancecomplianceenforcement/guidanceforindustry/ucm052658.pdf (accessed May 2016).

Anon. (2003), European Union, Regulation (Ec) No 1831/2003 Of The European Parliament and of The Council of 22 September 2003 on additives for use in animal nutrition, Official Journal of the European Union, October 10, 2003, L 268/29 http://faolex.fao.org/docs/pdf/eur40306.pdf (accessed April 2016).

Anon. (2014) WHO. Antimicrobial resistance: Global report on surveillance, http://apps.who.int/gb/ebwha/pdf_files/WHA67/A67_39-en.pdf (accessed April 2016).

Anon. (2015), Title 21, Code of Federal Regulation, Part 558 (21 CFR 558); https://www.federalregister.gov/articles/2015/06/03/2015-13393/veterinary-feed-directive (accessed May 2016); and Guidance for industry small entity compliance guide veterinary feed directive regulation questions and answers, GFI #120, U.S. Food and Drug Administration, Center for Veterinary Medicine, http://www.fda.gov/downloads/AnimalVeterinary/GuidanceComplianceEnforcement/GuidanceforIndustry/UCM052660.pdf (accessed May 2016).

Anon. (2016a) WHO. Evaluation of certain veterinary drug residues in food: Eighty-first report of the Joint FAO/Who Expert Committee on Food Additives, http://apps.who.int/iris/bitstream/10665/204670/1/9789240695504_eng.pdf?ua=1 (accessed July 2016).

Anon. (2016b), US fuel ethanol industry biorefineries and capacity and annual world ethanol production by country,' Renewable Fuel Association, Washington, DC, http://www.ethanolrfa.org/pages/statistics/(accessed April 2016).

Balizs, G. and Hewitt, A. (2003), Determination of veterinary drug residues by liquid chromatography and tandem mass spectrometry, Anal. Chim. Acta, 492, 105–31.

Barceló, D. and Bennett, J. (2016), Antibiotic resistance in the environment, Virtual Special Issues, *Science of the Total Environment*, http://www.journals.elsevier.com/science-of-the-total-environment/virtual-special-issues/antibiotic-resistance-in-the-environment/(accessed May 2016).

Bettiga, M., Hahn-Hägerdal, B. and Gorwa-Grauslund, M. (2008), Comparing the xylose reductase/xylitol dehydrogenase and xylose isomerase pathways in arabinose and xylose fermenting *Saccharomyces cerevisiae* strains. *Biotechnol. Biofuels*, 1, 16 (doi:10.1186/1754-6834-1-16).

Beyene, T. (2016), 'Veterinary drug residues in food-animal products: risk factors and potential effects on public health', *J. Veterinar. Sci. Technol.*, 7(1), http://www.omicsonline.org/open-access/veterinary-drug-residues-in-foodanimal-products-its-risk-factors-andpotential-effects-on-public-health-2157-7579-1000285.php?aid=66580 (accessed July 2016).

Bischoff, K., Zhang, Y. and Rich, J. (2016), Fate of virginiamycin through the fuel ethanol production process, *World J. Microbiol. Biotechnol.*, 32, 76–81.

Blasco, C., Picó, Y. and Torres, C. (2007), Progress in analysis of residual antibacterial in food, TrAC – Trends Anal. Chem., 26, 895–913.Boix, C, Ibáñez, M., Sancho, J, León, Yusá, V. and Hernández, F. (2014), 'Qualitative screening of 116 veterinary drugs in feed by liquid chromatography-high resolution mass spectrometry: potential application to quantitative analysis', *Food Chem.*, 160, 313–20.

Borras, S., Companyo, R., Granados, M., Guiteras, J., Perez-Vendrell, A., Brufau, J., Medina, M. and Bosch, J. (2011), Analysis of antimicrobial agents in animal feed, *TrAC – Trends Anal. Chem.*, 30, 1042–64.

Boscher, A., Guignard, C., Pellet, T., Hoffmann, L. and Bohn, T. (2010), Development of a multi-class method for the quantification of veterinary drug residues in feedingstuffs by liquid chromatography-tandem mass spectrometry, *J. Chromatogr. A*, 1217, 6394–6404.

Botsoglu, N. and Fletouris, D. (2001), Drug Residues in Food-Pharmacology, *Food Safety and Analysis*, Marcel Decker AG.

Brisaert, M., Heylen, M. and Plaizier-Vercammen, J. (1996), Investigation on the chemical stability of erythromycin in solutions using an optimization system, *Pharm. World Sci.*, 18, 182–6.

Careri, M., Bianchi, F. and Corradini, C. (2002), Recent advances in the application of mass spectrometry in food-related analysis, J. Chromatogr. A, 970, 3–64.

Cha, J., Yang, J. and Carlson, J. (2005), Rapid analysis of trace levels of antibiotic polyether ionophores in surface water by solid-phase extraction and liquid chromatography with ion trap tandem mass spectrometric detection, *J. Chromatogr. A*, 1065, 187–98.

Compart, P., Carlson, A., Crawford, G., Fink, R., Diez-Gonzalez, F., DiCostanzo, A. and Shurson, G. (2013), Presence and biological activity of antibiotics used in fuel ethanol and corn co-product production, *J. Anim. Sci.*, 91, 2395–2404.

Corrigan, M. and Mass, R. (2009), Utilisation of distillers' dried grains with solubles (DDGS) by cattle, *The Alcohol Textbook*, Nottingham University Press, pp. 323–35.

Cronly, M., Behan, P., Foley, Malone, E., Earley, S., Gallagher, M., Shearan, P. and Regan, L. (2010), Development and validation of a rapid multi-class method for the confirmation of fourteen prohibited medicinal additives in pig and poultry compound feed by liquid chromatography–tandem mass spectrometry, *J. Pharm. Biomed. Anal.*, 53, 929–38.

da Costa, P., Oliveira, M., Bica, A., Vaz-Pires, P. and Bernardo. F. (2007), Antimicrobial resistance in *Enterococcus* spp. and *Escherichia coli* isolated from poultry feed and feed ingredients, *Vet. Microbiol.*, 120, 122–31.

De Alwis, H. and Heller, D. (2010), Multiclass, multiresidue method for the detection of antibiotic residues in distillers grains by liquid chromatography and ion trap tandem mass spectrometry, *J. Chromatogr. A*, 121, 3076–84.

De Olivia Neto, P., Amorim Ferreira, M. and Yokoya, F. (2004), Screening for yeast with antibacterial properties from an ethanol distillery. *Bioresour. Technol.*, 92, 1–6.

Edrington, T., MacDonald, J., Farrow, R., Callaway, T., Anderson, R. and Nisbet. D. (2010), Influence of wet distiller's grains on prevalence of *Escherichia coli* O157:H7 and *Salmonella* in feedlot cattle and antimicrobial susceptibility of generic *Escherichia coli* isolates, *Foodborne Pathog. Dis.*, 7, 605–8.

Farré, M., Barceló, D. and Barceló, D. (2013), Analysis of emerging contaminants in food, TrAC – Trends Anal. Chem., 43, 240–53.Fink-Gremmels. J. (2012), *Animal Feed Contamination: Effects on Livestock and Food Safety*, Woodhead Publishing Limited, Cambridge, UK.

Fletcher, S. (2015), Understanding the contribution of environmental factors in the spread of antimicrobial resistance, *Environ. Health Prev. Med.* 20, 243–52.

Gomez-Perez, M., Plaza-Bolanos, P., Romero-Gonzalez, R., Martinez-Vidal, J. and Garrido-Frenich, A. (2012), Comprehensive qualitative and quantitative determination of pesticides and veterinary drugs in honey using liquid chromatography–Orbitrap high resolution mass spectrometry, *J. Chromatogr. A*, 1248, 130–8.

Gómez-Pérez, M., Romero-González, Martinez Vidal, J. and Garrido Frenich, A. (2015), 'Analysis of veterinary drug and pesticide residues in animal feed by high resolution mass spectrometry: comparison between time-of-flight and Orbitrap', *Food Addit. Contam Part A Chem. Anal Control Expo Risk Assess*, 32(10), 1637–46.

Guillarme, D., Schappler, J., Rudaz, S. and Veuthey, J. (2010), Coupling ultra-high-pressure liquid chromatography with mass spectrometry, TrAC – Trends Anal. Chem., 29, 15–27.

Hamdy, M., Toledo, R., Shieh, C., Pfannenstiel, M. and Wang, R. (1996), Effects of virginiamycin on fermentation rate by yeast, *Biomass Bioenerg.*, 11, 1–9.

Hurd, H., Doores, S., Hayes, D., Mathew, A., Maurer, J., Silley, P., Singer, R. and Jones, R. (2004), Public health consequences of macrolide use in food animals: A deterministic risk assessment, *J. Food Prot.* 67, 980–92.

Hynes, S., Kjarsgaard, D., Thomas, K. and Ingledew, W. (1997), Use of virginiamycin to control the growth of lactic acid bacteria during alcohol fermentation, *J. Ind. Microbiol. Biotechnol.*, 18, 284–91.

Ingledew, W. (2009), Yeasts: physiology, nutrition and ethanol production. In *The Alcohol Textbook*, Nottingham University Press, pp. 101–13.

Islam, M., Toledo, R. and Hamdy, M. (1999), Stability of virginiamycin and penicillin during alcohol fermentation, *Biomass Bioenergy*, 17, 369–85.

Jacob, M., Fox, J., Narayanan, S., Drouillard, J., Renter, D. and Nagaraja, T. (2008), Effects of feeding wet corn distillers grains with solubles with or without monensin and tylosin on the prevalence and antimicrobial susceptibilities of fecal foodborne pathogenic and commensal bacteria in feedlot cattle, *J. Anim. Sci.*, 86, 1182–90.

Juranek, P. and Duquette, P. (2007), Antibiotic regulatory considerations for distiller's grains, Distillers Grains Q., 4th quarter, BBI International Media, Grand Forks, ND.

Kaklamanos, G., Vincent, U. and von Holst, C. (2013a), Analysis of antimicrobial agents in pig feed by liquid chromatography coupled to Orbitrap mass spectrometry, *J. Chromatogr. A*, 1293, 60–74.

Kaklamanos, G., Vincent, U. and von Holst, C. (2013b), Multi-residue method for the detection of veterinary drugs in distillers grains by liquid chromatography–Orbitrap high resolution mass spectrometry, *J. Chromatogr. A*, 1322, 38–48.

Kantiani, L., Farre, M., Freixiedas, J. and Barceló, D. (2010), Development and validation of a pressurised liquid extraction liquid chromatography-electrospray-tandem mass spectrometry method for β-lactams and sulfonamides in animal feed, *J. Chromatogr. A*, 1217, 4247–54.

Kaufmann, A., Butcher, P., Maden, K. and Widmer, M. (2008), Quantitative multiresidue method for about 100 veterinary drugs in different meat matrices by sub 2-μm particulate high-performance liquid chromatography coupled to time of flight mass spectrometry, *J. Chromatogr. A*, 1194 66–79.

Kaufmann, A., Butcher, P., Maden, K., Walker, S. and Widmer, M. (2011a), Quantification of anthelmintic drug residues in milk and muscle tissues by liquid chromatography coupled to Orbitrap and liquid chromatography coupled to tandem mass spectrometry, *Talanta*, 85, 991–1000.

Kaufmann, A., Butcher, P., Maden, K., Walker, S. and Widmer, M. (2011b), Development of an improved high resolution mass spectrometry based multi-residue method for veterinary drugs in various food matrices, *Anal. Chim. Acta*, 700, 86–94.

Kellmann, M., Muenster, H., Zomer, P. and Mol, H. (2009), Full scan MS in comprehensive qualitative and quantitative residue analysis in food and feed matrices: how much resolving power is required, *J. Am. Soc. Mass Spectrom.*, 20, 1464–76.

Kemper, N. (2008), Veterinary antibiotics in the aquatic and terrestrial environment, *Ecol. Ind.*, 8, 1–13.

Kheirolomoom, A., Kazemi-Vaysari, A., Ardjmand, M. and Baradar-Khoshfetrat, A. (1999), The combined effects of pH and temperature on penicillin G decomposition and its stability modeling, *Process. Biochem.*, 35, 205–11

Le Bizec, B., Pinel, G. and Antignac, J. (2009), Options for veterinary drug analysis using mass spectrometry, J. Chromatogr. A, 1216, 8016–34.
Lehman, R. and Rosentrater. K. (2007), Microbial development in distiller's wet grains produced during fuel ethanol production from corn (*Zea mays*), *Can. J. Microbiol.*, 53, 1046–52.
Logothetis, S., Nerantzis, E., Gioulioti, A. and Walker, G., (2010), Influence of sodium chloride on wine yeast fermentation performance. *Int. J. Wine Res.*, 2, 35–42.
Lushia, W. and Heist, P. (2005), Antibiotic resistant bacteria in fuel ethanol fermentations, *Ethanol Producer Mag.*, 80–2.
Luther, M. (2012a), Report of FY 2008 nationwide survey of distiller's products for antibiotic residues, Center for Veterinary Medicine, FDA, Rockville, MD. http://www.fda.gov/AnimalVeterinary/Products/AnimalFoodFeeds/Contaminants/ucm331189.htm (accessed April 2016).
Luther, M. (2012b), Report of FY 2010 nationwide survey of distiller's products for antibiotic residues, Center for Veterinary Medicine, FDA, Rockville, MD. http://www.fda.gov/AnimalVeterinary/Products/AnimalFoodFeeds/Contaminants/ucm300126.htm (accessed April 2016).
Makanjuola, D., Tymon, A. and Springham, D. (1992), Some effects of lactic-acid bacteria on laboratory-scale yeast fermentations, *Enzyme Microb. Technol.*, 14, 350–7.
Malik, A., Blasco, C. and Picó, Y. (2010), Liquid chromatography–mass spectrometry in food safety, J. Chromatogr. A, 1217, 4018–40.
Marazuela, M. and Bogialli, D. (2013), chapter 17–Determination of Veterinary Drug Residues in Foods by Liquid Chromatography–Mass Spectrometry: Basic and Cutting-Edge Applications, Liquid Chromatogr., 455–76.
Marazuela, M. and Bogialli, S. (2009), A review of novel strategies of sample preparation for the determination of antibacterial residues in foodstuffs using liquid chromatography-based analytical methods, Anal. Chim. Acta, 645, 5–17.
McGlinchey, T., Rafter, P., Regan, F. and McMahon, G. (2008), A review of analytical methods for the determination of aminoglycoside and macrolide residues in food matrices, *Anal. Chim. Acta.*, 624, 1–15.
Meek, R., Vyas, H. and Piddock, L. (2015) Nonmedical uses of antibiotics: time to restrict their use?, *PLoS Biol.*, 13(10), e1002266. doi:10.1371/journal.pbio.1002266.
Mol, H., Plaza-Bolanos, P., Zomer, P., de Rijk, T., Stolker, A. and J. Mulder, P. (2008), Toward a generic extraction method for simultaneous determination of pesticides, mycotoxins, plant toxins, and veterinary drugs in feed and food matrixes, *Anal. Chem.*, 80, 9450–9.
Monceaux, D. and Kuehner, D. (2009), Dryhouse technologies and DDGS production, *The Alcohol Textbook,* Nottingham University Press, pp. 303–22.
Narendranath, N. (2003), Bacterial contamination and control in ethanol production, *The Alcohol Textbook,* Nottingham University Press, pp. 287–98.
Narendranath, N., Hynes, S., Thomas, K. and Ingledew, W. (1997), Effects of lactobacilli on yeast-catalyzed ethanol fermentations, *Appl. Environ. Microbiol.,* 63, 4158–63.
Ortelli, D., Cognard, E., Jan, P. and Edder, P. (2009), Comprehensive fast multiresidue screening of 150 veterinary drugs in milk by ultra-performance liquid chromatography coupled to time of flight mass spectrometry, *J. Chromatogr. B*, 877, 2363–74
Pedersen, C., Boersma, M. and Stein. H. (2007), Digestibility of energy and phosphorus in ten samples of distillers dried grains with solubles fed to growing pigs, *J. Anim. Sci.*, 85, 1168–76.
Peters, R., Bolck, Y., Rutgers, P., Stolker, A. and Nielen, M. (2009), Multi-residue screening of veterinary drugs in egg, fish and meat using high-resolution liquid chromatography accurate mass time-of-flight mass spectrometry, *J. Chromatogr. A,* 1216, 8206–16.
Peters, R., Stolker, A., Mol, J., Lommen, A., Lyris, E., Angelis, Y., Vona-parti, A., Stamou, M., Georgakopoulos, C. and Nielen, M. (2010), Screening in veterinary drug analysis and sports doping control based on full-scan, accurate-mass spectrometry, TrAC – Trends Anal. Chem., 29, 1250–68.
Phillips, I., Casewell, M., Cox, T., De Groot, B., Friis, C., Jones, R., Nightingale, C., Preston, R. and Waddell, J. (2004), Does the use of antibiotics in food animals pose a risk to human health? A critical review of published data, *J. Antimicrob. Chemother.* 53, 28–52.

Pilgrim, C. (2009), Status of the worldwide fuel alcohol industry, *The Alcohol Textbook*, Nottingham University Press, pp. 7–17.

Romero-Gonzalez, R., Aguilera-Luiz, M., Plaza-Bolanos, P., Garrido-Frenich, A. and Martinez-Vidal, J. (2011), Food contaminant analysis at high resolution mass spectrometry: Application for the determination of veterinary drugs in milk, *J. Chromatogr. A*, 1218, 9353–65.

Samanidou, V. and Evaggelopoulou, E. (2008), Chromatographic analysis of banned antibacterial growth promoters in animal feed, *J. Sep. Sci.*, 31, 2091–12.

Schenck, F. and Callery, P. (1998), Chromatographic methods of analysis of antibiotics in milk, *J. Chromatogr. A*, 812, 99–109.

Schroeder, T., Drouillard, J., Fox, J. and Strode, J. (2010), Distiller's grain market development and price: Risk management: Final report, https://www.ams.usda.gov/sites/default/files/media/KS0668.pdf (accessed April 2016).

Shurson, G., Spiehs, M., Wilson, J. and Whitney, M. (2003), Value and use of 'new generation' distiller's dried grains with solubles in swine diets, Proceedings of Alltech's 19th Annual Symposium. Nottingham, Nottingham, UK, pp. 223–35.

Skinner, K. and Leathers, T. (2004), Bacterial contaminants of fuel ethanol production, *J. Ind. Microbiol. Biotechnol.*, 31, 401–8.

Smyth, C. (2010), US and EU indicative bioethanol margins, *Biofuel Ind. News*, 41, November 24.

Stolker, A., Zuidema, T. and Nielen, M. (2007), Residue analysis of veterinary drugs and growth-promoting agents, TrAC – Trends Anal. Chem., 26, 967–79.

Stolker, A. and Brinkman, U. (2005), Analytical strategies for residue analysis of veterinary drugs and growth-promoting agents in food-producing animals- a review, *J. Chromatogr. A*, 1067, 15–53.

Stolker, A., Rutgers, P., Oosterink, E., Lasaroms, J., Peters, R., van Rhijn, J. and Nielen, M. (2008), Comprehensive screening and quantification of veterinary drugs in milk using UPLC–ToF-MS, *Anal. Bioanal. Chem.*, 391, 2309–22.

Sunde, M., Fossum, K., Solberg, A. and Sorum, H. (1998), Antibiotic resistance in *Escherichia coli* of the normal intestinal flora of swine, *Microb. Drug Resist.*, 4, 289–99.

Thomas, K. and Ingledew, W. (1992), Production of 21% (v/v) ethanol by fermentation of very high gravity (VHG) wheat mashes, *J. Ind. Microbiol.*, 10, 61–8.

van der Heeft, E., Bolck, Y., Beumer, B., Nijrolder, A., Stolker A. and Nielen, M. (2009) Full-scan accurate mass selectivity of ultra-performance liquid chromatography combined with time-of-flight and Orbitrap mass spectrometry in hormone and veterinary drug residue analysis, *J. Am. Soc. Mass Spectrom.*, 20, 451–63.

Villar-Pulido, M., Gilbert-Lopez, B., Garcia-Reyes, J., Martos, N. and Molina-Diaz, A. (2011), Multiclass detection and quantitation of antibiotics and veterinary drugs in shrimps by fast liquid chromatography time-of-flight mass spectrometry, *Talanta*, 85, 1419–27.

Walker, G. (2011), 125th Anniversary review: fuel alcohol: current production and future challenges, *J. Inst. Brew.*, 117(1), 3–22.

Wan, E., Ho, C., Sin, D. and Wong, Y. (2006), Detection of residual bacitracin A, colistin A, and colistin B in milk and animal tissues by liquid chromatography tandem mass spectrometry, *Anal. Bioanal. Chem.*, 385, 181–8.

Wang, J. (2009), Analysis of macrolide antibiotics, using liquid chromatography-mass spectrometry, in food, biological and environmental matrices, *Mass Spectrom. Rev.*, 28, 50–92.

Wang, L., Yang, H., Zhang, C., Mo, Y. and Lu, X. (2008), Determination of oxytetracycline, tetracycline and chloramphenicol antibiotics in animal feeds using subcritical water extraction and high performance liquid chromatography, *Anal. Chim. Acta*, 619, 54–8.

Yang, S. and Carlson, K. (2004), Solid phase extraction-high-performance liquid chromatography-ion trap mass spectrometry for analysis of trace concentrations of macrolide antibiotics in natural and waste water matrices, *J. Chromatogr. A*, 1038, 141–5.

Yang, S., Cha, J. and Carlson, K. (2006), Trace analysis and occurrence of anhydroerythromycin and tylosin in influent and effluent wastewater by liquid chromatography combined with electrospray tandem mass spectrometry, *Anal. Bioanal. Chem.* 385, 623–36.

Part 2

Ensuring safety at slaughter

Chapter 7

Beef carcass inspection systems

William James, formerly Food Safety and Inspection Service (FSIS)-USDA, USA

1 Introduction

Globally, foodborne diseases and zoonoses are important public health and animal health problems. Prevention and reduction of foodborne diseases and zoonoses by ante- and post-mortem meat inspection are a historical responsibility of Veterinary Services (see definition in Table 1). Veterinary Services in most advanced countries are responsible for designing and implementing ante- and post-mortem meat inspection programmes (Anon., 2016).

Some international organizations make reference to a 'central competent authority' to encompass the many other disciplines contributing to a successful meat inspection programme. However, many of the veterinarians leading their national meat inspection programmes are also formally educated in public health. They have spent their careers working at the nexus of animal health and public health, and every credible meat inspection programme in the world has veterinarians in top leadership positions (Anon., 2004a). Out of respect for their current and historical contributions to the art and science of meat inspection, this chapter will use the term Veterinary Services. This and other important terms used in this chapter are defined in Table 1.

This chapter begins by providing an overview of the purposes and importance of meat inspection programmes. Subsequent sections cover ante- and post-mortem inspection, and provide a review of some important procedures used by Veterinary Services to inspect cow heads, viscera and carcasses.

http://dx.doi.org/10.19103/AS.2016.0008.11

Table 1 Definition of terms

Beef	All parts of a bovine animal that are intended for, and have been determined to be safe and suitable for, human consumption (Anon., 2005)
Hazard	A biological, chemical or physical agent in, or condition of, food with the potential to cause an adverse health effect (Anon., 2005).
Inspector	A person who has the training, knowledge, skills and experience to perform an assigned task, and who is subject to requirements specified by Veterinary Services (Anon., 2005).
Safe (for human consumption)	Safe for human consumption according to the following criteria (Anon., 2005): • has been produced by applying all food safety requirements appropriate to its intended end use • meets risk-based performance and process criteria for specified hazards • does not contain hazards at levels that are harmful to human health
Suitable (for human consumption)	Suitable for human consumption according to the following criteria (Anon., 2005): • has been produced under hygienic conditions • is appropriate to its intended use • meets standards for specified diseases or defects as established by Veterinary Services
Veterinary Services	The official authority charged by a nation's government with the control of meat hygiene, including setting and enforcing regulatory meat hygiene requirements (Anon., 2005). Sometimes referred to as the 'central competent authority' in international guidelines.

2 Successful meat inspection programmes: an overview

The best meat inspection programmes are risk-based. Management systems should cover hazards (see definition in Table 1) to both human and animal health in the animals slaughtered, as determined by a country's Veterinary Services (Petersen et al., 1996). An inspection programme should be sufficiently flexible to inspection procedures that can be easily adjusted based on the emergence or disappearance of specific hazards (Walker et al., 2000).

Sound public health and animal health principles require that finite resources be employed to control hazards that are reasonably likely to occur, as determined by the Veterinary Services (Petersen et al., 1996). There are many theoretical hazards for which control measures can be developed. However, hazards that are unproven or exceedingly rare should be considered of negligible importance to the design of inspection systems. They rob resources from management of those hazards which will do the most good for the most people in a country.

All gross pathology of food animals is condemnable at inspection, but not all are hazardous to public health. So, not all diseases and conditions deserve the same level of attention. There are three basic categories of animal diseases and conditions in relation to their importance as zoonoses:

1 Those that are not transmissible to people: Most animal diseases and conditions have no effects on people. The affected tissues would be safe (see definition in Table 1)

to eat. For example, bruised muscle tissue seen at slaughter. However, bruises are trimmed and condemned because they are not considered suitable (see definition in Table 1) for consumption.

2 Those that are occupational health concerns: Occupational health problems arise from working with or inspecting animals or their products which have infectious diseases at slaughter that are dangerous to people. An example is brucellosis in cattle, sometimes called Malta fever in people.
3 Those that are food safety concerns: A few animal diseases with food safety implications are visible to the naked eye. Most of these are parasites like beef tapeworms. However, most food safety concerns in meat arise from sources that cannot be seen during traditional organoleptic inspection – bacteria.

Inspection programmes based on risk help to focus inspection activities where they will do the most good.

Many of the bacteria that cause foodborne illnesses in people are well adapted to their animal hosts. An inspector (see definition in Table 1) cannot detect signs of disease in animals from specific human pathogens. Inspectors are reduced to inspecting for evidence of poor processing, such as faecal contamination. This is why process control inspection is more important than historical carcass-by-carcass inspection.

In addition, traditional organoleptic inspection is primarily useful as a means of sorting abnormal tissues and parts from normal tissues and parts. Using sight, smell and touch, inspectors make dispositions rather than diagnoses. A disposition consists of one of three decisions:

1 Passed – The animal is passed for slaughter at ante-mortem inspection, or the product is passed for consumption at post-mortem inspection.
2 Suspect – The animal or product is held for closer examination and testing by a veterinarian.
3 Condemned – The animal or product is rejected for slaughter or human consumption.

Many of the signs of an unhealthy animal inspected at ante-mortem inspection cannot be attributed to a specific disease. Also, much of the pathology seen at post-mortem inspection does not lend itself to an explicit diagnosis. Such condemnations are assigned to general disease categories such as septicaemia with no certain underlying aetiology.

Dispositions should only be made by inspectors with a level of training, knowledge and skills required by Veterinary Services (Anon., 2005). Since an individual bovine at slaughter represents a significant investment by the producer, only a veterinarian should condemn an animal at ante-mortem inspection or an entire carcass at post-mortem inspection.

The information provided in this chapter targets specifically cattle and beef (see definition in Table 1) inspection, but the principles are generally capable of extending to inspection of other species and their products.

The functions of a nation's Veterinary Services include:

1 risk analysis
2 establishment of standards and procedures
3 management of the inspection programme
4 verification and certification of compliance with standards and procedures
5 dissemination of information throughout the meat production chain.

Veterinary Services should use risk assessment in the development of an inspection programme. Priority should be given to addressing microbiological contamination resulting from dressing procedures, as this has been found to be the most important source of hazards in beef (Petersen and James, 1998). Serological or other testing of cattle and beef as part of ante- and post-mortem inspection should be used to support surveillance for animal diseases. The information gathered should be linked to human disease data to allow an assessment of the effectiveness of various animal disease management options and evaluation of foodborne diseases (Anon., 2016).

Standards and procedures for ante- and post-mortem inspection should be designed to detect and remove hazards of public health and animal health significance from the beef production chain. This is achieved by the removal of live animals at ante-mortem inspection or by the condemnation of specific tissues at post-mortem inspection (Anon., 2016).

There are different effective methods of managing inspection. Veterinary Services in various countries allow flexibility in meat inspection service delivery. For instance, inspectors (see definition in Table 1) in some countries are required to be government employees trained and managed by Veterinary Services. However, in some countries, inspectors can be private contractors to the government, whose activities are managed by Veterinary Services. Also, some countries permit the companies engaging in slaughter of cattle to perform some functions traditionally associated only with Veterinary Services. These functions include sorting abnormal animals and products from normal animals and products under the verification of Veterinary Services.

In any inspection programme, inspectors or those performing sorting functions under verification should be educated or trained to meet the requirements established under the Veterinary Services' standards and procedures. Verification and certification of inspection and compliance activities is a vital function of Veterinary Services to assure the consistent application of these standards and procedures. Veterinary Services must provide full assurance for domestic consumers and international trading partners that beef produced under its inspection programme is safe and suitable for consumption.

Dissemination of information throughout the beef production chain enables Veterinary Services, farmers and ranchers, and slaughter establishments to evaluate the efficacy of their hazard controls. Animal identification and animal traceability systems should be developed to trace slaughtered animals back to their place of origin, and products derived from them forward through the meat production chain (Anon., 2016).

3 Ante-mortem inspection

The term 'ante-mortem' means 'before death'. Ante-mortem inspection is the inspection of live animals before slaughter.

3.1 Design of ante-mortem inspection

Ante-mortem inspection must be included as a fundamental element of a risk-based system for the production of beef. All animals presented for slaughter should be subjected to routine ante-mortem inspection on an individual basis (Anon., 2004b).

Ante-mortem inspection is most often conducted at the slaughter establishment. However, with proper facilities and equipment, some or all ante-mortem procedures can be conducted at another site such as the farm or ranch of origin. Such an arrangement can be beneficial in particular circumstances, for example, instances where control of an outbreak of an infectious disease is of concern (Anon., 2005).

Inspection should include confirmation that the animals are properly identified as to slaughter class, health status, and farm or ranch of origin. It is important that this information is available at ante-mortem inspection for relevant public health and animal health controls. For example, a relevant public health concern might exist if the farm or ranch of origin has a history of veterinary drug residue violations. An example of an animal health concern is the presence of a reportable disease in a country, such as foot-and-mouth disease (Anon., 2005).

Ante-mortem inspection should support post-mortem inspection by utilization of procedures or tests designed to detect signs of disease in the live bovine that cannot be observed at post-mortem inspection. Neurologic signs are an example of a condition or disease that cannot be observed at post-mortem inspection (Anon., 2005).

The slaughter establishment's responsibilities for maintaining the premises where ante-mortem inspection is to be conducted for cattle should include pens in satisfactory condition for ease of access and safety of the inspector performing ante-mortem inspection (Anon., 2013).

The lighting must be sufficient for inspection. The establishment must provide adequate areas for holding animals that are identified by Veterinary Services as suspect and condemned. An establishment should also provide a restraining device such as a chute or squeeze gate for restraining animals and taking temperatures during the examination of animals, when appropriate (Anon., 2013).

The establishment should have a system for the identification of animals presented for slaughter to enable an inspector to document that ante-mortem inspection has been performed before cattle are moved to slaughter. A common method is use of identification cards presented to an inspector before ante-mortem inspection is performed. An identification card contains spaces to record the date and time of inspection, pen or lot number, number and slaughter class of animals presented, and the inspector's signature or initials (Anon., 2013).

An inspector should count and record all cattle passed at ante-mortem inspection for correlation with the number of cattle presented for post-mortem inspection. The numbers should match exactly (Anon., 2013).

Information generated from ante-mortem inspection of importance to public health or animal health should be provided to the producer to enable improvement in husbandry methods of cattle presented for slaughter (Anon., 2005).

3.2 Implementation of ante-mortem inspection

Veterinary Services should determine how ante-mortem inspection is to be implemented, including identification of the components that may be applied at primary production rather than the slaughter establishment. Veterinary Services should establish the required knowledge, skill and experience of inspectors.

An inspector must perform assigned duties with caution, employing principles of safety. Working with cattle in tight quarters is dangerous.

Inspectors should verify that the establishment carries out its responsibilities in support of ante-mortem inspection (Anon., 2013):

1 application of identification systems for individual animals that document the findings of ante-mortem inspection throughout the slaughter process

2 presentation of reasonably clean animals
3 prompt removal of animals that have died in the pens, with permission of an inspector

Ante-mortem inspection should occur soon after the arrival of cattle (Anon., 2004b).

Ante-mortem inspection consists of two parts (Anon., 2013):

1 observe animals at rest
2 observe animals in motion

It is important to inspect the animals both at rest and in motion because some abnormal signs, such as laboured breathing, are easier to detect when the animals are at rest, while other abnormalities, such as lameness, may not be detected unless the cattle are in motion.

Among lots of cattle that contain high pathology, the animals should be observed from both sides to detect diseases such as epithelioma of the eye (Anon., 2013).

3.3 Results of ante-mortem inspection

Results of ante-mortem inspection should be recorded on an official form and made available to the inspector performing post-mortem inspection to inform a final disposition. An establishment should identify the carcass and parts of the animal from which they come (Anon., 2013).

The inspector performing ante-mortem inspection should hold a bovine in special facilities for more detailed inspection, diagnostic tests or treatment when indicated. Animals condemned as unsafe or unsuitable for human consumption should be immediately euthanized, or moved to a suitable location for treatment. The reason for condemnation should be recorded. This information should be provided to the producer.

Cattle in which no disease or condition has been identified that would make the meat unsuitable for food should be released for slaughter.

Cattle that exhibit signs indicating the animal might be affected by a disease or condition that would make the meat unsuitable for food should be designated as suspect. The bovine should be segregated in a suspect pen for closer inspection by a veterinarian.

After further examination of an animal in a suspect pen, the veterinarian might determine that the animal is not fit for human food, and that it must be condemned. Alternately, a veterinarian might determine that the suspect animal should be released for slaughter. Suspect cattle released for slaughter should be slaughtered separately for special evaluation during post-mortem inspection.

An animal that is condemned during ante-mortem inspection is not eligible for slaughter because it has been identified as having diseases or conditions making it unfit for human food. These cattle are sometimes referred to as 4-D animals (dead, dying, disabled or diseased).

Many countries require temperature checks of animals at ante-mortem inspection when there are presenting signs (e.g. mastitis). A bovine with too high a temperature cannot proceed to slaughter. In the United States, a bovine with a temperature of 105°F (40.6°C) must be condemned (Anon., 2013).

Since the establishment cannot slaughter a condemned bovine, the establishment should promptly and humanely kill the animal and immediately dispose of the carcass. A condemned carcass should be marked, denatured or destroyed in a manner that renders it obviously unsuitable for food (Anon., 2013).

Condemned cattle should be identified and the findings recorded. Identification data recorded should include (Anon., 2013):

1 tag numbers associated with the incoming bovine (e.g. back tag, ear tag)
2 breed (e.g. Hereford, Holstein)
3 slaughter class (e.g. steer, calf)
4 general ante-mortem findings (e.g. emaciation, fever)

4 Post-mortem inspection

The term 'post-mortem' means 'after death'. Post-mortem inspection is the inspection of carcasses of animals after slaughter.

4.1 Design of post-mortem inspection

All carcasses and other relevant parts should be subjected to post-mortem inspection, which should be part of an overarching risk-based system for the production of meat (Anon., 2005). After arrival at the slaughter establishment, the slaughter of the animals should not be delayed (Anon., 2004b).

Post-mortem inspection of cattle should utilize information from primary production and ante-mortem inspection, together with the findings from organoleptic inspection of the head, carcass and viscera, to make a disposition on the safety and suitability of parts intended for human consumption. When organoleptic inspection is insufficient to make a disposition of carcasses and other relevant parts as safe or suitable, the parts should be set aside and followed up with confirmatory inspection procedures or tests (Anon., 2005).

The establishment's procedures to prepare cattle for inspection must take place in sanitary conditions and must use sanitary procedures to prevent contamination of the carcasses and parts. During slaughter, the establishment must use sanitary dressing procedures to minimize contamination of edible tissues (Anon., 2004b).

Relevant information on the cattle population (e.g. breed, health status, origin) should be used in both the design and implementation of post-mortem inspection systems (Anon., 2005).

Where indicated by public health concerns, routine screening of carcasses and other relevant parts by methods other than organoleptic inspection may be required for suspected hazards, for example, testing for tuberculosis (Anon., 2005).

Characteristics of a risk-based post-mortem inspection programme include (Anon., 2005):

1 design and application of organoleptic procedures and tests that are relevant and proportional to meat-borne risks associated with grossly detectable abnormalities;
2 tailoring of procedures to the spectrum and prevalence of diseases and defects reasonably likely to be present in the particular slaughter population;
3 procedures that minimize cross-contamination through handling to the greatest extent practicable, and may include procedures that are limited to visual observation of carcasses and other relevant parts in the first instance if justified by risk assessment;
4 inspection of non-edible parts of animals where they may play an indicator role in the disposition of edible parts;

5 application of additional risk-based inspection procedures on a routine basis when live animals are positive to a diagnostic test, for example, tuberculin test;
6 use of laboratory tests for hazards that are unaddressed by organoleptic inspection, for example, veterinary drug residues;
7 return of information to the primary producer so as to seek continuous improvement in the safety and suitability status of animals presented for slaughter.

The establishment must also ensure that the carcasses are routinely presented for inspection in a specified manner. The organs must be displayed in a specified order so that the inspector does not have to spend time locating them before performing inspection procedures. Proper presentation helps to ensure consistent and accurate inspection (Anon., 2012).

4.2 Implementation of post-mortem inspection

Post-mortem inspection should occur immediately after slaughter. Inspection should take into account all relevant information from the level of primary production and ante-mortem inspection (Anon., 2005).

Veterinary Services should determine how post-mortem inspection is to be implemented, the training, knowledge, skills and experience required of personnel involved (including the role of any personnel not employed by Veterinary Services). The final responsibility for verifying that all post-mortem inspection and disposition requirements are met should lie with Veterinary Services (Anon., 2005).

Carcasses and other relevant parts condemned as unsafe or unsuitable for human consumption should be identified and handled in a manner that does not result in cross-contamination of meat from other carcasses and parts. The reason for condemnation should be recorded (Anon., 2005).

The following conditions must be provided by the establishment (Anon., 2012):

1 Adequate space for conducting inspection
2 Adequate lighting for conducting inspection
3 Hand-rinsing facilities to ensure that sanitary conditions are maintained
4 Condemned containers for disposal of condemned carcasses or parts

The post-mortem inspection process for cattle involves the following steps (Anon., 2012):

1 head inspection
2 viscera inspection
3 carcass inspection

In large establishments, inspectors are assigned to cover one of these areas and rotate to different sites according to a rotation pattern. At small establishments, the inspector may perform all of the post-mortem inspection procedures on each animal (Anon., 2012).

In general, when abnormalities are observed while performing inspection, the following actions must take place (Anon., 2012):

1 If the disease or condition of the head, organ or carcass is localized, the affected tissues may be trimmed.
2 If the disease or condition is generalized and affects the majority of the head, organ or carcass, it must be retained for veterinary disposition.

4.3 Results of post-mortem inspection

Veterinary Services and industry should record and disseminate the results of post-mortem inspection as appropriate. Notifiable human or animal health diseases and cases of non-complying residues or contaminants should be reported to Veterinary Services and to the owner of the animal(s). Analysis of the results of post-mortem inspection over time is the responsibility of Veterinary Services, and the results of such analyses should be made available to all interested parties (Anon., 2005).

Inspection results can be divided into two categories:

1 pathology findings
2 processing deficiencies

The causes of pathologic findings are manifested in many ways, and have a variety of aetiologies. Processing deficiencies manifest as contamination by faeces, ingesta or filth from poor dressing techniques; cross-contamination from one carcass to another; or metal or dirt from equipment or rails.

Post-mortem disposition of edible parts as safe and suitable for human consumption should primarily be based on foodborne risks to human health. Other risks to human health, for example, from occupational exposure or from handling of meat at home, are also an important consideration. Dispositions in relation to suitability characteristics of meat should reflect consumer acceptability preferences appropriate to intended end use (Anon., 2005).

Where the initial results of post-mortem inspection are insufficient to accurately make a disposition of edible parts as safe or suitable for human consumption, a provisional disposition should be followed up with more detailed inspection procedures and/or tests. Pending the outcome of more detailed inspection and/or diagnostic tests, all parts of the animal that are required for further investigation should be held under the control of an inspector (Anon., 2005).

Condemned product is a product that has been determined through inspection to be diseased or to have a condition that renders it unfit for human consumption. It is prohibited from entering commerce for use as human food (Anon., 2013).

Inedible product is any product that is adulterated, uninspected or not intended for use as human food. The term 'inedible' refers to product that by its nature is not handled as human food. Examples include the four Hs – horns, hair, hide and hooves (Anon., 2013).

Both condemned and inedible products are not fit for human consumption. Due to the edible appearance of some condemned materials, they must be controlled (Anon., 2013).

Maximum line speeds are governed by inspection or sanitary processing requirements. Sanitary processing is affected by conditions such as heavily soiled animals or larger breeds. Line speeds can be maximized on the slaughter or eviscerating line when optimum conditions of animal health and cleanliness exist. Line speeds must be reduced to accommodate inspection or sanitary processing when optimum conditions do not exist (Hogue et al., 1993).

Appropriate line speeds can be developed by examining the cleanliness of carcasses at the end of the line, and by performing ergonomic work measurements at various locations on the line (Anon., 2013). Every line has a processing point that serves as a governor of line speed.

The marks of inspection are usually applied just prior to a carcass and parts entering the coolers (Anon., 2013). Once the inspection legend has been applied to a carcass, the product should be safe and suitable for further processing.

Meat products derived from an inspection programme designed, implemented and the results of which are verified and recorded according to the guidelines covered in this chapter should be acceptable for consumption domestically and eligible for international trade.

5 Important inspection procedures

The following inspection procedures for cattle heads, viscera and carcasses are summarized from training materials for veterinarians in the U.S. Department of Agriculture's Food Safety and Inspection Service (Anon., 2012). Inspection procedures vary somewhat from country to country depending on hazards existing in cattle in specific countries.

5.1 Head inspection

1 Observe the outer surface of the head and eyes. Some diseases and conditions that may be detected during this step include contamination and epitheliomas.
2 Incise and observe the four pairs of lymph nodes – mandibular, parotid, lateral retropharyngeal and medial retropharyngeal. Some diseases and conditions that may be observed when performing this step include tuberculosis and abscesses.
3 Incise and observe the masticatory or cheek muscles. Some diseases and conditions seen during this step include cysticercosis and eosinophilic myositis.
4 Observe and palpate the tongue. Some diseases and conditions that can be detected by this step include actinobacillosis and foreign bodies, such as thorns.

5.2 Viscera inspection

1 Abscesses are frequently detected during the palpation and observation of the rumino-reticular junction. These abscesses are usually localized and require only that the viscera be condemned. If abscesses are found in other locations in addition to the abdominal viscera, it could be an indication of a generalized condition requiring condemnation of the entire carcass and its parts.
2 The mesenteric lymph nodes might show evidence of tuberculosis, neoplasms and, in some cases, pigmentation changes.
3 The small intestines sometimes appear dark red to purple, this would indicate enteritis. A determination whether the condition is acute or chronic must be made.
4 Evidence of extensive adhesions may be seen. If the condition appears localized and no further carcass or viscera involvement is observed, the abdominal viscera would be condemned and the carcass retained for trimming.
5 The inspection of the spleen is done by observation. When an abnormal spleen is detected, retain it as well as the carcass and all parts. The spleen may be helpful in making a final disposition on any carcass. Ensure that the spleen is included with the viscera whenever a carcass is retained for a disease condition.
6 Observe the oesophagus for *Cysticercus*, eosinophilic myositis and evidence of grub infestation. *Cysticercus* and eosinophilic myositis require retention. Grub infestation is usually a localized condition requiring that affected organs and areas be trimmed or condemned, but the carcass will usually be passed without retention.
7 Pneumonia and pleuritis are the most common abnormalities observed in the lungs. Acute pneumonia is characterized by enlarged, oedematous lymph nodes and/or

dark red to purple sections or spots in the lung tissue. A chronic pneumonia may be characterized by a localized abscess within the lungs, or evidence that the lung has become adhered to the pleura. Observe the rest of the viscera and carcass to look for evidence that the condition is generalized. The carcass may appear degenerated. There may be watery tissue or fat sloughing indicating a generalized condition.

8 The inspection of the heart involves opening it by an incision from the base to the apex. Common conditions detected while inspecting the heart are pericarditis and endocarditis.

9 When inspecting the liver, abscesses are commonly observed. Many intensively fed cattle have localized abscesses. Make as many incisions as necessary to search for abnormal conditions, but do not mutilate product unnecessarily. A liver containing an abscess is condemned.

5.3 Carcass inspection

Bovine slaughter today is performed under one of two of the following basic operations.

1 Bed dressing is most often used in small, low-volume establishments. After the head has been removed, the carcass is lowered to a skinning bed. The skinning bed may be a cradle or it may be the floor.
2 Rail dressing is designed for large, high-volume establishments. The animal is moved around the slaughter floor by means of a rail and instead of one employee dressing the entire animal, several people perform various dressing procedures as the carcass moves past them.

The establishment is responsible for assigning an employee prior to inspection to trim and remove all bruises, blood clots, grubs and other localized abnormalities. The establishment employee must not remove any abnormality that could affect the disposition of the carcass.

The following steps are followed when inspecting a carcass:

1 Observe the back of the skinned carcass.
2 Palpate the scrotal, or mammary, and medial iliac lymph nodes.
3 Observe body cavities.
4 Observe the cut surfaces of muscles and bones, the diaphragm's pillars and the peritoneum.
5 Observe and palpate the kidneys and diaphragm.
6 Observe the pleura, neck and carcass exterior.

If a disease or condition is observed that requires the carcass to be retained, the inspector must have the carcass tagged, and have the viscera and head retrieved and tagged.

Tissues and parts removed and condemned for various reasons should be placed in a designated container.

6 Summary

An increasing focus on food safety by Veterinary Services in developed nations has resulted in reductions of foodborne illnesses. They have employed evolving inspection systems to focus more fully on food safety diseases and conditions rather than those diseases and conditions which might be repugnant, but do not cause illnesses.

Progress in the design and implementation of inspection systems, as well as analyses of the results of inspection have enabled veterinarians to improve cattle husbandry and better control cattle diseases. Additionally, these systems have enabled Veterinary Services to better prevent conditions that might cause illnesses during the raising and slaughter of cattle.

As our understanding of the complex association between animal health and public health improves, public health veterinarians and other professionals in allied disciplines will continue to advance beef safety in the world.

7 Where to look for further information

The references provided at the end of this chapter serve as an excellent starting point for those wishing to find further information on this subject, especially the Code of Hygienic Practice for Meat by the Codex Alimentarius Commission. In addition to this chapter's referenced materials, nations exporting beef to countries of the European Union, North America and other developed countries have published regulatory requirements found to be equivalent among developed nations trading in beef.

8 References

Anon., 2004a. Regulation (EC) No 853/2004 of the European Parliament and of the Council of 29 April 2004.

Anon., 2004b. Regulation (EC) No 854/2004 of the European Parliament and of the Council of 29 April 2004.

Anon., 2005. Codex Alimentarius Commission, Code of Hygienic Practice for Meat, CAC/RCP 58-2005

Anon., 2012. U.S. Department of Agriculture, Food Safety and Inspection Service, Public Health Veterinarian Training Module, Post-Mortem Inspection, 27 January 2012.

Anon., 2013. U.S. Department of Agriculture, Food Safety and Inspection Service, Public Health Veterinarian Training Module, Ante-Mortem Inspection, 17 September 2013.

Anon., 2016. World Organization for Animal Health (OIE), Terrestrial Animal Health Code, chapter 6.2, Control of Biological Hazards of Animal Health and Public Health Importance through Ante- and Post-Mortem Meat Inspection, 13 June 2016.

Hogue, A. T., et al., 1993. Bacteria on beef briskets and ground beef: correlation with slaughter volume and antemortem condemnation. *Journal of Food Protection* 56(2): 110–13; 119.

Petersen, K. E., et al., 1996. Use of a priority rating process to sort meatborne zoonotic agents in beef. *Journal of Agromedicine* 3(1): 17–36.

Petersen, K. E. and William O. James, 1998. Agents, vehicles, and causal inference in bacterial foodborne disease outbreaks: 82 reports (1986–1995). *Journal of the American Veterinary Medical Association* 212: 1874–81, 15 June 1998.

Walker, H. L., et al., 2000. Relevance of carcass palpation in lambs to protecting public health, *Journal of Food Protection* 63(9): 1287–90.

Chapter 8

Maintaining the safety and quality of beef carcass meat

James S. Dickson, Iowa State University, USA and Gary R. Acuff, Texas A&M University, USA

1 Introduction

Contamination of animal carcasses during slaughtering procedures is undesirable, but unavoidable in the conversion of live animals to meat for consumption. Internal muscle tissues are essentially sterile, and most initial contamination of red meat carcasses is contributed by the hide during removal (Elmonssalami and Wassef, 1971; Gill and Penny, 1979; Gill et al., 1976). The exposed surface of the hide and the hair accumulate dust, dirt and faecal material, and this is the primary source of bacterial contamination during slaughter (Ayres, 1955; Shotts et al., 1961). The factors that affect the extent of this contamination are reviewed by Patterson (1969) and Grau et al. (1968). Much of the microflora transferred to the tissue surfaces, while aesthetically undesirable, is non-pathogenic; however, pathogens such as *Salmonella*, *Campylobacter* and pathogenic *Escherichia coli* can be present.

http://dx.doi.org/10.19103/AS.2016.0008.12

2 Process flow description

The slaughter process begins when the animals are unloaded from the trucks which bring them to the slaughter establishment. The animals are typically held in holding pens (lairage) for several hours to reduce transportation stress. Although the lairage pens are livestock pens and not sanitary food production areas, they still require routine removal of faecal material and uneaten food to prevent excessive contamination of the animals prior to the actual slaughter process. If the lairage pens are uncovered, they should be located in such a place that they drain well during rainy seasons. Although these steps are important in regard to animal welfare, they also have an impact on the safety and quality of the meat, in that they minimize additional contamination of the hide prior to slaughter. Since the hide is a significant source of carcass contamination, minimizing contamination of the hide can help minimize subsequent contamination of the carcass.

The slaughter process proceeds as the animals are moved from the lairage pens to the slaughter floor. The animals are moved with as little stress as required, to minimize stress-related quality defects in the meat. The next two steps – humane stunning and exsanguination – occur in rapid succession. The carcasses are typically allowed to bleed out for approximately 5 to 6 minutes. Other than maintaining the hygienic quality of the stick knives, there are no other processes which either contribute to or reduce safety or quality issues during these two steps.

The impact of sticking on contamination is not clear. A classic study by Jensen and Hess (1941) evaluated the process of sticking, and suggested that bacteria could enter the bloodstream during the sticking. These conclusions were based on the fact that fewer bacteria were found in the blood retained in the hearts of hogs 'sterilely' stuck, as compared to those that were septically stuck. However, they also noted that when cultures of *E. coli* were added to blood drawn from a live hog, the bacteria could not be recovered after two to five hours, and they attributed this to the bactericidal activity normally associated with blood. For bacterial pathogens to enter the bloodstream during the sticking operation, several events would have to occur. The first is that pathogens would have to be present at the exact point of the stick wound. Secondly, the bacteria would have to be carried into the bloodstream of the animal by the knife, either from previous contamination or from material at the site of the stick wound. Research has shown that salmonellae may be carried on improperly sterilized knives (Peel and Simmons, 1978). While there might be a reasonable probability of these events happening, the individual cells would be rapidly dispersed throughout the entire bloodstream, resulting in a rapid dilution of the initial population. This dilution, coupled with the documented bactericidal properties of the blood, suggests that the stick would not be a major source of salmonellae contamination of the muscle tissue. In addition, the site of the stick wound itself is normally trimmed out at a later point in the process, removing any bacteria which may have adhered to the tissue.

In some slaughter processes, the hides of the carcasses are sprayed or washed with water prior to further processing. The objective of this step is dependent upon current environmental conditions. Under dry environmental conditions, a light water spray may be applied to the hides to minimize dust during the hide removal operation. This reduces contamination of the meat by reducing the dust which may be generated by mechanical hide removal. The airborne dust may be contaminated with bacteria which may impact either the quality or safety of the carcass. In wet environmental conditions, the carcasses may be heavily contaminated with mud and manure. In this case, a more forceful washing system may be employed to remove some of this external contamination from the surface of the hide. Although the wash does not remove all of the contamination, an incremental decrease in hide contamination may result in less contamination of the carcass as it proceeds through the remainder of the slaughter process. Implementation of hide wash systems that apply antimicrobial solutions has been reported to reduce the presence of enteric pathogens on carcass surfaces (Arthur et al., 2007; Bosilevac et al., 2005), and Yang et al. (2015) reported that hide decontamination could reduce transfer of *E. coli* to carcass surfaces during dehiding processes.

The next step in the process is the removal of the hide. The hide may be removed by hand; however, in many large slaughter establishments hide removal is a mechanically assisted process. When the hide is removed from a carcass during slaughter/dressing procedures, the sterile surface protected by the hide is exposed and then subsequently contaminated with bacteria that most likely originate from the hide, gastrointestinal tract or slaughter plant environment. After hide removal, the head and viscera are removed, which may lead to additional contamination. Although good dressing practices will minimize this contamination, it is almost impossible to prevent microbial contamination from occurring during these steps. Most bacteria contaminating the surface will be exposed and vulnerable to chemical decontamination treatments; however, some may have been forced into small cuts just under the surface of the carcass, or flaps of tissue may hang after disconnecting from the removed hide such that the bacteria are covered. These cells may be positioned such that they are protected from exposure to chemicals used for decontamination of the surface. In addition, a newly exposed carcass surface is typically warm, soft, moist and tacky, but evolves after cooling to a hard, dry surface. Temperature of the carcass surface, presence of a moist surface or water film and solidification of fat surfaces during cooling all likely affect the ability to decontaminate. This chapter specifically addresses chilled carcass surfaces through fabrication and packaging, likely some of the more difficult surfaces to successfully apply an intervention. The meat surface is cool, bacteria have likely begun to attach and they may be embedded within solidified fatty tissues.

3 Bacterial attachment to meat surfaces

Bacterial attachment to meat tissue is a complex mechanism which may have a practical effect on the transfer of pathogens between carcasses, the effectiveness of sampling methods and the performance of treatments for carcass decontamination.

Meat decontamination is usually achieved by sprays or washes with water or antimicrobial solutions. Depending on the treatment applied, the reduction in microbial numbers on the meat surface will be due to physical removal of the microorganisms, a killing effect of the decontaminating solution or a combination of both factors. Among the factors influencing the effectiveness of carcass interventions, bacterial attachment onto the meat surface has attracted the attention of many researchers. The molecular basis for bacterial attachment has been reviewed by Hardin (1995). Fratamico et al. (1996) reported that, once bacteria are attached to meat, rinse solutions such as acetic acid or trisodium phosphate (TSP) were not effective in removing a large part of the contaminating bacteria. In contrast, Castillo et al. (1998a) and Hardin et al. (1995) did not find any differences in populations of *E. coli* O157:H7 or *S.* Typhimurium on beef carcass surfaces treated by hot water or organic acids immediately or 20–30 min after contaminating the beef surfaces. Since bacterial attachment on carcass surfaces has been reported to occur within 20 min of contact, Butler et al. (1979) concluded that bacterial attachment did not affect the antimicrobial effect of hot water or organic acid sprays against pathogens on beef.

Other factors, such as the surface fat characteristics of the carcass region, may be more important regarding effectiveness of antimicrobial sprays. Using transmission and scanning electron microscopy, Mattila and Frost (1988) described the attachment of *E. coli* to beef and chicken surfaces. The effect of the surface fat characteristics of the meat is important because fat, being hydrophobic, may interfere with the effectiveness of water washes to remove bacterial contamination. In contrast to the above reports, Dickson (1991a) found more attachment of *S.* Typhimurium and *L. monocytogenes* to lean tissue than to fat tissue. He also found organism-intrinsic factors, such as inoculum size, temperature at which the cells had been exposed before attaching and age of the cells, to have a significant effect on the attachment of these pathogens to beef surfaces. Bouttier et al. (1997) also studied the role of bacterial flagella on attachment to meat surfaces by determining the number of cells of *Salmonella* Choleraesuis adhering to lean or fat tissue after treating with antibodies to flagellar antigens. The count of attached cells was significantly lower for antibody-treated cells than that of control cells or cells treated with antibodies to somatic antigens. Although this would indicate that flagella have an effect on bacterial attachment, other experiments involving potential chemical receptors on beef tissues saturated by a suspension of flagella showed no effect of flagella on the attachment of *S.* Choleraesuis. This demonstrated that the surface of beef did not possess any receptors for the flagella of this species. Butler et al. (1979) reported greater attachment of Gram-negative, motile cells than Gram-positive or non-motile bacteria. Cells that are attached to meat surfaces have been shown to transfer between surfaces at a lower rate than unattached cells, although the highest transfer rate was observed from adipose to lean tissue (Dickson, 1990). This is of practical importance because carcasses are often in close contact for a lengthy time in coolers, and cross-contamination may occur during storage. The presence of organic matter may be another factor affecting the extent of bacterial attachment on meat surfaces. Dickson and MacNeil (1991) reported greater attachment of *S.* Typhimurium and *L. monocytogenes* to beef carcass surfaces when the inoculum had been diluted in phosphate buffer compared to cow manure. The nature of the inoculum needs to be taken into consideration when research is conducted involving inoculation of pathogens onto meat surfaces.

4 Decontamination methods

4.1 Knife trimming

Current USDA-FSIS regulations require that all faeces, ingesta and milk must be physically removed from beef carcasses by knife trimming or, when such contamination is less than one inch in its greatest dimension, by vacuuming with hot water or steam (Federal Register, 1996). The traditional approach of trimming contamination from beef carcasses has been evaluated for efficacy by several authors. Gorman at al. (1995b) reduced the aerobic plate counts (APC) and *E. coli* counts on inoculated beef brisket 2.0 to 2.5 $\log_{10}$ ($P<0.05$) by knife trimming without any other combined treatment. Application of a water wash subsequent to trimming achieved no additional reduction unless the water temperature was at least 66°C. Other authors have also reported reductions of bacterial counts through trimming (Gorman et al., 1995a; Hardin et al., 1995; Prasai et al., 1995a, b; Reagan et al.,

1996); however, in most instances the evaluations of decontamination by knife trimming have been performed under laboratory conditions. Prasai et al. (1995b) concluded that an observed large bacterial reduction demonstrated by trimming might have been due to the artificial conditions under which this operation was accomplished. In their work, the trimming samples were collected from locations that had been completely trimmed by making one cut using a sterile knife, a procedure not likely to be comparable to the trimming performed in plants during normal slaughter operations. In a study conducted within a beef packing plant, Gill et al. (1996) reported no differences in total bacterial and *E. coli* counts from carcasses sampled before and after trimming. These authors also found that a water wash treatment subsequent to trimming did not produce further reduction in bacterial counts, concluding that the reduction in bacterial numbers achieved by either trim or water wash is insufficient to enhance the safety of the meat. These reports may indicate that although trimming has been reported to significantly reduce bacterial counts under laboratory conditions, the circumstances under which slaughter plants typically perform this operation may not achieve the expected carcass decontamination. In addition, spread of bacterial contamination from non-visible areas of faecal contamination to other clean areas during trimming is a concern (Hardin et al., 1995).

4.2 Water wash

Washing of carcasses with water after slaughter is a common practice in beef slaughter, with the intent of removing visible contamination from carcasses and improving the visual quality of the meat. Under the USDA-FSIS zero-tolerance policy for faecal contamination, only knife trimming and, to some extent, steam vacuum treatments are allowed as a means to eliminate visible faecal contamination from carcasses. However, many researchers have studied the ability of water washes to reduce microbial contamination. Anderson et al. (1975) studied the effect of factors such as water volume and pressure, angle of droplet impact, droplet size, spray force and the speed at which the meat passed through the spray on the removal of the yeast *Rhodotorula rubra* on beef plate meat. Factors such as water pressure, water flow rate and speed of movement of meat through spray had significant effect on microbial removal, whereas mean droplet size was not a significant factor. The angle of droplet impact was not significant when the pressure was 28 kg/cm^2 but became significant as the pressure was decreased. Contrasting the findings of Anderson et al. (1975), Crouse et al. (1988) did not find any reduction in numbers of *Enterobacteriaceae* or APC on beef carcasses as affected by the spray pressure or the chain speed. In a similar study, DeZuniga et al. (1991) found no significant differences due to pressure for the reduction in APC or counts of *Enterobacteriaceae* on meat surfaces.

DeZuniga et al. (1991) also addressed the effect of high-pressure water wash on bacterial penetration into the meat. Using an insoluble dye (Blue Lake) with a particle size slightly smaller than most bacteria, these authors reported that the depth of the dye penetration after an automated water wash was directly proportional to the line pressure, and that the type of nozzle used had a significant effect on penetration of Blue Lake at pressures above 4,140 kPa. DeZuniga et al. (1991) concluded that bacteria might similarly penetrate into the meat as a result of high-pressure cabinet water wash treatment. Of significance is the possibility that if a sanitizer is used after washing for decontamination purposes, the solution might not reach bacteria implanted within the meat by a pressured wash.

It is generally accepted that treatments with hot water (74°C) will produce a sanitizing effect rather than a simple washing effect (Federal Register, 1996). Cabedo et al. (1996)

reported larger reductions of *E. coli* on beef brisket after spraying with 74°C water compared to 35°C. However, under sublethal temperature conditions, higher temperatures would likely have an effect on fat softening, which may also affect the ability to remove bacterial contamination. Gorman et al. (1995b) produced larger *E. coli* reductions on beef tissue by spraying water at 35°C than by spraying water at 16°C. However, there were no differences in *E. coli* reductions after spraying water at 35, 66, or 74°C. When sprays at 16, 35, or 74°C were followed by a second wash at 16°C, the reductions increased with the temperature of the first wash, possibly indicating that a warm carcass wash would be more helpful than a cold wash in removing microbial contamination, especially if the water wash is followed by a sanitizing step.

4.3 Trimming versus washing

Many studies evaluating trimming as a means of reducing bacterial contamination compare the reductions obtained by water washing, and indicate that trimming produces similar or larger reductions than those obtained by water wash (Gorman et al., 1995a; Hardin et al., 1995; Prasai et al., 1995b). The USDA-FSIS affirms that trimming, if performed properly, will effectively remove the visible contamination as well as any accompanying microbial contamination, whereas, if not properly conducted, may spread the contamination to other newly exposed areas (Federal Register, 1996). However, Gill et al. (1996) reported that numbers of *E. coli*, coliforms and aerobic bacteria that contaminate beef carcasses during dehiding and evisceration were not reduced by trimming, and were halved by washing. Conversely, Reagan et al. (1996) showed a significant superiority of trimming over water wash at reducing aerobic bacteria and *E. coli* biotype I on beef carcasses.

Whether trimming reduces more contamination than water wash or vice versa, neither procedure appears to be particularly effective in decontaminating carcasses. In studies where trimming and water wash are compared to sanitizing treatments for beef carcass decontamination, both water wash alone and trim alone have been reported to produce significantly smaller reductions than sanitizing agents such as hot water or organic acids (Gorman et al., 1995b; Hardin et al., 1995; Reagan et al., 1996). These studies indicate that both trimming and water washing of carcasses should not be practiced for decontamination purposes, but for carcass cleaning. Even a visually clean carcass may be contaminated with pathogenic bacteria at unsafe levels. In addition, both treatments are likely to spread pathogenic contamination to clean areas of the carcass. Therefore, trim, wash, or any other cleaning treatment, should be followed by a subsequent sanitizing treatment.

4.4 Steam vacuum

Application of hot water or steam combined with vacuuming is a commonly used carcass cleaning process that is allowed to be used instead of knife trimming to physically remove faecal contamination while sanitizing the contaminated area (Federal Register, 1996). A typical steam vacuum machine includes a vacuum wand with a build in hot water spray nozzle, which delivers water at 82 to 88°C. This internal nozzle is intended to sanitize the carcass surface as the vacuum removes the faecal material. Two external spray nozzles are positioned on the top and bottom of the wand to provide a continuous steam flow. This design would allow for steam to continuously keep the outside of the wand clean and sterile, while also helping in the carcass surface sanitation process. The steam vacuum machine has been designed to clean only small areas of contamination, and is not applicable to clean and sanitize the entire carcass surface. Because of this limitation, applying steam vacuum treatment for carcass cleaning is permitted only when the extent of the contamination is no larger than 6.25 cm^2 (1 in^2) (Federal Register, 1996).

Investigators have studied the efficacy of spot cleaning vacuum machines for reducing bacterial contamination on carcass surfaces. Phebus et al. (1997) observed no differences between a steam vacuum system and interventions such as steam pasteurization and knife trimming in reducing *E. coli* O157:H7, *S.* Typhimurium or *L. monocytogenes* on freshly slaughtered beef surfaces. However, these authors noted that steam vacuum is a localized cleaning device, whereas steam pasteurization is a full-carcass treatment. During in-plant evaluations of two steam-vacuuming units, Kochevar et al. (1997) found lower APC and coliform counts on carcass surfaces that had been treated with the steam vacuum unit, when compared to surfaces that had been knife trimmed. According to the results of their study, use of steam vacuum reduced microbiological contamination and improved visual appearance of carcasses for which otherwise knife trimming would have been required by the inspector. In a review on procedures for beef carcass decontamination, Dorsa (1997) described an in-plant testing of steam vacuum, designed to determine its efficacy under industrial use, indicating that a steam-vacuuming unit consistently reduced bacterial populations from contaminated areas of less than 2.5 cm. From these results, he concluded that a commercial steam vacuum system could perform better than knife trimming for removing bacterial contamination on beef carcasses.

5 Decontamination treatments: hot water and organic acids

5.1 Hot water

In contrast to regular water washes, sprays with water at temperatures above 74°C may be used as sanitizing interventions. The effect of applying hot water versus washing carcasses with warm water was reported by Smith (1992), finding an average 0.2-$\log_{10}/cm^2$ reduction of various pathogens on fresh meat after washing with 40°C water, but reporting a reduction of 3.1 $\log_{10}/cm^2$ after applying 80°C water. Multiple reports indicate that the application of hot water treatments can effectively reduce microbiological contamination on meat carcasses, and several studies also report that washing carcasses with water at temperatures greater than 80°C will not produce permanent discolouration of the carcass surface (Barkate et al., 1993; Patterson, 1969; Smith and Graham, 1978). In an early report on hot water decontamination, Patterson (1969) reported that beef carcasses treated with a steam and hot water spray (80–96°C) for 2 min contained significantly lower bacterial numbers than untreated carcasses. A volume of 18.9 L of water was sprayed on each carcass; however, the actual temperature at the carcass surface during the treatment was not provided. Smith and Graham (1978) reported that pouring hot water (80°C) on beef and lamb samples for 10 s destroyed more than 2 logs of *E. coli* and *Salmonella* inoculated at levels of 6.5 $\log_{10}/cm^2$. In a laboratory evaluation of a hot water cabinet, Davey and Smith (1989) obtained *E. coli* reductions of approximately 3 $\log_{10}/cm^2$ for artificially contaminated beef carcass sides treated with hot water that elevated the carcass surface temperature to 83.5°C for 20 s. Kelly et al. (1981) reported that lamb carcasses sprayed with hot water at temperatures above 80°C caused significant decreases (>1.0 $\log_{10}/cm^2$)

in APC. Dorsa et al. (1996b) and Gorman et al. (1995b) reported reductions in coliform or *E. coli* counts of approximately 3.0 $\log_{10}/cm^2$ when evaluating hot water treatments.

Proper design of a hot water treatment for carcass treatment is of paramount importance for obtaining effective reduction in bacterial populations, as loss of heat within the spray from the nozzle to the carcass surface may be so great that the carcass surface receives an insufficient temperature increase. Barkate et al. (1993) sprayed areas of hot beef carcass surfaces using 95°C water with the objective of raising the carcass surface temperature to 82°C for approximately 10 s and bacterial contamination on the carcass surface was reduced significantly. Problems reported in applying hot water included designing a water spray that would adequately attain a bactericidal temperature at the surface of the carcass. The volume of the spray and the size of the water droplets were reported to greatly affect the temperature of the water from the point of origin at the spray nozzle and prior to contacting the carcass surface. Using a type of nozzle which addressed the limitations reported by Barkate et al. (1993), Castillo et al. (1998a) sprayed hot water onto different hot carcass surface regions, obtaining average reductions of initial counts for *E. coli* O157:H7 and *S.* Typhimurium of 3.7 and 3.8 $\log_{10}/cm^2$. Corresponding reductions for APC and counts of coliforms and thermotolerant coliforms in their study were 2.9, 3.3 and 3.3 $\log_{10}/cm^2$, respectively. In this study, the hot water spray was combined with a previous water wash at 35°C, which significantly improved the visual quality of the carcass surfaces.

5.2 Organic acids

Among different organic acids, acetic and lactic acids were more extensively used for carcass decontamination. The effectiveness of these two acids, as well as other carcass interventions was reviewed by Dickson and Anderson (1992). In an early study on pork carcass decontamination, Biemuller et al. (1973) reported that spraying carcasses with acetic acid at pH 2.0 for 30 or 60 s resulted in large reductions of naturally occurring microflora and inoculated *S. enteritidis* on pork carcasses at a slaughter plant. Ockerman et al. (1974) also reported significant reductions in numbers of naturally occurring microorganisms on lamb carcasses by acetic and lactic acids. Additionally, these authors reported a small residual effect of the acids on the microbial numbers on the carcasses during 12 days of refrigerated storage, which was affected by the concentration and type of the acid. Anderson et al. (1977) produced reductions of 2.55 log on counts of viable microorganisms on meat by spraying 3% acetic acid. These reductions were significantly greater than the reductions obtained by spraying hypochlorite solution (200–250 ppm). Quartey-Papafio et al. (1980) sprayed beef strips with acetic acid; formic acid; and a mixture of acetic, formic and propionic acids. All treatments significantly reduced the bacterial counts with respect to untreated controls. However, the reductions were usually less than one log. Treatment with 1% formic acid produced the smallest reduction in viable counts (0.66 log), followed by a mixture of 0.5% acetic, 0.25% formic and 0.25% propionic acids (0.76 log); 3% acetic acid (0.89 log); and 2% formic acid (1.56 log). After 7 days of storage at 7°C, bacterial counts on the strips increased between 0.92 and 2.24 log, whereas the counts on untreated controls increased 4.66 log. They also reported that 5% ascorbic acid sprayed to prevent browning of meat treated with formic acid also enhanced the antimicrobial effect. Dickson (1992a) observed consistent reductions in populations of inoculated *S.* Typhimurium on lean and adipose beef tissue sprayed with 2% acetic acid, irrespective of the initial cell population. The acetic acid treatment had an immediate lethal effect on part of the population of *S.* Typhimurium, while another part was sublethally injured. In general, the reductions in counts of different pathogens on beef, as reported by different authors, vary between 2 and 4.3 log cycles after spraying 2% acetic acid (Dickson, 1991b; Dickson and Anderson, 1991; Hardin et al., 1995; Tinney et al., 1997). Variations in reductions obtained by different investigators may be due to differences in factors such as the temperature of the acid solution, which ranged from room temperature to 55°C in these reports. In a rare report on ineffectiveness of organic acid treatments to decontaminate beef tissue, Brackett et al. (1994) found that acetic, citric and lactic acid solutions at different concentrations were unable to reduce *E. coli* O157:H7 on beef sirloin pieces, regardless of the concentration and temperature of the acid solution. These authors explained the difference in these results from other studies can be attributed to differences in methodology. In several studies, the acid treatment is applied by dipping beef pieces in acid solutions, whereas these authors sprayed the acid solutions onto the beef pieces. However, pH data in this study indicates that the inability of organic acid solutions to reduce counts of *E. coli* O157:H7 on beef was most likely due to their failure to reduce the beef surface pH to antimicrobial levels. In other similar papers, Anderson and Marshall (1990a) reduced the pH of beef dipped in lactic acid solutions, from 5.6 (untreated meat) to 3.95, and Hardin et al. (1995) obtained surface pH values on beef carcass surfaces of 2.64 to 2.88 after spraying lactic acid, and of 3.14 to 3.47 after spraying acetic acid. Even though *E. coli* O157:H7 has been reported to be resistant to low pH environments, recent studies indicate that lactic or acetic acid sprays, when applied at 55°C, can effectively reduce levels of *Salmonella* or *E. coli* O157:H7 (Castillo et al., 1998b; Hardin et al., 1995).

Studies on carcass decontamination using lactic acid indicate that this acid shows a strong antibacterial capacity. Hardin et al. (1995) reported that lactic acid was more effective than acetic acid in reducing *E. coli* O157:H7, and as effective as acetic acid in reducing *S.* Typhimurium on beef carcass surfaces. Woolthuis et al. (1984) found immersing porcine livers for 5 min in a 0.2% lactic acid solution to be significantly more effective than immersing in hot water (65°C) for 15 s in reducing total bacterial counts and lactic acid bacteria, whereas *Enterobacteriaceae* counts were reduced at the same rate after both treatments were applied. In another study, mean *Enterobacteriaceae* counts of 1.8 $\log_{10}$ CFU/cm^2 were reduced to undetectable levels on calf carcasses by spraying 1.25% L-lactic acid. This treatment also reduced the APCs by 0.8 to 1.3 $\log_{10}$ CFU/cm^2 depending on the carcass region treated (breast or perineum) (Woolthuis and Smulders, 1985). Prasai et al. (1991) reduced the APC of beef carcasses at two slaughter plants by ca. 2 $\log_{10}$/cm^2 by spraying 1% lactic acid at 55°C after dehiding and eviscerating. However, APCs of vacuum packaged loins cut from these carcasses were not different from those of loins cut from non-treated carcasses, indicating that the quality of subprimals depends, to a large extent, on the degree of recontamination after applying carcass intervention. An European Food Safety Authority (EFSA) panel (2011) concluded, among other things, that, 'lactic acid was shown to reduce the prevalence of Salmonella and/or STEC/VTEC on carcasses, beef cuts and trimmings to varying degrees depending on study design and contamination level, but reductions were generally significantly higher compared to controls'.

Some researchers have reported on the impact of organic sprays on the sensory characteristics of meat. Bell et al. (1986) did not observe significant ($P<0.05$) discolouration of beef after dipping the meat in 1.2% v/v acetic acid for 1 min. When the treatment was extended to 10 min, a concentration of 0.6% lactic acid was enough to produce significant discolouration when compared to untreated controls. A mixture of 0.6% acetic acid and 0.046% formic acid was not different from 1.2% acetic acid in its antibacterial activity and did not produce discolouration or noticeable flavours in the meat. Garcia-Zepeda et al. (1994a) compared the changes in psychrotrophic counts and acceptability scores of chuck subprimals obtained from carcasses treated with 3% lactic acid, 200 ppm chlorine or water. Subprimals obtained from carcasses sprayed with lactic acid showed lower psychrotrophic counts but also lower acceptability scores than subprimals from carcasses treated with chlorine or water. In contrast, Goddard et al. (1996) found no differences in meat colour, fat colour or odour in beef strip loins treated with a mixture of lactic and acetic acids when compared to untreated controls. Acuff et al. (1987) found no differences in bacterial and shelf life of steaks cut from beef loins sprayed with different acid solutions and steaks cut from untreated loins. Similar results were obtained by Dixon et al. (1987) in strip loins packed in either polyvinyl chloride or high oxygen barrier films.

Different factors may impact the effectiveness of organic acid treatments for decontaminating carcasses. The ability of different organisms to acquire acid tolerance has been reviewed by Rowbury (1995). This author mentioned growth at high temperatures or acidic conditions as environmental factors inducing acid tolerance. Heat- or acid-induced acid tolerance in bacterial pathogens requires the cells remaining under specific environmental conditions for a time long enough to synthesize the outer membrane proteins required for the acquired acid tolerance (Wang and Doyle, 1998). Since the length of all treatments in carcass decontamination is reduced to few seconds, the impact of this acquired acid tolerance on the effectiveness of different carcass interventions should be evaluated. Dickson and Kunduru (1995) addressed the acid adaptation in different strains of *Salmonella* as a potential factor influencing the effectiveness of organic acid rinses on beef. All strains of acid-adapted *Salmonella* were at least as sensitive to the organic acid rinses as the non-adapted parent strains. This study also addressed the effect of acid adaptation on heat resistance in *Salmonella*. Again, no effect of acid adaptation on the heat resistance was observed for any strain of *Salmonella*. These investigators concluded that acid adaptation of salmonellae in the environment, if occurred, would not create a new hazard with the use of organic acid rinses on beef carcasses. Another factor that might affect the practical application and lethality evaluation of organic rinses against pathogens is sublethal injury. Van Netten et al. (1984) demonstrated that acid-stressed cells of various pathogenic Enterobacteriaceae can remain undetected during evaluations of organic acid rinses, so that pathogen reduction by these treatments may be overestimated.

Several studies were conducted to determine the effect of temperature and concentration of the acid solution on the reduction of meat-borne pathogens and spoilage bacteria on beef surfaces (Anderson et al., 1987, 1988, 1992; Anderson and Marshall, 1989, 1990a,b; Greer and Dilts, 1992). In all these studies the temperature of the acid solution was found to have a profound effect on the magnitude of the reductions in bacterial counts. The concentration of the acid in the sanitizing solution has generally been determined to be of minimal importance for the effectiveness of organic rinses when it is above 1% (Anderson and Marshall, 1990a; Greer and Dilts, 1992).

Other factors such as bacterial attachment, type of meat surface (lean vs. fat), rigour state, inoculating menstruum or level of inoculum have been studied in their effect on the efficacy of organic acid sprays for carcass decontamination. Dickson (1992b) did not find differences in populations of attached *S.* Typhimurium on pre- or post-rigour, lean or fat beef tissue. Two separate reports (Dickson, 1992b; Cutter et al., 1997) indicate that rigour state (pre- and post-rigour) does not affect the removal of pathogenic bacteria following treatment with 2% acetic acid. In general, the inoculating menstruum (buffer, tryptic soy broth, rumen fluid or faeces) had no effect on reduction of *S.* Typhimurium or *E. coli* O157:H7 after spraying acetic acid onto the inoculated beef tissues. However, Dickson (1992b) observed less reduction of *S.* Typhimurium populations on fat tissue after acid treatment when the inoculum menstruum was manure. The attachment rate and type of surface tissue do not seem to impact the effectiveness of treatments for carcass decontamination. Hardin et al. (1995) found no differences in reduction of *S.* Typhimurium and *E. coli* O157:H7 in the outside round, brisket, flank and clod carcass surface regions treated with acetic or lactic acids immediately and 20 to 30 min after inoculation. The reductions of both pathogens were significantly smaller on the inside round region, which shows a surface mostly composed of lean muscle. However, as with the other carcass surface regions, no effect of bacterial attachment was observed on bacterial reductions for outside round. Similar results were reported by Castillo et al. (1998a) for beef carcass surfaces treated with hot water. Cutter and Siragusa (1994a) also reported greater reduction rates for Gram-negative organisms on fat than on lean tissues. The inoculation menstruum has been shown to affect the attachment rate as well. Dickson and Macneil (1991) reported that *S.* Typhimurium and *Listeria monocytogenes* attached to beef carcass surfaces at a higher rate when the inoculum had been diluted in phosphate buffer compared to cow manure.

In addition to lactic and acetic acids, other organic acids have been tested for ability to reduce bacterial populations on beef. Podolak et al. (1996) found that fumaric acid solutions produced greater reductions in microbial populations and growth in ground beef. In another study, these same authors found fumaric acid at concentrations of 1% and 1.5% to be more effective than 1% lactic or acetic acids in reducing populations of *E. coli* O157:H7 and *L. monocytogenes* on beef lean muscle. In contrast, Anderson et al. (1992) found lactic acid to be more effective than acetic acid or a mixture of lactic, acetic, citric and L-ascorbic acids in reducing Gram-negative pathogens on lean meat. García-Zepeda et al. (1994b) compared gluconic acid (1.5 and 3.0%), 1.5% lactic acid, and combinations of gluconic and lactic acid as fresh beef decontaminants. Beef samples were inoculated with *Lactobacillus fermentum*, treated with the different acid solutions, vacuum packaged and stored at 1°C for up to 56 d. A mixture of 3% gluconic acid combined with 1.5% lactic acid produced the lowest psychrotrophic or lactobacilli counts. However, this mixture was detrimental to the colour characteristics of the meat. A 50:50 mixture of 1.5% each gluconic and lactic acids appeared to be beneficial for the colour characteristics of the meat at display, whereas lactic acid alone effectively reduced the bacterial counts but negatively affected the redness of the meat. Cutter and Siragusa (1994a) found no differences in log reductions of *E. coli* O157:H7 and *Pseudomonas fluorescens* on beef surfaces after spraying citric, acetic or lactic acids at equal concentrations. Reynolds and Carpenter (1974) used a 60:40 w/w mixture of acetic and propionic acids, which is used as fungicide in cereal storage, for pork carcass decontamination. By modifying the molarities of these two acids, they reduced bacterial populations by 2.0 $\log_{10}$. In general, concentrations below 2.15 M of each acid produced little effect on the visual quality of the carcasses, while achieving bacterial reductions similar to those obtained by applying the acid mixture with higher molarities. More research on the usefulness of other organic acids with reported antibacterial activity (Richards et al., 1995) might be necessary for offering alternatives for carcass decontamination.

6 Decontamination treatments: other interventions

TSP is a non-acid compound commonly used for carcass decontamination, and applications including TSP were patented for poultry decontamination (Bender and Brotsky, 1992). Dickson et al. (1994) obtained reductions of *S.* Typhimurium, *L. monocytogenes* and *E. coli* O157:H7 ranging from ca. 0.8 to 1.2 log_{10}/cm^2 by spraying TSP solutions (55°C) on lean beef muscle. On adipose tissue the reductions ranged from 1.2 to 2.5 log_{10}/cm^2. TSP concentration was not a significant factor in bacterial reduction by this chemical and, in general, greater reductions were observed when the temperature of the TSP solution was increased from 25 to 55°C. It is likely that the high pH of the TSP solution (ca. 13) was responsible for the bacterial reductions reported by Dickson et al. (1994). Antimicrobial effect of high-pH solutions on foodborne bacterial pathogens has been reported, and is apparently due to membrane disruption of the cells and an increase in the water solubility of the DNA at high pH (Mendonca et al., 1994).

The effectiveness of bacteriocins in beef decontamination was also evaluated. Cutter and Siragusa (1994b) applied nisin solution (5000 activity units/ml) to beef carcass tissue inoculated with various Gram-positive bacteria. Reductions in counts produced by this treatment ranged from 1.79 to 3.54 log_{10}/cm^2. Cutter and Siragusa (1996) enhanced the inhibition of *B. thermosphacta* on beef surfaces by immobilizing nisin in calcium alginate gels. Using this approach produced greater bacterial reductions on the beef immediately after treatment, and counts remained lower than those of beef treated with nisin during refrigerated storage for 7 d. After storage, the numbers of *B. thermosphacta* on the beef surfaces were 7.1 log_{10} CFU/cm^2 for controls, 6.45 CFU/cm^2 for beef treated with calcium alginate only, 5.26 CFU/cm^2 for beef treated with nisin only and 2.37 CFU/cm^2 for beef treated with alginate-immobilized nisin. Other bacteriocins have also been used for controlling pathogens on meat (Goff et al., 1996).

Additional chemicals have been investigated for use as carcass sanitizers. Cutter and Dorsa (1995) found chlorine dioxide at concentrations of up to 20 ppm to be no more effective than regular water in reducing bacteria of faecal origin on beef carcass tissue. Anderson et al. (1977) reported no differences in bacterial reductions on meat after spraying tap water or water added with 200 ppm sodium hypochlorite, but Kotula et al. (1974) found lower bacterial counts on beef carcasses treated with 200 ppm sodium hypochlorite compared to untreated carcasses. In addition, they reported some continued effect of chlorine during storage of the carcasses. Since treatment of carcasses with non-chorinated water was not included in their study, the reported effect of chlorinated water on bacterial counts may have involved both washing effect and an extended antimicrobial effect. After inoculating different pathogenic and non-pathogenic organisms onto lean and adipose beef tissue, Dickson (1988) applied washes with phosphate buffer, ethanol, sodium chloride, sodium hydroxide and potassium hydroxide. Phosphate buffer, ethanol and sodium chloride produced reductions of <1 log, while sodium and potassium hydroxide reduced the populations of inoculated bacteria by as much as 4 logs, with greater reductions in bacterial counts on adipose tissue than on lean tissue. Castillo et al. (1998d) observed reductions in *E. coli* O157:H7 and *S.* Typhimurium counts on beef carcass surfaces after the application of acidified sodium chlorite (ASC) solutions. When phosphoric acid was used to acidify sodium chlorite, the resulting ASC solution reduced populations of both pathogens by 3.8 to 3.9 log cycles, while when ASC solutions were prepared by acidifying with citric acid, the reductions obtained ranged from 4.5 to 4.6 log cycles. Bromine-containing compounds have also been shown to reduce microbial populations on beef carcasses, with Kalchayanand et al. (2009) reporting reductions in APC and Enterobacteriaceae populations up to 3.6 logs, similar to hot water washing.

Steam treatments of beef or sheep carcasses were attempted by various investigators with little success (Anderson et al., 1979; Dorsa et al., 1996a). Dorsa (1997) reviewed studies that led to the development of a steam pasteurization treatment, and in the late 1990s the process gained rapid popularity among meat processors (Wilson and Leising, 1994). Phebus et al. (1997) assembled an experimental steam pasteurization chamber, and reported reduced counts of *E. coli* O157:H7, *S.* Typhimurium and *L. monocytogenes* by 3.4 to 3.7 log cycles on surfaces of freshly slaughtered beef. However, steam pasteurization alone was not found to achieve greater reductions than other treatments such as knife trimming or steam vacuuming. Nutsch et al. (1997) conducted commercial evaluations of the steam pasteurization process in a beef processing plant in which carcasses were subjected to a preliminary water wash before passing through air blowers to eliminate excessive carcass surface humidity that would favour steam condensation. The carcasses then passed through a steam chamber followed by another section of the cabinet where cold water was applied. Applying this treatment, carcass APCs were reduced from 2.12–2.19 $\log_{10}$ CFU/cm^2 to 0.56–0.84 $\log_{10}$ CFU/cm^2. Counts of *E. coli* were also reduced from original counts of 0.60–1.53 $\log_{10}$ CFU/cm^2 to undetectable levels after steam treatment.

Chemical dehairing could have potential benefits for reducing contamination on meat carcasses. This process was developed by Bowling and Clayton (1992) based on the hair removal process used in the tanning of leather. Schnell et al. (1995) adapted the chemical dehairing process to slaughter operations in a commercial facility. These authors found no differences in APCs or coliform counts in samples from carcasses of dehaired cattle and those of conventionally slaughtered cattle. They concluded that dehairing enhanced the visual cleanliness of the carcasses but was not able to effectively reduce bacterial counts. This report did not include any determination of the extent to which bacterial counts were reduced on the hides of the slaughtered cattle by the dehairing process; therefore, the impact of this treatment on the prevention of faecal contamination from the hides to the carcass surfaces could not be determined. Adapting the chemical dehairing process to laboratory testing, Castillo et al. (1998c) found reductions in *E. coli* O157:H7, *S.* Typhimurium, *E. coli*, coliforms and APCs on artificially contaminated bovine skin, ranging from 3.4 to >4.8 $\log_{10}$/cm^2. The authors concluded that dehairing had commercial potential in controlling contamination during dehiding operations. More recently, Nou et al. (2003) evaluated dehairing of cattle hides in a commercial beef processing facility and concluded that hide-to-carcass contamination with pathogens was reduced.

Other agents such as ozonated water or hydrogen peroxide were also evaluated, with variable results when compared to treatments such as trimming or washing (Gorman et al., 1995b; Reagan et al., 1996). Meat irradiation was recently approved for decontamination of fresh frozen meat, and it was previously approved for treating poultry and pork. Many reports indicate that this process is quite effective for reducing pathogenic contamination from ground beef and poultry carcasses (Clavero et al., 1994; Lee et al., 1996; Tarté et al., 1996). A multiple hurdle approach, or use of sequential process interventions, might be necessary for reducing contamination with pathogens (Arthur et al., 2004; Castillo et al., 1998b; Phebus et al., 1997). Investigations were conducted on combinations of carcass decontamination procedures to determine possible levels of pathogen reduction during meat processing. Reported success is variable and sometimes contradictory. For example, in reviewing treatments at four beef packing plants, it was reported that spraying carcasses with 2% lactic acid, steam-vacuum or trimming was ineffective, and that using only steam or hot water treatments without the other treatments substantially reduced bacterial contamination (Gill and Landers, 2003). Gill and Badoni (2004) compared 0.02% peroxyacetic acid, acidified 0.16% sodium chlorite, 2% lactic acid and 4% lactic acid on chilled beef surfaces. They found that a negligible reduction of coliforms or *E. coli* resulted from exposure to peroxyacetic acid and ASC, and that both were less effective than 4% lactic acid. They surmised that varied results in evaluating antimicrobial treatments might be due to their application on different types of meat surfaces, and that the meat surface microflora likely reflects the effect of prior antimicrobial treatments. In addition, peroxyacetic acid concentrations up to 600 ppm were reported to be ineffective antimicrobial treatments when applied to chilled beef carcass surfaces that were inoculated (King et al., 2005).

Irradiation of carcasses to reduce or eliminate *E. coli* O157:H7 on beef surfaces was investigated by Arthur et al. (2005) and Maxim et al. (2014). Arthur et al. (2005) subjected beef surfaces to E-beam irradiation at 1 kGy and reported a 4-log/cm^2 reduction. The authors concluded that chilled carcasses could be irradiated to reduce surface contamination without affecting flavour or aroma of subsequently produced ground beef. Using a different approach, Maxim et al. (2014) developed a chamber designed to obtain uniform E-beam dose distribution on the surface of a carcass to obtain a more reliable and controlled treatment. The 'Maxim Chamber' was tested on rabbit carcasses and successfully obtained an even distribution of dose, providing a >5-log reduction of *E. coli* O157:H7.

Many approaches to decontaminate meat surfaces were investigated and implemented; however, none of the approved interventions is capable of eliminating the presence of pathogens. In an attempt to reduce risk, it is common for meat processors to apply several redundant pathogen reduction technologies, but elimination of bacterial contamination cannot be guaranteed (Barkocy-Gallagher et al., 2003; Elder et al., 2000). Proper handling of meat products by end users is still required to assure safety.

7 Processing operations: fabrication

Many beef carcasses are chilled for at least 18 hours prior to fabricating them into smaller portions. Although contamination during chilling is rarely addressed, it is certainly possible. Potential sources of contamination include direct contact by employees, contact between individual carcasses or environmental contamination by either air, water or physical contaminants, such as rail dust. Since the beef carcasses were processed to minimize contamination during the slaughter process, it is imperative to maintain hygienic conditions during further processing of these carcasses.

Fabrication is the process of breaking down the carcass into smaller and smaller portions. The process of fabrication is typically considered to begin when the carcasses exit the chillers. Although all components of fabrication are important, bacterial control can be categorized within four general areas:

a General hygienic practices
b Equipment sanitation
c Environmental controls
d Interventions

To assure the safety and quality of fresh beef, all of these categories must be effectively managed to reduce further contamination as well as minimize the growth of the existing microbial contamination.

General hygienic practices, often referred to as good manufacturing practices (GMPs) or good hygienic practices (GHPs), are the fundamental programmes required to produce safe and wholesome food. These may include personal hygiene, control of foreign materials (glass, metal, etc.), pest control and building maintenance. There are many sources of information on these types of programmes, including but not limited to the British Standards Institute PAS 220:2008, ISO/TS 22002-1:2009 and the Codex Alimentarius report entitled 'Guidelines for the Control of Nontyphoidal *Salmonella* spp. in Beef and Pork Meat' (http://www.fao.org/fao-who-codexalimentarius/sh-proxy/en/?lnk=1&url=https%253A%252F%252Fworkspace.fao.org%252Fsites%252Fcodex%252FStandards%252FCAC%2BGL%2B87-2016%252FCXG%2B087e.pdf). The important part of GMPs/GHPs is that they all strive to avoid unnecessary contamination. While contamination in and of itself is unavoidable, a significant amount of contamination can be minimized or eliminated by these practices. Processors will implement these programmes to various degrees, depending on the physical structure of the establishment and the geographic location, but every processor can use these programmes to instil a sense in all of their employees that they are not manufacturing random industrial items, but are in fact producing food which will be consumed by families like theirs.

Sanitation is usually thought of as a prerequisite programme, much like the GMPs/GHPs. However, the significance of sanitation, especially equipment sanitation, to the production of safe and wholesome beef cannot be overstated. Consider the results of poor sanitation. Meat residue remains on the equipment at temperatures conducive to rapid growth of microorganisms. The entire microbiome, including spoilage and pathogenic bacteria, reach high populations within a relatively short time. When production begins again, this meat residue with high microbial populations becomes an inoculum for all of the fresh meat with lower populations of bacteria. Improper sanitation contributes to both food safety and shelf life issue.

There are a variety of approaches to sanitize beef processing establishment, but all follow the same basic outline. The initial step is a 'dry' clean up, intended to remove large particles of meat residue and other miscellaneous soil from the equipment. This dry clean up is important, it reduces the amount of labour required later in the sanitation process, and also reduces the amount of sanitation chemicals required. Sanitation chemicals contribute to operational costs in two ways. First, there is the initial purchase of the chemical, and excessive chemical use requires more chemicals to be purchased. The second cost is often poorly understood, even by management. All of the sanitation chemicals go down the drain. Wastewater requires treatment, and excessive chemical use can contribute to increased waste treatment costs. Of equal importance, these waste chemicals enter the environment, and at some point may require additional treatment, if the waste is discharged into a water system which is used for drinking water by a community downstream. Minimizing chemical usage, within the boundaries of effective use, is both financially and environmentally responsible.

The second common step in equipment sanitation is a water rinse. In many cases, the water will be warmed to facilitate the removal of fat from the surfaces. However, the temperature should be below 57°C, 135°F, to avoid coagulating the meat protein on to the equipment surface. After rinsing, the cleaning chemicals are applied with some type of mechanical force (manual scrubbing, pressure wash). The commonly used cleaning chemicals are generally alkaline to facilitate the removal of lipids and proteins. The cleaning chemicals are then rinsed and the surface inspected for visual residue. The inspection step is extremely important, in that residual meat cannot be adequately sanitized, and residual cleaning chemicals will often neutralize sanitation chemicals.

The final step is sanitizing, which can be accomplished in a variety of physical or chemical methods. Hot water is a commonly used sanitizer, but as with any thermal process requires both an adequate temperature and a sufficient contact time to accomplish the intended purpose. Chemical sanitizers are widely used, with the most common being chlorine-based chemicals. Daily sanitation is essential, and additional sanitation steps may be required when the establishment may not have been operated for several days (national holidays, seasonal events, etc.).

Environmental controls may also be considered part of the GMP/GHP programme. The most common environmental control beyond sanitation is the temperature of the fabrication room. In the United States, the USDA-FSIS permits continuous operations over two shifts (nominally 16 hours) without a full sanitation programme if the room temperature is maintained below 10°C. The rationale for this is that most of the pathogens of concern, notably non-Typhoidal salmonellae, grow slowly or not at all below 10°C. By maintaining the room below this temperature, along with having the carcass typically chilled to less than 5°C prior to processing, the potential growth of non-Typhoidal salmonellae was effectively controlled. While the focus of this environmental control is primarily food safety, it also has the potential to impact spoilage and shelf life. Environmental control is one of the tools commonly used to extend the shelf life of fresh beef, especially for export purposes. As with any GMP/GHP, all of the individual components must work together to achieve the goal.

As food safety has become a larger concern for the beef industry, interventions similar to those included during the slaughter are becoming more common on the fabrication side. Peroxyacetic acid or ASC are commonly applied to subprimal cuts prior to fabrication to retail cuts (Kalchayanand et al., 2012; Liao et al., 2015; Ransom et al., 2003).

8 Packaging, storage and shelf life

8.1 Packaging

After the beef carcasses are cut into smaller portions, they are usually packaged in some form prior to entering commerce. The packaging depends upon the type and size of the cut, such as primals, being shipped for further cutting compared to pre-packaged product that is retail ready, and the intended use. The intended use may be as diverse as delivery to a local restaurant or retailer, or placing in a refrigerated container for international shipment. All of these factors are involved in the selection of the packaging method, and may be employed simultaneously on meat from the same lot of cattle.

Packaging serves two purposes. It protects the meat from further contamination, and also may serve as an environmental control for bacteria. The most widely used method of environmental control by packaging is vacuum packaging. Vacuum packaging simply removes the majority of the oxygen from the package, slowing the growth of bacteria. Many members of the spoilage microbiome are facultatively anaerobic, capable of growing with or without oxygen. However, in many cases anaerobic metabolism is much less efficient than aerobic metabolism, and the same bacteria simply have a slower growth rate in the absence of oxygen. Vacuum packaging is also dependent upon the packaging film, as using a film with a high oxygen transmission rate will negate the benefit of vacuum packaging.

A variation of vacuum packaging is modified atmosphere packaging. In this case, after a vacuum is pulled to remove the oxygen, the package is back-flushed with a gas mixture. Back-flushing with an oxygen-free gas removes the very small amount of residual oxygen left by the vacuum, and therefore results in an environment with even less residual oxygen. The gas may be an inert gas such as nitrogen or carbon dioxide, or may be a mixture of several gasses. Carbon monoxide was added in a small percentage to the gas, because it helps to maintain the colour of the meat pigment.

Some works documented the inclusion of antimicrobials in the package itself or incorporated into the packaging film. Much of this research has been conducted with processed meats, primarily to address concerns with *L. monocytogenes*. The antimicrobials typically are organic acids, or the salts of organic acids, or natural plant-derived materials. Although they have been shown to be effective, in many cases they are not permitted to be used with fresh beef. These compounds, in this specific application, are considered additives and must therefore be listed on the label. In the United States, the standard of identity for fresh beef prohibits additives, and the products could not be labelled as fresh beef. Many other regulatory restrictions exist, both within and between different countries, and so at this time these additives are not employed to improve the safety and shelf life of fresh beef.

8.2 Storage and shelf life

The factors which influence shelf life are no different from the environmental factors previously discussed. The primary environmental factor is temperature, as at this point the beef should be packaged in a way to prevent further contamination. In 2014, EFSA published a detailed report on the importance of maintenance of the cold chain during storage and distribution. This document is perhaps the most recent and comprehensive review of the subject, and certainly a valuable reference. Their conclusions included:

a Carcass surface temperature is a more relevant indicator of the effect of chilling on bacterial growth than core temperature.
b If there is equivalent or less bacterial growth, there is no additional risk for the consumer. Total bacterial growth is affected by the continuum of chilling in the slaughter plant, during transport, deboning, storage, retail and catering/domestic refrigeration.
c It is possible to have different combinations of slaughterhouse–transportation time–temperature chilling scenarios that result in equivalent or less bacterial growth than those obtained using the currently mandated chilling requirements (chilling to a core temperature of 7°C in the slaughterhouse chillers before transportation for a maximum of 48 hours).

While these recommendations are specific to the European Union, the fundamental principles of bacteriology do not differ throughout the world, and so the same principles are applicable to all fresh beef.

9 Conclusions

Many programmes and processes contribute to the safety and quality of fresh beef. These programmes overlap, so that a partial failure in one does not lead to an overall processing failure. The concept was best described by Leistner (2000) as the 'hurdle' concept, where multiple interventions were applied so that ultimately the majority of the pathogenic microorganisms would be eliminated. Although the hurdle concept was originally applied to pathogenic bacteria, the same concept applies to spoilage bacteria as well. Ultimately, the safety and quality of fresh beef re-lives on an integrated system which starts with the production of healthy live cattle, free of chemical residues and ends with the presentation of the beef to the consumer, ready to eat.

10 References

Acuff, G. R., Vanderzant, C., Savell, J. W., Jones, D. K., Griffin, D. B., and Ehlers, J. G. 1987. Effect of acid decontamination of beef subprimal cuts on the microbiological and sensory characteristics of steaks. *Meat Sci.* 19: 217–26.

Anderson, M. E., Huff, H. E., Naumann, H. D., Marshall, R. T., Damare, J. M., Pratt, M., and Johnston, R. 1987. Evaluation of an automated beef carcass washing and sanitizing system under production conditions. *J. Food Prot.* 50: 562–6.

Anderson, M. E., Huff, H. E., Naumann, H. D., and Marshall, R. T. 1988. Count of six types of bacteria on lamb carcasses dipped or sprayed with acetic acid at 25° or 55°C and stored vacuum packaged at 0°C. *J. Food Prot.* 51: 874–7.

Anderson, M. E., Marshall, R. T., Naumann, H. D., and Stringer, W. C. 1975. Physical factors that affect removal of yeasts from meat surfaces with water sprays. *J. Food Sci.* 40: 1232–5.

Anderson, M. E., Marshall, R. T., Stringer, W. C., and Naumann, H. D. 1977. Combined and individual effects of washing and sanitizing on baceterial counts of meat - A model system. *J. Food Prot.* 40: 688–70.

Anderson, M. E., Marshall, R. T., Stringer, W. C., and Naumann, H. D. 1979. Microbial growth on plate beef during extended storage after washing and sanitizing. *J. Food Prot.* 42: 389–92.

Anderson, M. E. and Marshall, R. T. 1989. Interaction of concentration and temperature of acetic acid solution on reduction of various species of microorganisms on beef surfaces. *J. Food Prot.* 52: 312–15.

Anderson, M. E. and Marshall, R. T. 1990a. Reducing microbial populations on beef tissues: Concentration and temperature of lactic acid. *J. Food Saf.* 10: 181–90.

Anderson, M. E. and Marshall, R. T. 1990b. Reducing microbial populations on beef tissues: Concentration and temperature of an acid mixture. *J. Food Sci.* 55: 903–5.

Anderson, M. E., Marshall, R. T. and Dickson, J. S. 1992. Efficacies of acetic, lactic, and two mixed acids in reducing numbers of bacteria on surfaces of lean meat. *J. Food Saf.* 12: 139–47.

Arthur, T. M., Bosilevac, J. M., Nou, X., Shackelford, S. D., Wheeler, T. L., Kent, M. P., Jaroni, D., Pauling, B., Allen, D. M. and Koohmaraie, M. 2004. *Escherichia coli* O157 prevalence and enumeration of aerobic bacteria, *Enterobacteriaceae*, and *Escherichia coli* O157 at various steps in commercial beef processing plants. *J. Food Prot.* 67: 658–65.

Arthur, T. M., Bosilevac, J. M., Brichta-Harhay, D. M., Kalchayanand, N., King, D. A., Shackelford, S. D., Wheeler, T. L. and Koohmaraie, M. 2007. Effects of a minimal hide wash cabinet on the levels and prevalence of *Escherichia coli* O157:H7 and *Salmonella* on the hides of beef cattle at slaughter. *J. Food Prot.* 70: 1076–9.

Arthur, T. M., Wheeler, T. L., Shackelford, S. D., Bosilevac, J. M., Nou, X. and Koohmaraie, M. 2005. Effects of low-dose, low-penetration electron beam irradiation of chilled beef carcass surface cuts on *Escherichia coli* O157:H7 and meat quality. *J. Food Prot.* 68: 666–72.

Ayres, J. C. 1955. Microbiological implications in the handling, slaughtering, and dressing of meat animals. *Adv. Food Res.* 6:109–61.

Barkate, M. L., Acuff, G. R., Lucia, L. M., and Hale, D. S. 1993. Hot water decontamination of beef carcasses for reduction of initial bacterial numbers. *Meat Sci.* 35: 397–401.

Barkocy-Gallagher, G. A., Arthur, T. M., Rivera-Betancourt, M. Nou, X., Shackelford, S. D., Wheeler, T. L. and Koohmaraie, M. 2003. Seasonal prevalence of Shiga toxin-producing *Escherichia coli*, including O157:H7 and non-O157 serotypes, and *Salmonella* in commercial beef processing plants. *J. Food Prot.* 66: 1978–86.

Bell, M. F., Marshall, R. T. and Anderson, M. E. 1986. Microbiological and sensory tests of beef treated with acetic and formic acids. *J. Food Prot.* 49: 207–10.

Bender, F. G. and Brotsky, E. 1992. Process for treating poultry carcass to control salmonellae growth. United States Patent 5,143,739.

Biemuller, G. W., Carpenter, J. A. and Reynolds, A. E. 1973. Reduction of bacteria on pork carcasses. *J. Food Sci.* 38: 261–3.

Bosilevac, J. M., Nou, X., Osborn, M. S., Allen, D. M. and Koohmaraie, M. 2005. Development and evaluation of an on-line hide decontamination procedure for use in a commercial beef processing plant. *J. Food Prot.* 68: 265–72.

Bouttier, S., Linxe, C., Ntsama, C., Morgant, G., Bellon-Fontaine, M. N. and Fourniat, J. 1997. Attachment of *Salmonella choleraesuis* to beef muscle and adipose tissues. *J. Food Prot.* 60: 16–22.

Bowling, R. A. and Clayton, R. P. 1992. Method for Dehairing Animals. United States Patent 5,149,295. Monfort Inc. Greeley, CO.

Brackett, R. E., Hao, Y.-Y., and Doyle, M. P. 1994. Ineffectiveness of hot acid sprays to decontaminate *Escherichia coli* O157: H7 on beef. *J. Food Prot.* 57: 198–203.

Butler, J. L., Stewart, J. C., Vanderzant, C., Carpenter, Z. L. and Smith, G. C. 1979. Attachment of microorganisms to pork skin and surfaces of beef and lamb carcasses. *J. Food Prot.* 42: 401–6.

Cabedo, L., Sofos, J. H. and Smith, G. C. 1996. Removal of bacteria from beef tissue by spray washing after different times of exposure to fecal material. *J. Food Prot.* 59: 1284–7.

Castillo, A., Lucia, L. M., Goodson, K. J., Savell, J. W. and Acuff, G. R. 1998a. Use of hot water for beef carcass decontamination. *J. Food Prot.* 61: 19–25.

Castillo, A., Lucia, L. M., Goodson, K. J., Savell, J. W. and Acuff, G. R. 1998b. Comparison of water wash, trimming, and combined hot water and lactic acid treatments for reducing bacteria of fecal origin on beef carcasses. *J. Food Prot.* 61: 823–8.

Castillo, A., Dickson, J. S., Clayton, R. P., Lucia, L. M. and Acuff, G. R. 1998c. Chemical dehairing of bovine skin to reduce pathogenic organisms and bacteria of fecal origin. *J. Food Prot.* 61: 623–5.

Castillo, A., Lucia, L. M., Kemp, G. K. and Acuff, G. R. 1998d. Reduction of *Escherichia coli* O157:H7 and *Salmonella* Typhimurium on beef carcass surfaces using acidified sodium chlorite. *J. Food Prot.* 61: *In Press.*

Clavero, M. R. S., Monk, J. D., Beuchat, L. R., Doyle, M. P. and Brackett, R. E. 1994. Inactivation of *Escherichia coli* O157: H7, salmonellae, and *Campylobacter jejuni* in raw ground beef by gamma irradiation. *Appl. Environ. Microbiol.* 60: 2069–75.

Crouse, J. D., Anderson, M. E. and Naumann, H. D. 1988. Microbial decontamination and weight of carcass beef as affected by automated washing pressure and length of time of spray. *J. Food Prot.* 51: 471–4.

Cutter, C. N. and Dorsa, W. J. 1995. Chlorine dioxide spray washes for reducing fecal contamination on beef. *J. Food Prot.* 58: 1294–6.

Cutter, C. N. and Siragusa, G. R. 1994a. Efficacy of organic acids against *Escherichia coli* O157: H7 attached to beef carcass tissue using a pilot scale model carcass washer. *J. Food Prot.* 57: 97–103.

Cutter, C. N. and Siragusa, G. R. 1994b. Decontamination of beef carcass tissue with nisin using a pilot scale model carcass washer. *Food Microbiol.* 11: 481–9.

Cutter, C. N. and Siragusa, G. R. 1996. Reduction of *Brochothrix thermosphacta* on beef surfaces following immobilization of nisin in calcium alginate gels. *Lett. Appl. Microbiol.* 23: 9–12.

Cutter, C. N., Dorsa, W. J. and Siragusa, G. R. 1997. Parameters affecting the efficacy of spray washes against *Escherichia coli* O157: H7 and fecal contamination on beef. *J. Food Prot.* 60: 614–18.

Davey, K. R. and Smith, M. G. 1989. A laboratory evaluation of a novel hot water cabinet for the decontamination of sides of beef. *Int. J. Food Sci. Tech.* 24: 305–16.

DeZuniga, A. G., Anderson, M. E., Marshall, R. T. and Iannotti, E. L. 1991. A model for studying the penetration of microorganisms into meat. *J. Food Prot.* 54: 256–8.

Dickson, J. S. 1988. Reduction of bacteria attached to meat surfaces by washing with selected compounds. *J. Food Prot.* 51: 869–73.

Dickson, J. S. 1990. Transfer of *Listeria monocytogenes* and *Salmonella typhimurium* between beef tissue surfaces. *J. Food Prot.* 53: 51–5.

Dickson, J. S. 1991a. Attachment of *Salmonella typhimurium* and *Listeria monocytogenes* to beef tissue: effects of inoculum level, growth temperature and bacterial culture age. *Food Microbiol.* 8: 143–51.

Dickson, J. S. 1991b. Control of *Salmonella typhimurium, Listeria monocytogenes,* and *Escherichia coli* O157: H7 on beef in a model spray chilling system. *J. Food Sci.* 56: 191–3.

Dickson, J. S. 1992a. Acetic acid action on beef tissue surfaces contaminated with *Salmonella typhimurium. J. Food Sci.* 57: 297–301.

Dickson, J. S. 1992b. Contamination of beef tissue surfaces by artificially inoculated low-moisture menstraa. *J. Food Safety.* 12: 181–90.

Dickson, J. S. and Anderson, M. E. 1991. Control of *Salmonella* on beef tissue surfaces in a model system by pre- and post-evisceration washing and sanitizing, with or without spray chilling. *J. Food Prot.* 54: 514–18.

Dickson, J. S. and Anderson, M. E. 1992. Microbiological decontamination of food animal carcasses by washing and sanitizing systems. A review. *J. Food Prot.* 55: 133–40.

Dickson, J. S., Cutter, C. J. N. and Siragusa, G. R. 1994. Antimicrobial effect of trisodium phosphate against bacteria attached to beef tissue. *J. Food Prot.* 57: 952–5.

Dickson, J. S. and Kunduru, M. R. 1995. Resistance of acid-adapted salmonellae to organic acid rinses on beef. *J. Food Prot.* 58: 973–6.

Dickson, J. S. and Macneil, M. D. 1991. Contamination of beef tissue surfaces by cattle manure inoculated with *Salmonella typhimurium* and *Listeria monocytogenes. J. Food Prot.* 54: 102–4.

Dixon, Z. R., Vanderzant, C., Acuff, G. R., Savell, J. W. and Jones, D. K. 1987. Effect of acid treatment of beef strip loin steaks on microbiological and sensory characteristics. *Int. J. Food Microbiol.* 5: 181–6.

Dorsa, W. J. 1997. New and established carcass decontamination procedures commonly used in the beef-processing industry. *J. Food Prot.* 60: 1146–51.

Dorsa, W. J., Cutter, C. N. and Siragusa, G. R. 1996a. Effectiveness of a steam-vacuum sanitizer for reducing *Escherichia coli* O157: H7 inoculated to beef carcass surface tissue. *Lett. Appl. Microbiol.* 23: 61–3.

Dorsa, W. J., Cutter, C. N., Siragusa, G. R. and Koohmaraie, M. 1996b. Microbial decontamination of beef and sheep carcasses by steam, hot water spray washes, and a steam-vacuum sanitizer. *J. Food Prot.* 59: 127–35.

Elder, R. O., Keen, J. E., Siragusa, G. R., Barkocy-Gallagher, G. A., Koohmaraie, M. and Laegreid, W. W. 2000. Correlation of entrohemorrhagic *Escherichia coli* O157 prevalence in feces, hides, and carcasses of beef cattle during processing. *Proc. Natl. Acad. Sci.* USA 97: 2999–3003.

Elmonssalami, E. and N. Wassef. 1971. Penetration of some microorganisms in meat. *Zbl. Vet. Med. B* 18: 329–36.

European Food Safety Authority (EFSA). 2011. Scientific Opinion on the evaluation of the safety and efficacy of lactic acid for the removal of microbial surface contamination of beef carcasses, cuts and trimmings. *EFSA J.* 9(7): 2317.

European Food Safety Authority (EFSA). 2014. Scientific Opinion on the public health risks related to the maintenance of the cold chain during storage and transport of meat. *EFSA J.* 12(3): 3601–82.

Federal Register. 1996. Notice of Policy Change; Achieving the Zero Tolerance Performance Standard for Beef Carcasses by Knife Trimming and Vacuuming with Hot Water or Steam; Use of Acceptable Carcass Interventions for Reducing Carcass Contamination Without Prior Agency Approval. Food Safety and Inspection Service, USDA. *Fed. Regist.* 61: 15024–7.

Fratamico, O. M., Schultz, F. J., Benedict, R. C., Buchanan, R. L. and Cooke, P. H. 1996. Factors influencing attachment of *Escherichia coli* O157:H7 to beef tissues and removal using selected sanitizing rinses. *J. Food Prot.* 59: 453–9.

Garcia-Zepeda, C. M., Kastner, C. L., Kenney, P. B., Campbell, R. E. and Schwenke, J. R. 1994a. Aroma profile of subprimals from beef carcasses decontaminated with chlorine and lactic acid. *J. Food Prot.* 57: 674–8.

García-Zepeda, C. M., Kastner, C. L., Willard, B. L., Phebus, R. K., Schwenke, J. R., Fijal, B. A. and Prasai, R. K. 1994b. Gluconic acid as a fresh beef decontaminant. *J. Food Prot.* 57: 956–62.

Gill, C. O. and M. Badoni. 2004. Effects of peroxyacetic acid, acidified sodium chlorite or lactic acid solutions on the microflora of chilled beef carcasses. *Int. J. Food Microbiol.* 91: 43–50.

Gill, C. O. and N. Penny. 1979. Survival of bacteria in carcasses. *Appl. Environ. Microbiol.* 37: 667–9.

Gill, C. O., Badoni, M. and Jones, T. 1996. Hygienic effects of trimming and washing operations in a beef carcass dressing process. *J. Food Prot.* 59: 666–9.

Gill, C. O. and Landers, C. 2003. Microbiological effects of carcass decontaminating treatments at four beef packing plants. *Meat Sci.* 65: 1005–11.

Gill, C. O., Penny, N. and Nottingham, P. M. 1976. Effect of delayed evisceration on the microbial quality of meat. *Appl. Environ. Microbiol.* 31: 465–8.

Goddard, B. L., Mikel, W. B., Conner, D. E. and Jones, W. R. 1996. Use of organic acids to improve the chemical, physical, and microbial attributes of beef strip loins stored at −1°C for 112 days. *J. Food Prot.* 59: 849–953.

Goff, J. H., Bhunia, A. K. and Johnson, M. G. 1996. Complete inhibition of *Listeria monocytogenes* on refrigerated chicken meat with pediocin AcH bound to heat-killed Pediococcus acidilactici cells. *J. Food Prot.* 59: 1187–92.

Gorman, B. M., Morgan, J. B., Sofos, J. N. and Smith, G. C. 1995a. Microbiological and visual effects of trimming and/or spray washing for removal of fecal material from beef. *J. Food Prot.* 58: 984–9.

Gorman, B. M., Sofos, J. N., Morgan, J. B., Schmidt, G. R. and Smith, G. C. 1995b. Evaluation of hand-trimming, various sanitizing agents, and hot water spray-washing as decontamination interventions for beef brisket adipose tissue. *J. Food Prot.* 58: 899–907.

Grau, F. Ft., Brownlie, L. E. and Roberts, T. A. 1968. Effect of some preslaughter treatments on the Salmonella population in the bovine rumen and faeces. *J. Appl. Bacteriol.* 31: 157–63.

Greer, G. G. and Dilts, B. D. 1992. Factors affecting the susceptibility of meatborne pathogens and spoilage bacteria to organic acids. *Food Res. Int.* 25: 355–64.

Hardin, M. D. 1995. Comparison of methods for contamination removal from beef carcass surfaces. Doctoral dissertation. Texas A and M University.

Hardin, M. D., Acuff, G. R., Lucia, L. M., Oman, J. S. and Savell, J. W. 1995. Comparison of methods for contamination removal from beef carcass surfaces. *J. Food Prot.* 58: 368–74.

Jensen, L. B. and W. R. Hess. 1941. A study of ham souring. *J. Food Sci.* 6: 273–326.

Kalchayanand, N., Arthur, T. M., Bosilevac, J. M., Brichta-Harhay, D. M., Guerini, M. N., Shackleford, S. D., Wheeler, T. L. and Koohmaraie, M. 2009. Effectiveness of 1,3-Dibromo-5,5 Dimethylhydantoin on Reduction of Escherichia coli O157:H7– and Salmonella-Inoculated Fresh Meat. *J. Food Protect.* 72(1): 151–6.

Kalchayanand, N., Arthur, T. M., Bosilevac, J. M., Schmidt, J. W., Wang, R., Shackelford, S. D. and Wheeler, T. L. 2012. Evaluation of commonly used antimicrobial interventions for fresh beef inoculated with Shiga toxin-producing *Escherichia coli* serotypes O26, O45, O103, O111, O121, O145, and O157:H7. *J. Food Prot.* 75: 1207–12.

King, D. A., Lucia, L. M., Castillo, A., Acuff, G. R., Harris, K. B. and Savell, J. W. 2005. Evaluation of peroxyacetic acid as a post-chilling intervention for control of *Escherichia coli* O157:H7 and *Salmonella* Typhimurium on beef carcass surfaces. *Meat Sci.* 69:401–7.

Kelly, C. A., Dempster, J. F. and Mcloughlin, A. J. 1981. The effect of temperature, pressure and chlorine concentration of spray washing water on numbers of bacteria on lamb carcases. *J. Appl. Bacteriol.* 51: 415–24.

Kochevar, S. L., Sofos, J. H., Bolin, R. R., Reagan, J. O. and Smith, G. C. 1997. Steam vacuuming as a pre-evisceration intervention to decontaminate beef carcasses. *J. Food Prot.* 60: 107–13.

Kotula, A. W., Lusby, W. R., Crouse, J. D. and De Vries, B. 1974. Beef carcass washing to reduce bacterial contamination. *J. Anim. Sci.* 39: 674–9.

Lee, M., Sebranek, J. G., Olson, D. G. and Dickson, J. S. 1996. Irradiation and packaging of fresh meat and poultry. *J. Food Prot.* 59: 62–72.

Leistner, L. 2000. Basic aspects of food preservation by hurdle technology. *Int. J. Food. Microbiol.* 55: 181–6.

Liao, Y.-T., Brooks, C. J., Martin, J. N., Echeverry, A., Loneragan, G. H. and Brashears, M. M. 2015. Antimicrobial interventions for O157:H7 and Non-O157 Shiga toxin-producing *Escherichia coli* on beef subprimal and mechanically tenderized steaks. *J. Food Prot.* 78: 511–17.

Mattila, T. and Frost, A. J. 1988. Colonization of beef and chicken muscle surfaces by *Escherichia coli*. *Food Microbiol.* 5: 219–30.

Maxim, J. E., Neal, J. A. and Castillo, A. 2014. Development of a novel device for applying uniform doses of electron beam irradiation on carcasses. *Meat Sci.* 96: 373–8.

Mendonca, A. F., Amoroso, T. L. and Knabel, S. J. 1994. Destruction of Gram-negative food-borne pathogens by high pH involves disruption of the cytoplasmic membrane. *Appl. Environ. Microbiol.* 60: 4009–14.

Nou, X., Rivera-Betancourt, M., Bosilevac, J. M., Wheeler, T. L., Shackelford, S. D., Gwartney, B. L., Reagan, J. W. and Koohmaraie, M. 2003. Effect of chemical dehairing on the prevalence of *Escherichia coli* O157:H7 and the levels of aerobic bacteria and *Enterobacteriaceae* on carcasses in a commercial beef processing plant. *J. Food Prot.* 66: 2005–9.

Nutsch, A. L., Phebus, R. K. Riemann, M. J., Schafer, D. E., Boyer, J. E., Wilson, R. C., Leising, J. D. and Kastner, C. L. 1997. Evaluation of a steam pasteurization process in a commercial beef processing facility. *J. Food Prot.* 60: 485–92.

Ockerman, H. W., Borton, R. J., Cahill, V. R., Parrett, N. A. and Hoffman, H. D. 1974. Use of acetic and lactic acid to control the quantity of microorganisms on lamb carcasses. *J. Milk Food Technol.* 37: 203–4.

Patterson, J. T. 1969. Hygiene in meat processing plants 4. Hot-water washing of carcasses. *Rec. Agric. Res. Minist. Agric. N. I.* 18: 85–7.

Peel, B. and Simmons, G. C. 1978. Factors in the spread of Salmonellas in meatworks with special reference to contamination of knives. *Austral. Vet. J.* 54: 106–10.

Phebus, R. K., Nutsch, A. L., Schafer, D. E., Wilson, R. C., Riemann, M. J., Leising, J. D., Kastner, C. L., Wolf, J. R. and Prasai R. K. 1997. Comparison of steam pasteurization and other methods for reduction of pathogens on surfaces of freshly slaughtered beef. *J. Food Prot.* 60: 476–84.

Podolak, R. K., Zayas, J. F., Kastner, C. L. and Fung, D. Y. C. 1996. Reduction of bacterial populations on vacuum packaged ground beef patties with fumaric and lactic acids. *J. Food Prot.* 59: 1037–40.

Prasai, R. K., Acuff, G. R., Lucia, L. M., Hale, D. S., Savell, J. W. and Morgan, J. B. 1991. Microbiological effects of acid decontamination of beef carcasses at various locations in processing. *J. Food Prot.* 54: 868–72.

Prasai, R. K., Campbell, R. E., Vogt, L. R., Kastner, C. L. and Fung, D. Y. C. 1995a. Hot-fat trimming effects on the microbiological quality of beef carcasses and subprimals. *J. Food Prot.* 58: 990–2.

Prasai, R. K., Phebus, R. K., Garcia Zepeda C. M., Kastner, C. L., Boyle, A. E. and Fung, D. Y. C. 1995b. Effectiveness of trimming and/or washing on microbiological quality of beef carcasses. *J. Food Prot.* 58: 1114–17.

Quartey-Papafio, E. A., Marshall, R. T. and Anderson, M. E. 1980. Short-chain fatty acids as sanitizers for beef. *J. Food Prot.* 43: 168–71.

Ransom, J. R., Belk, K. E., Sofos, J. N., Stopforth, J. D., Scanga, J. A. and Smith, G. C. 2003. Comparison of intervention technologies for reducing Escherichia coli O157:H7 on beef cuts and trimmings. *Food Prot. Trends* 23: 24–34.

Reagan, J. O., Acuff, G. R., Buege, D. R., Buyck, M. J., Dickson, J. S., Kastner, C. L., Marsden, J. L., Morgan, J. B., Nickelson, R. Smith, C. G. and Sofos, J. N. 1996. Trimming and washing of beef carcasses as a method of improving the microbiological quality of meat. *J. Food Prot.* 59: 751–6.

Reynolds, A. E. and Carpenter, J. A. 1974. Bactericidal properties of acetic and propionic acids on pork carcasses. *J. Animal Sci.* 38: 515–19.

Richards, R. M. E., Xing, D. K. L. and King, T. P. 1995. Activity of *p*-aminobenzoic acid compared with other organic acids against selected bacteria. *J. Appl. Microbiol.* 78: 209–15.

Rowbury, R. J. 1995. An assessment of environmental factors influencing acid tolerance and sensitivity in *Escherichia coli, Salmonella* spp. and other enterobacteria. *Lett. Appl. Microbiol.* 20: 333–7.

Schnell, T. D., Sofos, J. N., Littlefield, V. G., Morgan, J. B., Gorman, B. M., Clayton, R. P. and Smith, G. C. 1995. Effects of postexsanguination dehairing on the microbial load and visual cleanliness of beef carcasses. *J. Food Prot.* 58: 1297–302.

Shotts, E. B., Martin, W. T. and Galton, M. M. 1961. Further studies on Salmonella in human and animal food and in the environment of processing plants. *Proc. 65th. Annu. Mtg*, U.S. Livestock Sanitary Assoc, Minneapolis, MN., October, p. 309.

Smith, M. G. 1992. Destruction of bacteria on fresh meat by hot water. *Epidemiol. Infect.* 109: 491–6.

Smith, M. G. and Graham, A. 1978. Destruction of *Escherichia coli* and salmonellae on mutton carcasses by treatment with hot water. *Meat Sci.* 2: 119–28.

Tarté, R. R., Murano, E. A. and Olson, D. G. 1996. Survival and injury of *Listeria monocytogenes, Listeria innocua* and *Listeria ivanovii* in ground pork following electron beam irradiation. *J. Food Prot.* 59: 596–600.

Tinney, K. S., Miller, M. F., Ramsey, C. B., Thompson, L. D. and Carr, M. A. 1997. Reduction of microorganisms on beef surfaces with electricity and acetic acid. *J. Food Prot.* 60: 625–8.

Van Netten, P., Van Der Zee, H. and Mossel, D. A. A. 1984. A note on catalase enhanced recovery of acid injured cells of Gram negative bacteria and its consequences for the assessment of the lethalityh of L-lactic acid decontamination of raw meat surfaces. *J. Appl. Bact.* 57: 169–73.

Yang, X., Badoni, M., Tran, F. and Gill, C. O. 2015. Microbiological effects of a routine treatment for decontaminating hide-on carcasses at a large beef packing plant. *J. Food Prot.* 78: 256–63.

Wang, G. and Doyle, M. P. 1998. Heat shock response enhances acid tolerance of *Escherichia coli* O157:H7. *Lett. Appl. Microbiol.* 26: 31–4.

Wilson, R. C. and Leising, J. D. 1994. Method and apparatus for steam pasteurization of meats. United States Patent appl. 08/335,437.

Woolthuis, C. H. J., Mossel, D. A. A., Van Logtestijn, J. G., Dekruijf, J. M. and Smulders, F. J. M. 1984. Microbial decontamination of porcine liver with lactic acid and hot water. *J. Food Prot.* 47: 220–6.

Woolthuis, C. H. J. and Smulders, F. J. M. 1985. Microbial decontamination of calf carcasses by lactic acid sprays. *J. Food Prot.* 48: 832–7.

Chapter 9

Optimizing the microbial shelf life of fresh beef

Declan J. Bolton, Teagasc Food Research Centre (Ashtown), Ireland

1 Introduction

Beef is a nutritious but highly perishable food that is readily spoiled by the action of microbes. Shelf-life is the length of time a food remains acceptable to the user and in the case of beef products is often the time between packaging and the product becoming unpalatable or unfit for human consumption. It includes appearance and odour as well as colour, texture, flavour and nutritive value (Singh and Singh, 2005). For beef, appearance and colour are particularly important and the shelf-life can be limited to the 'display-life' or period the product retains the bright, cherry-red colour that consumers expect, although the more widespread use of vacuum packaging for retail cuts is changing consumer expectations. Despite the beef being perfectly safe for consumption, changes in myoglobin resulting in a browning of the meat may render the product undesirable and therefore 'spoiled' by customers (Brooks, 2007). From the microbiology perspective, beef is considered spoiled when the total viable count reaches 10^{7-8} CFU/cm^2 (Nychas et al., 2008).

Beef carcasses are placed in the chilling unit immediately after slaughter where they usually remain for 24–72 hours before being moved to the boning hall. Current legislation in the European Union, Regulation EC 853/2004 (Regulation, 2004), requires that beef carcasses be immediately chilled after *post-mortem* inspection to ensure a

http://dx.doi.org/10.19103/AS.2016.0008.13

core temperature of not more than 7°C in the case of meat and not more than 3°C for offal. However, the biochemical processes and structural changes that occur in beef during the first 24 hours *post-mortem* are critical in determining quality and palatability. Thus, the temperature profiles used typically ensure that the core temperature does not decrease below 10°C in the first 10 hours to ensure cold shortening (covered in Section 5 'Chilling') does not occur (EFSA, 2014). In the boning hall the carcasses are cut into primals and sub-primals which are vacuum packaged and the meat allowed to mature.

Beef is usually delivered to retailers as vacuum-packaged primals in cardboard boxes or in display ready packages. In the butcher shop, vacuum-packaged beef is opened and cut/minced into roasts, steaks and ground product. These may be displayed unpackaged at the 'butchers counter' or packaged in trays with overwrap film that is typically oxygen permeable. However, in recent years some retail stores have used vacuum packaging for the more valuable cuts. The type of packaging system greatly influences shelf-life. The other key determinant of shelf-life at this stage in the beef chain is the storage temperature, which should be sufficiently low to inhibit the growth or spoilage and pathogenic bacteria. However, even when bacterial growth is inhibited, prolonged storage of meat results in rancidity (oxidation of lipids) and other biochemical changes that make the product unacceptable to consumers.

Extending the shelf-life of beef has many commercial advantages, including reduced costs associated with wastage and facilitating a broader geographical distribution, thus opening new market opportunities. However, maximizing shelf-life is dependent on minimizing microbial contamination of beef carcasses and primals in the slaughter and boning halls, maintaining chilled temperatures along the beef chain and using packaging systems that inhibit the growth of spoilage organisms.

2 The factors affecting beef spoilage and microbial shelf life

As with other foods, intrinsic factors such as pH and water activity (a_w) and external factors such as storage temperature and atmosphere, affect the spoilage of fresh beef. At slaughter, beef muscle is typically pH 7.0 decreasing to pH 5.3–5.8 over the first 40 hours (Smulders et al., 1992). Dark Firm Dry (DFD) meat, which has a pH of 5.9–6.8, occurs when pre-slaughter stress depletes glycogen reserves to below 0.6%. DFD meat can occur in all meat species but is more common in beef. The higher pH of DFD beef promotes the growth of spoilage bacteria. Fresh meat has an a_w of approximately 0.99 (ICMSF, 1998) which decreases during chilling to approximately 0.95 (Prendergast and Sheridan, 2008). During commercial beef carcass chilling, the mean surface and core temperatures decrease from approximately 17°C and 37°C, respectively, to approximately 0°C (Bolton, unpublished data). Meat is usually stored at temperatures of around 2°C (James and James, 2004), with the exception of meat being transported long distances where a temperature of −1.5°C is recommended (Jeremiah and Gibson, 2001). Small changes in temperature can significantly affect the shelf-life. Thus increasing the temperature from −1.5°C to 0°C, 2°C or 5°C will decrease the time to spoilage by approximately 30%, 50% and 70%, respectively (Sumner and Jensen, 2011). Different packaging systems may be used for fresh beef products. While beef primals are usually stored in vacuum packages under anaerobic conditions, fresh retail beef cuts may be stored in a range of atmospheres depending on the packaging system used; air, vacuum packaging or modified atmosphere packaging (MAP), that usually contains carbon dioxide and oxygen balanced with nitrogen. The time to spoilage increases in the order: air, high-oxygen MAP, vacuum, no-oxygen MAP and 100% carbon dioxide (Borch et al., 1996). Packaging systems are discussed in more detail later in this chapter.

3 Microbial spoilage of fresh beef

Although lipid peroxidation is a major cause of deterioration in organoleptic properties and nutritional value, beef spoilage is primarily caused by the action of bacteria and is characterized by discolouration, strong off-odours and the development of slime. While indigenous enzymes may also be involved, their contribution is considered to be negligible as compared to bacterial action (Tsigarida and Nychas, 2001). These bacteria, originating from the animal (bovine and intestines) and the abattoir environment, represent a range of different genera, including *Achromobacter, Acinetobacter, Aeromonas, Alcaligenes, Alteromonas, Arthrobacter, Bacillus, Campylobacter, Carnobacterium, Citrobacter, Clostridium, Corynebactenum, Enterobacter, Escherichia, Flavobacterium, Hafnia, Klebsiella, Kluyverqa, Kocuria, Kurthia, Lactobacillus, Lactococcus, Leuconostoc, Listeria, Microbacterium, Micrococcus, Moraxella, Paenibacillus, Pantoea, Proteus, Providencia, Pseudomonas, Swewenella, Staphylococcus, Streptococcus, Vibrio, Weissella* and *Yersinia* (Nychas et al., 2007). The main spoilage bacteria, associated meat products and spoilage defects are shown in Table 1.

The dominant spoilage bacteria on beef carcasses, stored under chilled aerobic conditions, are psychrophilic and psychrotrophic bacteria, usually dominated by *Pseudomonas* spp. (Stanbridge and Davis, 1998; Koutsoumanis et al., 2006). Three

Table 1 Beef products, the main spoilage bacteria and associated spoilage defects

Beef product	Causal bacteria	Defect
Aerobically stored/ packaged fresh beef	*Pseudomonas, Lactobacillus, Enterococcus, Weissella & Brochothrix*	Slime
	Weisella, Leuconostoc, Enterococcus & Lactobacillus	Hydrogen peroxide greening
Vacuum-packed fresh beef	*Shewanella & Clostridium*	Hydrogen sulphide greening
	Clostridium & Hafnia	Sulphide odour
	Brochothrix thermosphacta	Cheesy or dairy odour
	Lactic acid bacteria, Enterococcus, Micrococcus, Bacillus & Clostridium	Souring
Whole meats	*Clostridium & Enterococcus*	Bone taint

Source: Adopted from Nychas et al., 2008.

species of *Pseudomonas*, *Ps. fragi*, *Ps. fluorescens* and *Ps. Lundensis*, are primarily responsible for the formation of slime and off-odours, usually when their population reaches 10^{7-8} CFU/cm^2 (Nychas et al., 2008). *Enterobacteriaceae*, especially cold-tolerant species such as *Hafnia alvei*, *Serratia liquefaciens* and *Pantoea agglomerans*, are also found on fresh beef but rarely contribute to spoilage, unless there is temperature abuse (Nychas et al., 1998). Lactic acid bacteria (LAB) and *Brochothrix thermosphacta* are oxygen-tolerant anaerobes that are also commonly detected on aerobically stored chilled meat. However, they are not major spoilage contributors on aerobically stored beef (Holzapfel, 1998).

When beef carcasses are deboned and cut into primals, which are typically stored at 0–2°C under anaerobic conditions in vacuum packages, the bacterial population shifts towards organisms capable of growth under these conditions. LAB become the dominant spoilage bacteria and anaerobic spoilage is usually characterized by souring rather than putrefaction, which typically occurs when LAB numbers reach 10^8 CFU/cm^2 (Jones, 2004). Vacuum-packaged beef primals may also be spoiled by a range of psychrotolerant/psychrophilic *Clostridium* spp. This occurs after 2–4 weeks and is characterized by a putrid smell (H_2S) with a metallic sheen on the meat with or without gas production. *Clostridium algidicarnis*, *Clostridium frigoris*, *Clostridium bowmanii*, *Clostridium frigidicarmis* and *Clostridium ruminantium* have been associated with spoilage without gas production (Broda et al., 1999, 2000; Adam et al., 2010; Cavill et al., 2011), while other species such as *C. estertheticum* and *C. gasigenes* produce large volumes of gas, primarily carbon dioxide (Moschonas et al., 2010; Yang et al., 2010). The packs inflate and eventually burst, thus this type of spoilage is often referred to as 'blown pack spoilage' (Bolton et al., 2015).

4 Prerequisite actions for increased shelf life of beef

The initial microbial burden on the beef carcass or primals will greatly influence the shelf-life. Hygienic slaughter and dressing is reliant on the development and implementation of an effective prerequisite programme (PRP) and, if required, hazard analysis and critical control point (HACCP) system. PRP activities cover such areas as the premises, personnel, process and product. Although detailed descriptions of a beef slaughter plant, boning hall and butcher retail PRPs are beyond the scope of this chapter, they may be summarized as follows: premises should be designed to prevent cross-contamination and facilitate cleaning. Beef contact surfaces should be composed of non-absorbent, non-toxic and non-corrosive materials and services (water treatment, drainage flow, storage of chemicals, cleaning programme, pest control, maintenance/calibration of equipment and waste disposal) should be carefully monitored and controlled. Personnel should be well trained, maintain good personal hygiene and cleanliness (including wearing protective clothing) and subject to a communicable diseases and injuries programme.In the beef slaughter plant, microbial contamination on the carcasses (product) should be minimized though control of the slaughter process, including the application of an animal cleanliness policy, slaughter-line standard operating procedures (SOP, including sanitation standard operating procedures (SSOP)) and product recall, withdrawal and traceability procedures.

5 Chilling: ensuring palatability while reducing microbial spoilage of beef

After slaughter and dressing, beef carcasses are at the optimum temperature for the growth of spoilage organisms and it is essential that they be chilled to temperatures that retard growth. However, the primary consideration in beef carcass chilling is palatability and not microbial spoilage. Thus beef carcasses, typically chilled in air at 2°C to 4°C with an air velocity of less than 1 m s^{-1} and a relative humidity greater than 80%, are not chilled to below 10°C before the pH reaches 6.2 (usually 10 hours) to prevent cold shortening. During cold shortening, calcium ions are released from the sarcoplasmic reticulum of muscle fibres, which triggers muscle contraction in a process aided by adenosine triphosphate and a toughening of the meat. This typically occurs when the muscle temperature is reduced to less than 14°C to 19°C while in the pre-rigor state. Data provided by a commercial export beef plant [Bolton, personal communication] showed that the ambient temperature in the chillers was approximately 11°C for the first 10 hours, thereafter decreasing to approximately 1.5°C to 3°C. During this regime, the surface temperature decreased from 25.2°C to 11.4°C after 10 h, to 2.9°C after 24 h and to 1.6°C after 36 h, thereafter remaining at that temperature. The corresponding deep round temperatures were 39.2°C upon entering the chiller, 21.7°C after 10 h; 8.6°C after 24 h; 4.57°C after 36 h and 2.9°C after 48 h. Chilled beef carcasses may be aged at 0 to 2°C for 5–21 days. Beef boning halls/cutting rooms typically operate at temperatures of up to 12°C. Primals are usually vacuum packed and chilled by using air chilling systems at temperatures ranging between 0 and 2°C.

While beef carcasses are usually chilled by using air, spray chilling systems may also be applied (Brown and James, 1992; Brown et al., 1993) and have been used in the initial stages of beef carcass refrigeration in the United States since the 1980s. Applying water to the carcasses substantially increases the rate of carcass temperature decline, especially on the surface, as evaporation increases the rates of heat transfer (Jones and Robertson, 1988; Lee et al., 1990; James, 1996). Spray chilling also prevents carcass shrinkage as the water applied replaces water lost through evaporation (Gigiel et al., 1989). In contrast *post-mortem* shrinkage of up to 2% has been reported during the initial 24 h of conventional air chilling of beef carcasses (Greer and Jones, 1997). This advantage may, however, be lost as there is some evidence of increased purge loss from spray-chilled beef sides after 15 days of vacuum-pack storage (Allen et al., 1987).

The bacterial load on carcasses may increase, decrease or remain the same as a result of air chilling (Lenahan et al., 2010). The effect of other chilling systems is similarly unclear as there are conflicting reports on the effect of air versus spray chilling on the microbial status of red meat carcasses. Greer and Dilts (1988) observed increased bacterial growth with spray systems, but Hamby et al. (1987) and Kinsella et al. (2006) suggest that water sprays do not affect the total viable count (TVC), the total *Enterobacteriaceae* count (TEC) or the total coliform count (TCC). From the perspective of shelf-life, decreasing the surface temperature of beef carcasses as quickly as possible to below the temperature capable of supporting the growth of spoilage bacteria would optimize microbial quality. This may be achieved by using rapid chilling, also referred to as 'ultra-rapid', 'fast/very fast', 'extreme', 'blast' and 'accelerated air' chilling. Temperatures as low as −35°C are used to achieve a carcass temperature of −1°C within 5 h *post-mortem* (Aalhus et al., 2002). Although rapid chilling minimizes microbial growth as well as drip and evaporative loss (Mallikarjunan and Mittal, 1996), it may result in considerable toughening of the beef (Joseph, 1996).

Beef carcasses may also be boned hot or warm, without chilling. Hot/warm boning has many advantages, including increased yield, reduced costs, less requirement for chiller space, less repetitive strain injuries and increased productivity (Heinz, 1975; Taylor et al., 1980; Cain et al., 1984; Pisula and Tyburcy, 1996; Rees, Trout and Warner, 2002; Seyfort et al., 2004). Bowater (2011) recently estimated that the use of hot/warm boning would increase yield by 4%, representing a €2.7 million cost saving for a plant processing 500 cattle per day for 250 days per annum. Despite the economic and other advantages, the beef industry has been slow to adopt this technology (Spooncer, 1993; Zhou et al., 2010; Yang et al., 2010). This may be attributed to the fact that hot-boned and vacuum-packaged beef may have a surface temperature of up to 25°C for several hours, supporting the proliferation of spoilage organisms (Sheridan and Sherington, 1982).

Freezing beef for transport to distant markets has also been investigated, although the adverse effects on the quality of meat is still a significant problem. As the temperature drops, water within and between the muscle fibres freezes, causing an increase in the concentration of proteins, carbohydrates, lipids, vitamins and minerals, which disrupts cell membranes resulting in toughening of the beef (Lawrie, 1998; Fellows, 2000). Although the link between the rate of freezing and moisture loss has been well documented, the effects of freezing and thawing on beef sensory attributes, including tenderness, are less well understood. Moreover, during extended frozen storage, oxidation of myoglobin, lipids and proteins results in a further degradation. While vitamin E may be used to inhibit this oxidation and MAP to preserve colour, more research is needed on how best to preserve the quality of beef, before freezing can be applied as a commercially viable preservation strategy (Leygonie et al., 2012).

6 Packaging for beef products

A variety of beef packaging systems are available, ranging from short shelf-life (up to 6 days) aerobic overwrap systems to MAP, bulk-gas flushing and vacuum packaging for longer-term storage (up to 6 months). Packaging films may be composed of low-density polyethylene (PE-LD), high-density polyethylene (PE-HD), polyamide (PA), polypropylene (PP), polyesters (PET), polyvinylidene chloride (PVC), polystyrene (PS) or ethylene/vinyl acetate (EVAC), or a combination thereof (Jan et al., 2005; Marsh and Bugusu, 2007). Each film has its own properties conferring different advantages and disadvantages in terms of extending shelf-life, environmental impact and cost. Moreover, a single type of plastic composed of a single polymer type does not have the properties required for beef packaging applications (e.g. low gas and moisture transmission), so most beef packaging films incorporate a variety of polymer resins in a multilayer construction. The main characteristics considered when selecting a packaging material include oxygen and carbon dioxide permeability, water transmission rates, seal reliability, anti-fogging properties and resistance to puncture. Anti-fogging agents, such as glycerol esters, polyglycerol esters, sorbitan esters, alcohol ethoxylates and nonyl phenol ethoxylates, may also be externally applied to the polymer surface by dip coating, spraying or blending into the packaging film polymer.

Beef cuts were traditionally packaged in air by using polyvinyl chloride high-permeability packaging systems, which allowed the development of oxymyoglobin, responsible for the red meaty colour expected by consumers. However, beef packaged in this way had a relatively short shelf-life as compared to anaerobic packages (Cole, 1986) as the aerobic conditions facilitated the growth of spoilage bacteria such as pseudomonads.

6.1 Modified atmosphere packaging

Maintaining a chilled temperature when storing packaged beef is critical to preventing microbial growth and chemical reactions within the meat, but higher temperatures may alter packaging atmospheres, negating the positive effects of packaging systems such as MAP (Limbo et al., 2010). MAP may be used to enhance the stability of bulk beef during short journeys, for example, from the slaughter plant to the cutting hall or from the cutting hall to retail. Bulk beef can be stored for 9–15 weeks at −1 to −2°C in atmospheres saturated with carbon dioxide and devoid of oxygen, and these systems have been used during international transport (Tewari et al., 1999). MAP is also an effective system for extending the shelf-life of beef cuts and ground product (Limbo et al., 2010). The gaseous composition typically includes a combination of oxygen, carbon dioxide and nitrogen.

Oxygen is included in MAP beef to promote oxymyoglobin formation that is responsible for the bright red colour consumers typically associate with fresh meat (Farber, 1991). However, oxygen also stimulates the growth of aerobic spoilage bacteria, such as pseudomonads. Moreover, the oxygen levels must be carefully controlled as high levels promote the formation of metmyoglobin (Mancini and Hunt, 2005), as well as off-odours and off-flavours associated with lipid oxidation (Limbo et al., 2010).

Carbon dioxide is water- and lipid-soluble and has both bacteriostatic and fungistatic properties that inhibit microbial growth, especially during the early lag phase. Carbon dioxide may dissolve to form carbonic acid (H_2CO_3) in beef tissue. Although the exact mechanism of action is not fully understood, it is thought that carbonic acid does not significantly increase the antimicrobial properties. It is more likely that the observed bacteriostatic effect is due to a range of metabolic interferences whereby carbon dioxide penetrates the microbial cell and adversely affects various enzymatic and biochemical pathways (Daniels et al., 1985). Regardless, the preservative effect of carbon dioxide is influenced by a range of parameters, including concentration, particle pressure, volume of headspace gas, storage temperature, pH and water activity of the beef (Church and Parsons, 1995). Of these, storage temperature is especially important as the solubility of carbon dioxide increases with decreasing temperature.

Nitrogen is an inert gas filler which displays no antimicrobial properties. As it has low water and lipid solubility, it is particularly useful in products where high levels of carbon dioxide could result in pack collapse if excessive carbon dioxide absorption occurs. Moreover, in beef products, it can displace oxygen, thereby inhibiting oxidative rancidity and the growth of aerobic spoilage organisms (Faber, 1991).

Carbon monoxide is also effective in maintaining the red colour of meat due to the formation of carboxymyoglobin which has a stable red colour similar to that of oxymyoglobin and eliminates the need for oxygen to maintain the fresh red colour expected by consumers (Sorheim et al., 1997). Moreover, lipid oxidation is reduced (Hunt et al., 2004; Brooks et al., 2008). However, the highly toxic nature of carbon monoxide restricts it use, although it is applied in countries such as Norway, in gaseous mixtures typically containing 60% carbon dioxide, 0.3–0.5% carbon monoxide and the remainder nitrogen (Sorheim et al., 1997).

The exact combination of gases used in MAP depends on a range of factors, including packaging materials and storage temperature, but all packs must contain sufficient headspace to allow the gas to interact with the product. MAP meat products typically use 60–85% oxygen and 15–40% carbon dioxide, with the remainder being nitrogen. High-oxygen MAP systems (e.g. 80% oxygen) will keep beef a bright red colour during prolonged storage as the depth of the oxymyoglobin layer is increased by 3–5 mm. Microbial growth is inhibited by using carbon dioxide in the remainder of the gas composition. Beef stored at −1.5°C using 80:20 oxygen:carbon dioxide has a shelf-life of 2 weeks (Gill and Jones, 1994). Large retailers may use this approach, often as part of a master-pack system. The master-pack system is a variation of the MAP system that consists of 4–8 packages of product (packed in air-permeable films ready for retail display) placed in a master pack composed of materials impermeable to oxygen or moisture. The master pack is then evacuated to remove the air and back-flushed with the desired MAP gas mixture, usually 80:20, carbon dioxide:nitrogen or 100% carbon dioxide. When required, the master pack is opened, the air-permeable packages are exposed to the air, and the beef blooms usually within 30 minutes of exposure to atmospheric oxygen. Beef products in the master pack have a 10–14-day storage life followed by a 2–7-day shelf-life in retail display (Buffo and Holley, 2005). A similar shelf-life may be achieved in high-oxygen MAP retail packs, which compares favourably with the 3–7 days obtained for PVC-packaged meats. The disadvantages of high-oxygen MAP include faster lipid oxidation, off-flavour development (Jakobsen and Bertelsen, 2002; Jayasingh et al., 2002) and premature browning during cooking (John et al., 2005).

6.2 Low-oxygen packaging

Despite the name, low-oxygen packaging (also referred to as CO-MAP) usually contains no oxygen, with a typical gas composition of 20–30% carbon dioxide, 0.4% carbon monoxide and the remainder nitrogen. This offers several advantages over aerobic or high-oxygen packaging, including decreased growth of spoilage and pathogenic bacteria (Cornforth and Hunt, 2008), red meat colour stability for up to 28 days for ground beef and 35 days for meat cuts (Jayasingh et al., 2001; Krause et al., 2003; Hunt et al., 2004), protection against lipid oxidation (Jayasingh et al., 2002), no premature browning during cooking (Seyfort et al., 2004; John et al., 2004, 2005) and increased tenderness (Lund et al., 2007; Grobbel et al., 2007). Yang et al. (2016) investigated the effects of four MAP systems including CO-MAP (0.4%:30%:69.6% $CO:CO_2:N_2$), 80%:20% $O_2:CO_2$, 50%:30%:20% $O_2:CO_2:N_2$ and vacuum packaging on the shelf-life of heavily marbled beef steaks at 4°C. This research found that the anaerobic systems (CO-MAP and vacuum packaging) extended the microbial shelf-life by up to 12 days while achieving lower oxidation and better colour stability. Similar studies on ground beef using the same MAP systems, but replacing 50%:30%:20% $O_2:CO_2:N_2$ with 30%:70% $CO_2:N_2$, also found that the anaerobic packaging systems, especially CO-MAP, were most effective at inhibiting microbial growth while maintaining desirable sensory characteristics (Rogers et al., 2014).

6.3 Vacuum packaging

Vacuum packaging, commonly used for beef primal cuts, provides the anaerobic conditions that extend both the microbiological and oxidative shelf-life of beef (Strydom and Hope-Jones, 2014). Vacuum-packaged primals stored under commercial chilling conditions (usually 0–4°C) have a storage life of 70 days which may be extended to 77 days if a temperature of 1°C is maintained (Bell and Garout, 1994), after which discolouration and off-flavours may occur. While most vacuum packaging is done by using shrink bags, an alternative vacuum-packaging system using vacuum skin-packaging may further extend the shelf-life (Lagerstedt et al., 2011; Strydom and Hope-Jones, 2014). Steaks cut from vacuum-packaged primals recover any loss of their fresh red colour after up to 72 hours of aerobic display (Gill and Jones, 1994).

Retail beef products are usually wrapped in PVC materials and have a shelf-life of 5–7 days maximum, the limiting factor being the formation of metmyoglobin resulting in browning of the meat (Greene et al., 1971). Browning typically occurs when oxygen levels are above 0.15% and can be prevented by packaging the beef products in vacuum packs. However, until recently the purplish hue associated with the formation of deoxymyoglobin was unacceptable to most retail consumers, and vacuum packaging was rarely used for retail cuts.

7 Active and intelligent packaging for beef products

Active packaging incorporates additives, often into the packaging material, including oxygen scavengers, carbon dioxide scavengers/emitters, chlorine dioxide generators, moisture control agents and antimicrobial compounds. Intelligent packaging uses sensor technology such as time-temperature indicators (TTI) and radio frequency identification (RFID) that interact with microbial growth metabolites, changes in gaseous composition and/or temperature changes to indicate freshness and/or microbial quality. Yoon et al. (1994), for example, demonstrated a positive correlation between oxidative stability and a phospholipid/phospholipase-based TTI in meat. However, although freshness indicators based on target metabolites associated with microbial deterioration offer great potential for beef products, intelligent package indicators are currently unavailable.

7.1 Oxygen scavengers

Although oxygen scavengers may also reduce drip loss (Payne et al., 1998) in addition to inhibiting the growth of spoilage bacteria, their value in the beef industry is not universally accepted, not least because beef tissue is inherently a very efficient oxygen scavenger (Tewari et al., 2002). Regardless, oxygen scavenging is usually achieved by including iron-based oxygen scavenger sachets into the package. Several oxygen scavengers have been validated and are commercially available for use in beef packs. Allen et al. (1996) demonstrated improved colour stability of *M. Longissimus dorsi* and *M. psoas major* using oxygen scavengers in master packs flushed with 50:50 carbon dioxide:nitrogen. Tewari et al. (2001) used two different oxygen scavengers to prevent discolouration of *M. psoas major* steaks stored at 1°C. The same technology may also be used to prevent discolouration of ground beef (Gill and McGinnis, 1995). The oxygen scavengers may also be incorporated into the film polymers instead of using a sachet delivery system. Commercially available oxygen scavenging films have been demonstrated to reduce oxygen concentrations in MAP systems from 2% to less than 0.1% in 3 days (Coma, 2008); a capacity that is considered low as compared to what can be achieved with iron-based sachets.

7.2 Carbon dioxide scavengers/emitters

Carbon dioxide is lost from beef packs through the film or absorbed into the beef tissue, necessitating the application of carbon dioxide emitters that typically convert oxygen to carbon dioxide to maintain the desired concentration (Ozdemir and Floros, 2004). These are usually based on ferrous carbonate or a mixture of sodium bicarbonate and ascorbic acid in a sachet, label or incorporated into the packaging material. Chen and Brody (2013) found that packaging films with CO_2 generators inhibited the growth of total aerobic bacteria, *Enterobacteriaceae* and *Listeria monocytogenes* in meat systems. However, not all bacteria are adversely affected by carbon dioxide. LAB (the main spoilage bacteria under anaerobic conditions), for example, are stimulated by carbon dioxide as is growth and toxin production by non-proteolytic *Clostridium botulinum* (Sivertsvik, 2003; Lovenklev et al., 2004).

7.3 Chlorine dioxide generators

Chlorine dioxide is also used to inhibit bacterial growth and may have potential in the beef sector. The gaseous form is generated from active packaging materials by reacting the solid compound (commercially available from Bernard Technologies, USA as 'Microspheres' and delivered in a sachet incorporated into the packaging material) with moisture. The concentrations of gas released into the beef pack usually range from 1 ppm to 100 ppm, and this can be achieved over a period of up to 6 weeks.

7.4 Moisture control agents

Excessive moisture promotes bacterial growth and absorbers and dessicants may be applied to extend shelf-life. Moisture drip absorbent pads, composed of two layers of microporous plastic film around a superabsorbent polymer (composed of polyacrylate salts, carboxymethyl cellulose or starch copolymers), are placed under fresh beef cuts to absorb drip exudate.

7.5 Antimicrobial compounds

Although the application of antimicrobial compounds to extend the shelf-life of beef products is currently limited by legislation and consumer acceptability (Quintavalla and Vicini, 2002), this area of research is showing potential. As microbial contamination occurs primarily on the surface of beef products, antimicrobial compounds may be applied to the contact surface of the packaging film to inhibit microbial growth (Han, 2005). These antimicrobial compounds may be included in a sachet, incorporated into the film, coated in a carrier matrix onto the inner surface of the film or applied directly to the beef (Cooksey, 2001). The film polymer (e.g. cationic amino-polysaccharides) may also have inherent antimicrobial properties. The first three approaches function by releasing the antimicrobial directly onto the surface of the food (Appendini and Hotcjkiss, 2002; Buonocore et al., 2003). Nadarajah et al. (2005), for example, demonstrated that allyl isothiocyanate, a volatile aliphatic sulphur containing compound, could be used to reduce *Escherichia coli* O157:H7 in fresh minced beef during refrigerated or frozen storage when incorporated into filter paper in a multilayer packaging system. Other antimicrobial substances that may be used include organic acids (e.g. propionic and sorbic acid), polysaccharides, alcohols, acid anhydride, bacteriocins (e.g. nisin), triclosan, essential oils and silver. Siragusa et al. (1999) reported reduced *Lactobacillus helveticus* and *Brochothrix thermosphacta* counts on beef, vacuum packaged using a polyethylene-based plastic film incorporating nisin, when stored at 4–12°C for 20 days. Triclosan (2,4,4′-trichloro-2′-hydroxydiphenyl-ether) may be combined with most of the major polymers to form effective antimicrobial films, although their effectiveness under refrigerated conditions has yet to be demonstrated (Kerry et al., 2006). Essential oils, rich in flavonoids and phenolic compounds, also have antimicrobial properties (Matan et al., 2006; Han et al., 2007). Seydim and Sarikus (2006) incorporated oregano and garlic into whey-protein based films that were effective against *Salmonella enteritidis, L. monocytogenes, E. coli, Staphylococcus aureus* and *Lactobacillus plantarum*. Although beef studies are limited, Kerry et al. (2006) used oregano essential oil on filter paper to extend the shelf-life of beef products.

Silver, a potent antimicrobial, is available commercially and can be incorporated into the packaging film or applied as a coating on the surface of the packaging material. The antimicrobial properties may be the result of extracellular binding to the cell wall and/or active transport into bacterial cells allowing binding to DNA and/or electron donor groups and/or ribosomes inhibiting cellular functions such as protein synthesis (Silvestry-Rodriguez et al., 2007; Lalueza et al., 2011). Although several studies have demonstrated antimicrobial affects *in vitro* (Dogan et al., 2009; Fernández et al., 2010; Laluza et al., 2011; Boschetto et al., 2012), relatively few studies have validated the use of silver as an antimicrobial in real food systems. Recent work by Kuuliala et al. (2015) investigating the efficacy of silver-containing packaging films against beef spoilage bacteria found that the shelf-life was not affected by these packaging films, although the same beef spoilage microbiota were inhibited *in vitro* by using a nanoscale silver coating. The authors attributed these observations to different release dynamics of silver ions on meat surfaces and/or interactions between the silver ions and amino acids. Moreover, cross-sectional imaging showed that some of the films containing the silver particles were covered by a thin polyethylene film preventing direct contact with the meat. Thus, the migration of silver or other antimicrobial compounds from a packaging film into the product may not be sufficient to inhibit the growth of spoilage bacteria, and the delivery matrix/film should also be carefully designed. Fernández et al. (2010), for example, found that silver zeolite composites prepared by solvent casting were more effective at inhibiting bacterial growth than those prepared by melt mixing.

7.6 Combination treatments

Different packaging systems may be used in combination to extend the shelf-life and maintain fresh beef meat quality. Lyu et al. (2016), for example, demonstrated that pre-treating beef with carbon monoxide in combination with ozone by using MAP for 1.5h before vacuum packaging and storage at 0°C for 46 days significantly ($p < 0.05$) decreased the bacterial counts at the end of storage while maintaining desirable sensory characteristics. Interestingly, drip loss was also reduced. Limbo et al. (2013) used a master bag system with oxygen scavengers and CO_2/N_2 MAP to extend the shelf-life of *Semitendinous* beef for 7 to 21 days.

8 Conclusions

Beef-quality parameters include colour, texture, flavour and nutritive value. Although oxidation and enzymatic autolysis may cause beef degradation, spoilage is primarily caused by bacteria, and maximizing shelf-life is dependent on a multiple hurdle approach to generate conditions that inhibit microbial growth. These include low-temperature chilling throughout the beef chain and using atmospheres within the packs, such as MAP or vacuum packaging, that inhibit the main spoilage bacteria, *Pseudomonas* spp. Anaerobic spoilage bacteria, such as LAB, can then be inhibited by using active and/or intelligent packaging systems. Although a range of these packaging technologies are currently commercially available, the emergence of new, geographically distant markets requires more research and the development of innovative beef shelf-life extension technologies based on a better understanding of the relationship between beef spoilage microbiota, biochemical factors and sensory components.

9 Where to look for further information

Further information on 'optimizing the microbial shelf-life of fresh beef' is available in a range of publications including: *Food Processing Technology - Principles and Practice* (Woodhead Publishing Limited, Cambridge, UK); *Innovations in Food Packaging* (Elsevier Academic Press Amsterdam), *Innovations in Food Packaging* (Elsevier Academic Press, Amsterdam) and *Novel Food Packaging Techniques* (Woodhead Publishing Limited, Cambridge, UK).

10 References

Aalhus, J. L., Robertson, W. M., Dugan, M. E. R. and Best, D. R. (2002), 'Very fast chilling of beef carcasses', *Can. J. Animal Sci.*, 82, 56–67.

Adam, K. H., Flint, S. H. and Brightwell, G. (2010), 'Psychrophilic and psychrotrophic clostridia: sporulation and germination processes and their role in the spoilage of chilled, vacuum-packed beef, lamb and venison', *Int. J. Food Sci. Technol.*, 45, 1539–44.

Allen, D. M., Hunt, M. C., Luchiari Filho, A., Danler, R. J. and Goll, S. J. (1987), 'Effects of spray chilling and carcass spacing on beef carcass cooler shrink and grade factors', *J. Animal Sci.*, 64, 165–70.

Allen, P., Doherty, A. M., Buckley, D. J., Kerry, J., O'Grady, M. N. and Monahan, F. J. (1996), 'Effect of oxygen scavengers and vitamin E supplementation on colour stability of MAP beef', In *Proceedings 42nd International Congress of Meat Science and Technology*, 1–6 September 1996, Lillehammer, Norway, pp. 88–9.

Appendini, P. and Hotchkiss, J. H. (2002), 'Review of antimicrobial food packaging. Innovative', *Food Sci. Emerg. Technol.*, 3, 113–26.

Bell, R. G. and Garout, A. M. (1994), 'The effective product life of vacuum packaged beef imported into Saudi Arabia by sea, as assessed by chemical, microbiological, and organoleptic criteria', *Meat Sci.*, 36, 381–96.

Bolton, D. J., Carroll, J. and Walsh, D. (2015), 'A four year survey of blown pack spoilage *Clostridium estertheticum* and *Clostridium gasigenes* on beef primal cuts', *Lett. Appl. Microbiol.*, 61(2), 153–7.

Borch, E., Kant-Muermans, M. L. and Blixt, Y. (1996), 'Bacterial spoilage of meat and cured meat products', *Int. J. Food Microbiol.*, 33, 103–20.

Boschetto, D. L., Lerin, L., Cansian, R., Pergher, S. B. C. and Di Luccio, M. (2012), 'Preparation and antimicrobial activity of polyethylene composite films with silver exchanged zeolite-Y', *Chem. Eng. J.*, 204–6, 210–16.

Bowater, F. J. (2011), 'Rapid carcass chilling plants compared to conventional systems'. The International Institute of Refrigeration. Available at http://www.fjb.co.uk/wp-content/themes/fjb/publications/Rapid%20Carcass%20Chilling%20vs%20Conventional.pdf. (accessed 20 February 2012).

Broda, D. M., Lawson, P. A., Bell, B. R. and Musgrave, D. R. (1999), '*Clostridium frigidicarnis* sp. A psychrotolerant bacterium associated with "blown pack" spoilage of vaccum-packaged meats', *Int. J. Sys. Evol. Microbiol.*, 49, 1539–50.

Broda, D. M., Saul, D. J., Bell, G. R. and Musgrave, D. R. (2000), '*Clostridium algidixylanolyticum* sp. Nov., a psychrotolerant, xylan-degrading, spore-forming bacterium', *Int. J. Sys. Evol. Microbiol.*, 50, 623–31.

Brooks, C. (2007), 'Beef packaging'. Beef Facts, product enhancement. Available at http://www.beefresearch.org/factsheets1/aspx.

Brooks, J. C., Alvarado, M., Stephens, T. P., Kellermeier, J. D., Tottor, A. W. and Miller, M. F. (2008), 'Spoilage and safety characteristics of ground beef packaged in traditional and modified atmosphere packages', *J. Food Prot.*, 71, 293–301.

Brown, T. and James, S. J. (1992), 'Process design data for pork chilling', *Int. J. Refrig.*, 15, 281–9.

Brown, T., Chourouzidis, N. and Gigiel, A. J. (1993), 'Spray chilling of lamb carcasses', *Meat Sci.*, 34, 311–25.

Buffo, R. A. and Holley, R. A. (2005), 'Centralized packaging systems for meats', in J. H. Han (ed.), *Innovations in Food Packaging*, pp. 227–36. Elsevier Academic Press, Amsterdam.

Bugusu, K. B. (2007), 'Food packaging – roles, materials, and environmental issues', *Food Sci.*, 72(3), R39–R55.

Buonocore, G. G., Del Nobile, M. A., Panizza, A., Corbo, M. R. and Nicolais, L. (2003), 'A general approach to describe the antimicrobial agent release from highly swellable films intended for food packaging applications', *J. Control. Release.*, 90, 97–107.

Cain, B. P., Powell, V. H., McPhail, N. G. and Anderson, J. (1984), 'A comparison of the yield of saleable meat from hot and cold boned beef carcasses', CSIRO Meat Research Report 6/84. CSIRO Division of Food research, Cannon Hill, Queensland, Australia.

Cavill, L., Renteria-Monterrubio, A. L., Helps, C. R. and Corry, J. E. L. (2011), 'Detection of cold-tolerant clostridia other than *Clostridium estertheticum* in raw vacuum-packed chill-stored meat', *Food Microbiol.*, 28, 957–63.

Chen, J. and Brody, A. L. (2013), 'Use of active packaging structures to control the microbial quality of a ready-to-eat meat product', *Food Control*, 30, 306–10.

Church, I. and Parsons, A. (1995), 'Modified atmosphere packaging technology: a review', *J. Sci. Food Agr.*, 67, 5113–21.

Cole, A. B. Jr (1986), 'Retail packaging systems for fresh red meat cuts'. Reciprocal Meat Conference. *Proc. Am. Meat Sci. Assoc.*, 39, 106–11.

Coma, V. (2008). 'Bioactive packaging technologies for extended shelf-life of meat based products', *Meat Sci.*, 78, 90–103.

Cooksey, K. (2001), 'Antimicrobial food packaging', *Food Cos. Drug Pack.*, 24(7), 133–7.

Cornforth, D. and Hunt, M. (2008), 'Low-Oxygen Packaging of Fresh Meat with Carbon Monoxide, Meat Quality, Microbiology and Safety', American Meat Science Association, White Paper Series No. 2, pp. 1–10.

Daniels, J. A., Krishnamurthy, R. and Rizvi, S. S. H. (1985), 'A review of effects of CO_2 on microbial growth and food quality', *J. Food Prot.*, 48, 32–537.

Dogan, H., Koral, M. and Inan, T. Y. (2009), 'Ag/Zn zeolite containing antibacterial coating for food-packaging substrates', *J. Plastic Film and Sheeting*, 25(3–4), 207–20.

EFSA (2014), 'Scientific Opinion on the public health risks related to the maintenance of the cold chain during storage and transport of meat. Part 1 (meat of domestic ungulates)', *EFSA J.*, 12(3): 3601, p. 81. doi:10.2903/j.efsa.2014.3601.

Farber, J. M. (1991), 'Microbiological aspects of modified atmosphere packaging technology – a review', *J. Food Prot.*, 54(1), 58–70.

Fellows, P. (2000), 'Freezing', *Food Processing Technology, Principles and Practice*, pp. 418–40, 2nd ed. Cambridge, UK, Woodhead Publishing Limited.

Fernández, A., Soriano, E., Hernández-Muñoz, P. and Gavara, R. (2010), 'Migration of antimicrobial silver from composites of polylactide with silver zeolites', *J. Food Sci.*, 75(3), E186–E193.

Gigiel, A. J., Collett, P. and James, S. J. (1989), 'Fast and slow beef chilling in a commercial chiller and the effect of operational factors on weight loss', *Int. J. Refrig.*, 12, 338–49.

Gill, C. O. and Jones, T. (1994), 'The display life of retail packaged beef steaks after their storage in master packs under various atmospheres', *Meat Sci.*, 38, 385–96.

Gill, C. O. and McGinnis, J. C. (1995), 'The use of oxygen scavengers to prevent the transient discoloration of ground beef packaged under controlled, oxygen depleted atmospheres', *Meat Sci.*, 41, 19–27.

Greene, B. E., Hsin, I. M. and Zipser, M. W. (1971), 'Retardation of oxidative colour changes in raw ground beef', *J. Food Sci.*, 36, 940–2.

Greer, G. G. and Dilts, B. D., (1988), 'Bacteriology and retail case life of spray-chilled pork', *Can. Inst. Food Sci. Technol. J.*, 3, 295–9.

Greer, G. C. and Jones, S. D. M. (1997), 'Quality and bacteriological consequences of beef carcass spray-chilling: effects of spray duration and boxed beef storage temperature', *Meat Sci.*, 45, 61–73.

Grobbel, J. P., Dikeman, M. E., Hunt, M. C. and Milliken, G. A. (2007), 'Different packaging environments alter tenderness and sensory traits from non-enhanced and injection-enhanced beef', In *Proceedings 53rd International Congress of Meat Science and Technology*, 5–10 August, Beijing, China, pp. 527–8.

Han, J. H. (2005), 'Antimicrobial packaging systems', in Juan H. Han (ed.), *Innovations in Food Packaging*, pp. 81–107. Amsterdam: Elsevier Academic Press.

Han, X., Shen, T. and Lou, H. (2007), 'Dietary polyphenols and their biological significance', *Int. J. Mol. Sci.*, 8, 950–88.

Hamby, P. L., Savell, J. W., Acuff, G. R., Vanderzant, C. and Cross, H. R. (1987), 'Spray-chilling and carcass decontamination systems using lactic and acetic acid', *Meat Sci.*, 21, 1–14.

Heinz, G. (1975), 'Frozen hot beef from overseas for use in manufacturing frankfurters', *Die Fleischwirtschaft V.*, 55, 910.

Holzapfel, W. H. (1998), 'The Gram-positive bacteria associated with meat and meat products', in R. G. Board and A. R. Davies (eds), *The Microbiology of Meat and Poultry*, pp. 35–84. Blackie Academic and Professional, London, UK.

Hunt, M. C., Mancini, R. A., Hachmeister, K. A., Kropf, D. H., Merriman, M. and DelDuca, G. (2004), 'Carbon monoxide in modified atmosphere packaging affects colour, shelf life, and microorganisms of beef steaks and ground beef', *J. Food Sci.*, 69, C45–C52.

ICMSF (International Commission on Microbiological Specifications for Foods) (1998), 'Micro-organisms in foods 6', *Microbial Ecology of Food Commodities*. Blackie Academic and Professional, London, p. 615.

Jakobsen, M. and Bertelsen, G. (2002), 'The use of CO2 in packaging of fresh red meats and its effect on chemical quality changes in the meat: a review', *J. Muscle Foods*, 13, 143–68.

James, S. (1996), 'The chill chain "from carcass to consumer"', *Meat Sci.*, 42(S), S203–S216.

James, C. and James, S. J. (2004), 'Refrigeration and freezing technology/equipment', *Encyclopedia of Meat Sciences*, pp. 1144–50.

Jan, J. H., Zhang Y. and Buffo R. (2005), 'Surface chemistry of food, packaging and biopolymer materials', in J. H. Han (ed.), *Innovations in Food Packaging*, pp. 45–59. Elsevier Academic Press, Amsterdam.

Jayasingh, P., Cornforth, D. P., Carpenter, C. E. and Whittier, D. (2001), 'Evaluation of carbon monoxide (CO) treatment in modified atmosphere packaging or vacuum packaging to increase colour stability of fresh beef', *Meat Sci.*, 59, 317–24.

Jayasingh, P., Cornforth, D. P., Brennand, C. P., Carpenter C. E. and Whittier D. R. (2002), 'Sensory evaluation of ground beef stored in high-oxygen modified atmosphere packaging', *J. Food Sci.*, 67 (9), 3493–6.
Jeremiah, L. and Gibson, L. (2001), 'The influence of storage temperature and storage time on color stability, retail properties and case-life of retail-ready beef', *Food Res. Int.*, 34, 815–26.
John, L., Cornforth, D. P., Carpenter, C. E., Sorheim, O., Pettee B. and Whittier, D. R. (2004), 'Comparisons of colour and thiobarbituric acid values of cooked hamburger patties after storage of fresh beef chubs in modified atmospheres', *J. Food Sci.*, 69(8), C608–C614.
John, L., Cornforth, D. P., Carpenter, C. E., Sorheim, O., Pettee, B. and Whittier, D. R. (2005), 'Colour and thiobarbituric acid values of cooked top sirloin steaks in modified atmospheres of 80% oxygen, or 0.4% carbon monoxide, or vacuum', *Meat Sci.*, 69, 441–9.
Jones, S. D. M. and Robertson, W. M. (1988), 'The effects of spray chilling carcasses on the shrinkage and quality of beef', *Meat Sci.*, 24, 177–88.
Jones, R. (2004), 'Observations on the succession dynamics of lactic acid bacteria populations in chill-stored vacuum-packaged beef', *Int. J. Food Microbiol.*, 90, 273–82.
Joseph, R. L. (1996), 'Very fast chilling of beef and tenderness – A report from an EU concerted action', *Meat Sci.*, 43(S), 217–27.
Kerry, J. P., O'Grady, M. N. and Hogan, S. A. (2006), 'Past, current and potential utilization of active and intelligent packaging systems for meat and muscle based products: a review', *Meat Sci.*, 74, 113–30.
Kinsella, K. J., Sheridan, J. J., Rowe, T. A., Butler, F., Delgado, A., Quispe-Ramirez, A., Blair, I. and McDowell, D. A. (2006), 'Impact of a novel spray-chilling system on surface microflora, water activity and weight loss during beef carcass chilling', *Food Microbiol.*, 23, 483–90.
Koutsoumanis, K., Stamatiou, A., Skandamis, P. and Nychas, G. J. E. (2006), 'Development of a microbial model for the combined effect of temperature and pH on spoilage of ground meat and validation of the model under dynamic temperature conditions', *Appl. Environ. Microbiol.*, 72, 124–34.
Krause, T. R., Sebranek, J. G., Rust, R. E. and Honeyman, M. S. (2003), 'Use of carbon monoxide packaging for improving the shelf life of pork', *J. Food Sci.*, 68, 2596–603.
Kuuliala, L., Pippuri, T., Hultman, J., Auvinen, S. K., Kolppo, K., Nieminen, T., Karp, M., Bjorkroth, J., Kuusipalo, J. and Jaaskelainen, E. (2015), 'Preparation and antimicrobial characterization of silver-containing packaging materials for meat', *Food Pack. Shelf-life*, 6, 53–60.
Lagerstedt, A., Ahnstrom, M. L. and Lundstrom, K. (2011), 'Vacuum skin pack of beef – a consumer friendly alternative', *Meat Sci.*, 88, 391–6.
Lalueza, P., Monzón, M., Arruebo, M. and Santamaría, J. (2011), 'Bactericidal effects of different silver-containing materials', *Mat. Res. Bull.*, 46(11), 2070–6.
Lawrie, R. A. (1998), Lawrie's meat science. In Anonymous (ed.), pp. 1–336, 6th ed. Lancaster, PA, Technomic Publishing Inc.
Lee, L. M., Hawrysh, Z. J., Jeremiah, L. E. and Hardin, R. T. (1990), 'Shrouding, spray-chilling and vacuum-packaged aging effects on processing and eating quality attributes of beef', *J. Food Sci.*, 55, 1270–3.
Lenahan, M., O'Brien, S. B., Kinsella, K., Sweeney, T. and Sheridan, J. J. (2010), 'Assessment of lamb carcass hygiene before and after chilling at five Irish abattoirs', *Food Con.*, 21, 313–18.
Leygonie, C., Britz, T. J. and Hoffman, L. C. (2012), 'Impact of freezing and thawing on the quality of meat: review', *Meat Sci.*, 91, 93–8.
Limbo, S., Torri, L., Sinelli, N., Franzetti, L. and Casiraghi, E. (2010), 'Evaluation and predictive modelling of shelf-life of minced beef stored in high oxygen modified atmosphere packaging at different temperatures', *Meat Sci.*, 84, 129–36.
Limbo, S., Uboldi, E., Adobati, A., Iametta, S., Bonomi, F., mascheroni, E., Santagostino, S., Powers, T. H., Franzetti, L. and Piergiovanni, L. (2013), 'Shelf-life of case-ready beef steaks (Semitendinosus muscle) stored in oxygen-depleted master bag system with oxygen scavengers and CO_2/N_2 modified atmosphere packaging', *Meat Sci.*, 93, 477–84.

Lovenklev, M., Artin, I., Hagberg, O., Borch, E., Holst, E. and Radstrom, P. (2004), 'Quantitative interaction effects of carbon dioxide, sodium chloride, and sodium nitrite on neurotoxin gene expression in nonproteolytic *Clostridium botulinum* type B', *Appl. Environ. Microbiol.*, 70, 2928–34.

Lund, M. N., Hviid, M. S. and Skibsted, L. H. (2007), 'The combined effect of antioxidants and modified atmospheric packaging on protein and lipid oxidation in beef patties during chill storage', *Meat Sci.*, 76, 226–33.

Lyu, F., Shen, K., Ding, Y. and Ma, X. (2016), 'Effect of pretreatment with carbon monoxide and ozone on the quality of vacuum packaged beef meats', *Meat Sci.*, 117, 137–46.

Mallikarjunan, P. and Mittal, G. S. (1996), 'Selection criteria for beef carcass chilling', *Food Res. Int.*, 29(7), 661–6.

Mancini, R. A. and Hunt, M. C. (2005), 'Current research in meat colour', *Meat Sci.*, 71, 100–21.

Matan, N., Rimkeeree, H., Mawson, A. J., Chompreeda, P., Harutharthanasan, V. and Parker, M. (2006), 'Antimicrobial activity of cinnamon and clove oils under modified atmosphere conditions', *Int. J. Food Microbiol.*, 107, 180–5.

Moschonas, G., Bolton, D. J., Sheridan, J. J. and McDowell, D. (2010), 'The effect of storage temperature and inoculum concentration on the time of onset of blown pack spoilage', *J. Appl. Microbiol.*, 102(92), 532–9.

Nadarajah, D., Han, J. H. and Holley, R. A. (2005), 'Inactivation of *Escherichia coli* O157:H7 in packaged ground beef by allyl isothiocyanate', *Int. J. Food Microbiol.*, 99, 269–79.

Nychas, G. J. E., Drosinos, E. H. and Board, R. G.(1998), 'Chemical changes in stored meat', in R.G. Board and A. R. Davies (eds), *The Microbiology of Meat and Poultry*, pp. 288–326. Blackie Academic and Professional, London, UK.

Nychas, G. J. E., Marshall, D. and Sofos, J. (2007), 'Meat poultry and seafood', in M. P. Doyle, L. R. Beuchat and T. J. Montville (eds), *Food Microbiology Fundamentals and Frontiers*, Chapter 6, ASM press.

Nychas, G. J. E., Skandamis, P. N., Tassou, C. and Koutsoumanis, K. P. (2008), 'Meat spoilage during distribution', *Meat Sci.*, 78, 77–89.

Ozdemir, M. and Floros, J. D. (2004), 'Active food packaging technologies', *Crit. Rev. Food Sci. Nut.*, 44, 185–93.

Payne, S. R., Durham, C. J., Scott, S. M. and Devine, C. E. (1998), 'The effects of non-vacuum packaging systems on drip loss from chilled beef', *Meat Sci.*, 49, 277–87.

Pisula, A. and Tyburcy, A. (1996), 'Hot processing of Meat', *Meat Sci.*, 43, S125–S135.

Prendergast, D. and Sheridan, J. J. (2008), 'Indicator organisms to determine the use of chilling as a critical point in beef slaughter HACCP', The National Food Centre Research Report No. 100. Dublin, Teagasc, ISBN 1841705330.

Quintavalla, S. and Vicini, L. (2002), 'Antimicrobial food packaging in meat industry', *Meat Sci.*, 62, 373–80.

Rees, M. P., Trout, G. R. and Warner, R. D. (2002), 'Tenderness of pork *longissimus dorsi et lumborum* after accelerated boning', Part 1. Effect of temperature conditioning. *Meat Sci.*, 61, 205–14.

Rogers, H. B., Brooks, J. C., Martin, J. N., Tittor, A., Miller, M. F. and Brashears, M. M. (2014), 'The impact of packaging system and temperature abuse on the shelf-life characteristics of ground beef', *Meat Sci.*, 97, 1–10.

Scetar, M., Kurek, M. and Galic, K. (2010), 'Trends in meat and meat products packaging-a review', *Croat. J. Food Sci. Technol.*, 2(1), 32–48.

Seydim, A. C. and Sarikus, G. (2006), 'Antimicrobial activity of whey protein based edible films incorporated with oregano, rosemary and garlic essential oils', *Food Res. Int.*, 39, 639–44.

Seyfort, M., Hunt, M. C., Mancini, R. A., Hachmeister, K. A., Krouf, D. H. and Unruh, J. A. (2004), 'Accelerated chilling, modified atmospheric packaging and injection enhancement affect colour and colour stability of beef round muscles', *Meat Sci.*, 68, 209–19.

Sheridan, J. J. and Sherington, J. (1982), 'The microbiology of hot and conventionally deboned vacuum packaged beef', *Meat Sci.*, 7, 245–58.

Silvestry-Rodriguez, N., Sicairos-Ruelas, E. E., Gerba, C. P. and Bright K. R. (2007), 'Silver as a disinfectant', *Rev. Environ. Contam. Toxic.*, 191, 23–45.

Singh, R. K. and Singh, N. (2005), 'Quality of packaged foods', in J. H. Han (ed.), *Innovations in Food Packaging*, pp. 22–4. Elsevier Academic Press, Amsterdam.

Siragusa, G. R., Cutter, C. N. and Willett, J. L. (1999), 'Incorporation of bacteriocin in plastic retains activity and inhibits surface growth of bacteria on meat', *Food Microbiol.*, 16, 229–35.

Sivertsvik, M. (2003), 'Active packaging in practice: fish', in R. Ahvenainen (ed.), *Novel Food Packaging Techniques*, pp. 384–400. Woodhead Publishing Ltd., Cambridge, UK..

Smulders, F. J. M., Toldra, F., Flores, J. and Prieto, M. (1992), 'New technologies for meat and meat products', *Audet Tijdschriften*, pp. 186–8, 182. Utrecht, The Netherlands.

Sorheim, O., Aune, T. and Nesbakken, T. (1997), 'Technological, hygienic and toxicological aspects of carbon monoxide use in modified-atmosphere packaging of meat', *Trend. Food Sci. Technol.*, 8, 307–12.

Spooncer, W. F. (1993), 'Options for hot boning', *Meat Update*, CSIRO, Australia.

Stanbridge, L. H. and Davis, A. R. (1998), 'The microbiology of chill-stored meat', in R. G. Board and A. R. Davies (eds), *The Microbiology of Meat and Poultry*, pp. 174–219. Blackie Academic and Professional, London, UK.

Strydom, P. E. and Hope-Jones, M. (2014), 'Evaluation of three vacuum packaging methods for retail beef loin cuts', *Meat Sci.*, 98, 689–94.

Sumner, J. and Jensen, I. (2011), 'The effect of storage temperature on shelf life of vacuum-packed lamb shoulders', *Food Aus.*, 63(6), 249–51.

Taylor, A. A., Shaw, B. G. and MacDoughall, D. B. (1980), 'Hot deboning beef with and without electrical stimulation', *Meat Sci.*, 5, 109–23.

Tewari, G., Jayas D. S. and Holley R. A. (1999), 'Centralized packaging of retail meat cuts: A review', *J. Food Prot.*, 62, 418–25.

Tewari, G., Jayas, D. S., Jeremiah, L. E. and Holley, R. A. (2001), 'Prevention of transient discoloration of beef', *J. Food Sci.*, 66, 506–10.

Tewari, G, Jeremiah, L. E., Jayas, D. S. and Holley, R. A. (2002), 'Improved use of oxygen scavengers to stabilize the colour of retail-ready meat cuts stored in modified atmospheres', *Int. J. Food Sci. Technol.*, 37, 199–207.

Tsigarida, E. and Nychas, G. J. E. (2001), 'Ecophysiological attributes of a *Lactobacillus* spp. and a *Pseudomonas* spp. on sterile beef fillets in relation to storage temperature and film permeability', *J. Appl. Microbiol.*, 90, 696–705.

Yang, X., Balamurugan, S. and Gill, C. O. (2010), 'Effects on the development of blown pack spoilage on the initial numbers of *Clostridium estertheticum* spores and *Leuconostoc mesenteroides* on vacuum packed beef', *Meat Sci.*, 88(3), 361–7.

Yang, X., Zhang, Y., Zhu, L., Han, M., Gao, S. and Luo, X. (2016), 'Effect of packaging atmospheres on storage quality characteristics of heavily marbled beef longissmus steaks', *Meat Sci.*, 117, 50–6.

Yoon, S. H., Lee, C. H., Kim, D. Y., Kim, J. W. and Park, K. H. (1994), 'Time-temperature indicator using a phospholipids-phospholipase system and application to storage of frozen pork', *J. Food Sci.*, 59, 490–3.

Zhou, G. H., Xu, X. L. and Liu, Y. (2010), 'Preservation technologies for fresh meat – a review', *Meat Sci.*, 86, 119–28.

Chapter 10

Ensuring beef safety through consumer education

Janet M. Riley, North American Meat Institute, USA

1 Introduction

For centuries, people consumed well-done meat, often boiled. This method of preparation stemmed from ancient Greeks and Romans, who recommended cooking in certain ways in accordance with their humoral theory of medicine. For instance, according to Hippocratic teachings, beef 'will agree best with those who use it well-boiled', and this belief lasted for centuries.

2 Beef preparation and food safety advice: before 1993

The introduction of the roasting thermometer in the twentieth century allowed cooks to prepare meat consistently, from occasion to occasion, without the guess work. As they experimented with less well-done meat, US beef consumers began to like it. This trend was helped along, in part, by cooks like Julia Child, who extolled the virtue of beef less done.

http://dx.doi.org/10.19103/AS.2016.0008.14

In a 1982 *New York Times* article (Fabricant, 1982), food writer Florence Fabricant noted the trend and observed that Julia Child recommended an internal temperature of 140 °F. for 'rare' in *Mastering the Art of French Cooking*, published in 1961. By 1978, however, she advised 125 °F not only for beef, but also for lamb in *Julia Child & Company.*

Given that whole-muscle cuts are internally sterile and external heat destroys bacteria, rare steaks and roasts were of little concern. However, as the popularity of the burger exploded at the same time that 'pink beef' was coveted, safety risks increased and a crisis would soon unfold.

In the early 1980s, a new pathogen arrived on the scene in the beef industry: *Escherichia coli* O157:H7. This pathogen exhibited a virulence never before seen, and required thorough cooking to destroy it. The first *E. coli* O157 outbreak was reported in 1982 when 47 people in Oregon and Michigan contracted illnesses characterised by haemorrhagic colitis (Riley et al., 1983). The outbreak ultimately was linked to hamburgers prepared by a fast-food chain.

By 1985, USDA had established its Meat and Poultry Hotline (USDA, 2016), an 800-strong experts trained to answer food safety questions. In 1995, the Hotline began answering questions by email. Beginning in 2002, the toll-free Hotline extended its service to callers whose first language is Spanish. In 2004, the Hotline launched 'Ask Karen', an online, 'virtual' food safety representative that offers advice and recommendations about safely handling, storing and preparing food to prevent illness, under all kinds of environments. The database of more than 1500 questions and answers is maintained by the Hotline's food safety experts. In the 30 years since its launch, the Hotline has answered more than two million calls as part of USDA's consumer food safety education efforts.

Three years before the 1993 outbreak in the Pacific Northwest, in 1990, USDA provided detailed, written information about safe handling in a publication called *Preventing Foodborne Illness*. The piece, a follow-up to the agency's previous *Safe Food Book*, referenced *E. coli* O157:H7 and advised consumers to cook ground beef to 160 °F. In 1991, the US Food and Drug Administration (FDA) released its own *Preventing Food-Borne Illness* document and recommended 140 °F as the minimal temperature for cooking beef generally. It did not acknowledge a higher temperature for ground beef. Years later, in 1997, the publication was revised, but a cooking temperature for ground beef was not added, a fact that underscores the difficulty that seemed to be occurring in delivering consistent, accurate information to consumers.

Meanwhile, in the early 1990s, state food codes were inconsistent. Different states recommended different practices and different end-point temperatures for hamburgers. Like a deadly case of déjà vu, the problem with this patchwork came to bear in 1993. Washington state law required burgers to be cooked to 155°F (68°C), the temperature necessary to completely kill *E. coli*. Jack-in-the-Box restaurants followed the federal standard of 140°F (60°C). Had the state cooking standard been followed in 1993, the *E. coli* outbreak might have been prevented, according to court documents and experts from the Washington State Health Department. (Of note, by 2015, all 50 states had adopted the FDA Food Code, with some adopting more recent versions than others.)

Following the outbreak, USDA's Food Safety and Inspection Service (FSIS) held a series of listening sessions across the country where the consuming public, industry and others could speak about the issue of beef safety. Emotions ran high as victims and parents of victims told tragic stories in painful detail. They demanded change, and they wanted pathogenic bacteria out of their ground beef.

3 Responses to the 1993 *E. coli* outbreak: 1993–2000

The American Meat Institute (AMI), representing meat packers and processors, and the Food Marketing Institute (FMI), representing retail grocers, quickly developed camera-ready art for two brochures that could be reproduced easily for use in retail stores with consumers and in foodservice operations to educate food preparers. Available in English and Spanish, the brochures were released in March 1993 and one represented the earliest efforts to spread comprehensive consumer handling information.

Notably, the brochures advise:

> Cook it safely. The center of patties and meat loaf should not be pink and the juices should run clear. Crumbled ground meats should be cooked until no pink color remains. Ground meat patties and loaves are safe when they reach 160 degrees F. in the center; ground poultry patties and loaves, 165 degrees F.

Missing in this early education piece is any reference to a thermometer. In the aftermath of the 1993 tragedy, industry statements that advised cooking hamburgers thoroughly were met with resistance from consumers and the organisations who prepared them. They called statements suggesting thorough cooking 'blaming the consumer' and argued that pathogens should not be there in the first place. Their reactions signalled a need for industry to tread carefully in any consumer education efforts lest they appear to be shifting blame. During this time, Safe Tables Our Priority (STOP) was formed by the families of victims and it became a powerful and influential voice in shaping public policy and consumer education efforts.

Soon after, the National Restaurant Association Education Foundation designated September as National Food Safety Education Month in 1994 'to heighten the awareness about the importance of food safety education. Each year a new theme and free training activities and posters are created for the restaurant and foodservice industry to help reinforce proper food safety practices and procedures'. The designation has continued and affords a hook for media and consumer outreach about a variety of safe food handling information.

The urgent need for efforts like the new Food Safety Education Month was underscored when FSIS in 1994 surveyed consumers and found that only 50% owned a food thermometer and most used it only for a holiday turkey. In the same year, the agency also announced that it would require mandatory safe handling labels on all packages of meat. Initial labels under consideration contained key messages instructing:

Keep refrigerated or frozen. Thaw in refrigerator.
Keep raw meat separate from other foods. Wash cutting surfaces, utensils and hands after touching raw meat.
Cook thoroughly.
Chill leftovers within two hours.

Research Triangle Institute (RTI) was contracted to conduct focus groups in three cities (a total of nine groups) to gauge consumer reaction to label options (Teague, 1995). They found consumers to be highly engaged in discussing the labels and consumers showed strong preference for labels that combined words, simple graphics and a 'why' statement that explained why safe handling was important. RTI found that consumers who cared for the elderly and the very young demonstrated greater motivation to embrace safe handling practices and greater desire for more information. Generally, however, the shorter the information, the more likely it was to be read.

Of particular interest in the paper that resulted from the RTI research was this:

> Most participants felt the short message 'cook thoroughly' was not enough; however, everyone thought that the using internal temperatures for hamburgers is absurd because 'no one sticks a meat thermometer in a burger'.

Most focus groups preferred the statement that hamburgers should be cooked 'until the juices are clear or the meat is no longer pink'.

After these messages were finalised, in 1995, a study by Kansas State University (Hunt, 1995) indicated that ground beef might turn brown before it reached a safe internal temperature, sufficient to destroy potentially dangerous bacteria. With that information in hand, FSIS commissioned the USDA's Agricultural Research Service to examine the colour of ground beef as it relates to doneness.

In 1997, AMI and FMI determined that extended efforts were needed and joined together to launch the Partnership for Food Safety Education (PFSE) (Hodges, 2016) in cooperation with USDA's FSIS and the FDA. The campaign resulted in part from a White Paper entitled *Food Safety from Farm to Table: A National Food Safety Initiative, Report to the President*. The Report ordered, 'FDA, USDA, CDC, and the Department of Education will work with the food industry, consumer groups and the states to launch a food safety public awareness and education campaign. The partnership will develop, disseminate, and evaluate a single food safety slogan and several standard messages. Industry has pledged $500,000 to date to support the partnership's activities and plans to raise additional funds'. USDA, FDA and CDC also pledged their expertise and resources to the new initiative.

With an infusion of financial and intellectual resources, the groups asked leading public relations agencies to propose campaigns that could reach and motivate consumers. The communications firm Hill + Knowlton Strategies was hired based upon their recommendation for a campaign called Fight BAC! The campaign featured an ugly, scary green germ named 'BAC', short for 'bacteria', cast as the enemy. The campaign enlisted educators whom they called 'BAC fighters' to help spread the message. They commissioned a costume that could be used at events. Over time, the PFSE expanded and eventually employed a full-time executive director. The partnership continues to this day and operates out of FMI offices.

4 Determining reliable indicators of beef safety: 2000–present

Just as the 1995 Kansas State University research raised concerns about the scientific underpinnings of the consumer messages used in the United States, in 2000, new USDA-commissioned research (Lyon, 2000) confirmed that emphasis on visual cues to determine the doneness of burgers might have been misguided. Agricultural Research Service's scientists cooked ground beef patties on a gas grill and used a thermometer to determine when the burgers reached temperatures of 135, 151 and 160 °F. They also cooked supermarket-purchased burgers until the meat colour turned brown without using a thermometer: some were shaped into patties and cooked immediately, some were shaped into patties and frozen, and some were frozen.

The scientists found that a burger with a brown centre is not necessarily cooked and safe to eat. Premature browning was not evident in frozen patties that were thawed and then cooked. However, beef that was frozen in bulk, thawed, formed into patties and immediately cooked showed brown colour at unsafe temperatures.

As expected, the longer the burger was cooked, the less pink the centre portion, but burgers removed from the grill with pink centres continued to brown for several minutes. Ground beef cooked to 135 °F and allowed to rest for about four minutes looked the same as a burger cooked to 160 °F.

This prompted FSIS to launch a new campaign aimed at consumers, urging them not to rely on colour as an indicator of doneness, but rather use a thermometer to verify that a burger was thoroughly cooked.

In May 2000, FSIS launched the 'Thermy the Thermometer' campaign that centred on a character that looked like an instant-read thermometer with a smiling face on top who proclaimed via his slogan, 'It's safe to Bite When the Temperature is Right'.

The education campaign included televised public service announcements and informational materials for consumers, educators and health professionals. A number of grocery chains around the country and the thermometer industry also participated in this public effort.

During this period, USDA approved a protocol for irradiation for fresh beef products in December 1999, and in May 2000, Huisken Meats of Minnesota became the first company to market it commercially in a frozen form. Notably, while the opportunity to irradiate ground beef was welcomed by many, no comprehensive consumer education programme accompanied USDA approval. After an initial burst of awareness and interest, as well as major opposition from a handful of vocal non-governmental organisations like the Organic Consumers Association and Public Citizen, irradiated ground beef never gained widespread acceptance and today comprises just a fraction of the food supply.

As the tenth anniversary of the 1993 outbreak approached, media and groups like STOP shone the spotlight on food safety issues again and assessed what progress had been made. USDA FSIS continued efforts to reach out to consumers and in 2003, FSIS launched Food Safety Mobile aimed at educating consumers through mobile marketing.

In 2004, the agency partnered with Michigan State University to launch a social marketing campaign, 'Is it done yet?' This effort targeted mothers of children under 10, as they seemed the most likely to be receptive to the information and motivated to change.

USDA, CDC and FDA partnered with the Ad Council to launch Food Safe Families campaign in 2011. The campaign represented a cross-federal effort to raise awareness of the risks of food poisoning and to motivate consumers, particularly parents, to take specific actions to reduce their risk and keep their families healthy. The campaign, still in progress in 2016, includes English- and Spanish-language television, radio, print and web advertising, as well as creative pieces designed specifically for in-store placement and an integrated PR and social media programme.

Public service announcements were released in 2011, 2013, 2014 and 2015. Before the launch of the PSA campaign in June 2011, FoodSafety.gov received an average of 92,000 visits per month. As a result, in part, of sustained campaign marketing efforts since then, USDA reports that FoodSafety.gov now receives an average of 860 000 visits each month. According to USDA, data collected before and after the launch of the PSA campaign shows that significantly more parents and caregivers believe that using a food thermometer is an important way to prevent food poisoning (61% in 2015 vs 34% in 2011), and significantly more parents and caregivers report using a food thermometer 'every time' they prepare meals at home (27% in 2015 vs 15% in 2011).

5 Global food safety education

The World Health Organization (WHO) has been a central repository of consumer food safety education materials and campaigns.

In the early 1990s, WHO developed the *Ten Golden Rules for Safe Food Preparation,* which were widely translated and reproduced. It was later determined that something simpler was needed. After consulting with experts in food safety and risk communication, WHO introduced the Five Keys to Safer Food poster in 2001. The new poster incorporated all the messages of the *Ten Golden Rules for Safe Food Preparation* under simpler headings that are more easily remembered and also provided more details on the reasoning behind the suggested measures.

The five core messages include the following:

Keep clean;
Separate raw and cooked;
Cook thoroughly;
Keep food at safe temperatures; and
Use safe water and raw materials.

Notably, WHO's more detailed 'cook' message is slightly different than the US message and reminiscent of some messaging from the early 1990s: 'Bring foods like soups and stews to boiling to make sure they have reached 70 °C (158 °F). For meat and poultry, make sure that juices are clear, not pink. Ideally, use a thermometer.' By 2016, the poster was available in 87 languages. WHO also made food safety the focus of World Health Day 2015. A campaign toolkit was provided with the theme 'How safe is our food? From farm to plate, make food safe.' The stated goal of the campaign was to spur governments to improve food safety through public awareness campaigns and to encourage consumers to take steps to ensure food safety.

In 1997, Canada followed the United States' lead and launched a Canadian Partnership for Food Safety Education (CPFSE). CPFSE used the character 'BAC' as the villainous germ, but also featured a positive character named Mrs. Cookwell, who was created to be a helpful resource for people.

Also, in 1997, Australia launched its own Food Safety Campaign Group, later renamed The Food Safety Information Council in 2001. The Council 'aims to reduce the number of Australians getting sick from food poisoning by providing simple, easy-to-follow consumer information on the handling, storage and preparation of food.' The Council sponsors a Food Safety Week annually that shifts themes each year to focus on different aspects of safe handling and food preparation. Among the Council's efforts is a poster bearing the slogan 'How you cook can make you (and others) crook.' Crook, in this context, is an Australian expression for 'sick'.

More recently, WHO and the Pan American Health Organization (PAHO) made food safety the focus of its 2015 World Health Day, celebrated each year on April 7. The centrepiece of the effort continued to be the Five Keys introduced in 2001. PAHO Director

Carissa F. Etienne wrote, 'While industry and regulatory systems are primarily responsible for these processes, we all have a role to play in ensuring food safety. Even food prepared at home can be a source of food-borne illness, if it is not handled properly. We need to adhere to the same food safety standards at home that we expect everyone else along the food chain to practice. Remember, food safety is everyone's responsibility.'

6 Awareness and behaviour changes in the United States

Consumer perceptions and behaviour related to food-borne illness changed little during the five-year period between 1988 and 1993 (Fein, 1995). Consumers believed incorrectly that food-borne illness was a minor illness that did not cause fever and that likely occurred within 24 hours of exposure to a pathogen via food. A 1992 study (Williamson, 1992) found that one-third of consumers thought that food safety problems occurred during manufacturing and that only 16% thought home was the most likely place for food mishandling. This perception may have caused consumers to believe that they could have little impact in preventing food-borne illnesses, although experts believed that small, sporadic cases of food-borne illness caused by preparation errors at home were more common than larger, well-publicised outbreaks.

In 1994, USDA began to test raw ground beef in meat packing plants and at retail stores for the presence of *E. coli* O157:H7. Industry expressed concern that test results only applied to the actual sample tested and not to the lot from which it was derived. Organisations like AMI were concerned that the much-publicised testing programme might mislead consumers into believing that the US meat supply was now somehow 'tested' and that they could relax their safe handling practices. After a failed legal challenge (Burros, 1994) by the Institute, the testing proceeded and continues to this day.

Many different efforts, from brochures to characters to mobile marketing, have resulted in some incremental progress in changing food safety behaviour since the early 1990s.

A national telephone survey conducted in 1996 found that of 1620 English-speaking US residents one-third reported unsafe handling practices. Unsafe practices were reported more often by men and by occasional food preparers than by women, persons over age 30 and frequent food preparers. Two-thirds of respondents reported using safe handling practices.

A 1998 mail survey (Bruhn, 1998) revealed that consumers considered themselves well informed about food safety and most reported taking action to minimise their risks, but most of these actions reported were related to nutritional changes. Twenty per cent of those surveyed said they did not know how to reduce microbial hazards. The study showed the most common food safety errors reported related to temperature control and handling leftovers. Consumers appeared unaware that food that looks and tastes good can contain pathogens. Bruhn and Schultz concluded that consumers need more information about technologies like irradiation, which could protect consumers when errors revealed to be so common, occur during preparation.

To tackle the underutilisation of thermometers, USDA-commissioned focus group research conducted by Macro International in January 1998, assessed 'Barriers that Limit Consumers' Use of Thermometers When Cooking Meat and Poultry Products.' According to the final report, consumers viewed thermometers as a tool used for large items like turkeys and roasts. They also told researchers they had cooked for years without an issue and did not see a need to begin using an instant-read thermometer. Costs, the time involved and firmly ingrained habits that did not involve instant-read thermometers were barriers. The report's authors said that parents of young children who were at greater risk showed more motivation to use a thermometer. The authors recommended emphasising thermometer use during daily meal preparation as opposed to holiday or special meal preparation, where they were already being used. They also noted that the possibility that thermometer use improved taste and quality of the meat served would likely motivate more effectively than safety alone.

In 2000, a study appeared in the *Journal of Food Protection* (Yang, 2016) that assessed awareness of the safe handling label. The study revealed that 79% of the 14 262 respondents to the survey reported that they had seen the safe handling label. Thirty-seven per cent claimed to have changed their raw meat preparation methods in response to the label. According to the researchers, 'Both label awareness and risky behaviors increased with education and income, suggesting that safe food handling labels have limited influence on consumer practices.' The researchers concluded that the labels may be more effective in discouraging cross-contamination than in encouraging thorough cooking practices. They argued that the label should be viewed as just one component among many food safety education programmes needed to educate and motivate change.

FSIS research found that consumer ownership of thermometers increased from 46% in 1998 to 60% in 2001, but the percentage of cooks using a thermometer for hamburgers increased from 3 to 6%. The disparity between ownership and actual use of thermometers for hamburgers spoke to the formidable challenge of rapid behaviour change and the need for more sustained food safety education efforts. FSIS again stepped up efforts to increase thermometer usage, this time targeting a specific population that they believed would move quickly to adoption. The agency chose the 'Boomburgs', defined as upscale, two-earner families in newer communities where the newest technological gadgets are frequently sought. The agency supported the choice by arguing that Boomburgs are major influencers of mass culture, setting trends and disseminating new ideas.

After extensive research, the agency arrived at the slogan, 'Is it done yet? You can't tell by looking. Use a food thermometer to be sure.' The agency conducted a pilot test of the programme that involved appearances by agency staff with the food safety mobile at festivals, schools and other events, distribution of free thermometers and placement of ads in newspapers and conveying campaign messages through radio.

Post-pilot results showed that thermometer use for cooking and grilling (sometimes, most times and all of the time) among the targeted group increased from 44 to 48%. However, use of a thermometer with hamburger patties specifically increased from 10.5% before the pilot test to 9.7% after the test.

Research by Utah State University researchers released in 2004 (Anderson, 2004) stunned the food safety education world with its use of cameras and 'glow germ' to depict vividly the rapid cross-contamination that could occur in a kitchen if proper safe handling procedures were not followed. 'Overall, subjects did not follow the Fight BAC! recommendations for safe food handling. Handwashing was inadequate. The average time spent washing hands was significantly lower than the 20-second recommendation. Only one-third of subjects' hand washing attempts were with soap. Surface cleaning was inadequate with only one-third of surfaces thoroughly cleaned,' the authors wrote. They continued:

> Moreover, one-third of subjects did not attempt to clean surfaces during food preparation. Nearly all subjects cross-contaminated raw meat, poultry, seafood, eggs, and/or unwashed vegetables with ready-to-eat foods multiple times during food preparation.

> Unwashed hands were the most common cross-contamination agent. Many subjects undercooked the meat and poultry entrees.

In fact, the study revealed that 76% of research participants attempted to check doneness of the meat, poultry or seafood entrée by using a knife or another utensil to cut or poke the entrée to evaluate changes in colour and texture rather than use a thermometer. In addition, nearly half of the participants were unaware of the safe internal temperatures for meat products.

In 2008, AMI launched a series of short videos in a series called 'Ask the Meat Scientist.' The video featured a staff scientist in a white lab coat addressing issues like the need to cook ground beef to 160 °F and ground turkey to 165 °F, how to use an instant-read thermometer and the need to refrigerate foods promptly after purchase. The videos have been viewed thousands of times since their release. The Institute also mailed thermometers to TV chefs to promote their use and to call their attention to the videos, but the effectiveness of the effort was not measured.

By 2009, with updated consumer education programmes firmly established, scientists again analysed consumer understanding of food safe handling practices and corresponding behaviours (Abbott et al., 2009).

Researchers surveyed 153 students at a major university about their food safety knowledge. In addition, the researchers observed participants in a laboratory setting and visited their kitchens. According to the study, participants followed less than half of the safe handling practices that were evaluated and correctly answered only two-thirds of the food safety knowledge questions. Knowledge of necessary practices was far higher than compliance when preparing foods safely. Highest compliance was with separation of raw and cooked foods while lowest compliance involved proper cooking. Ninety-seven per cent of participants did not use a thermometer to ensure doneness.

The Academy of Nutrition and Dietetics has tracked changes in consumer knowledge about food safety practices over time. Between 1999 and 2011, self-reported thermometer use declined from 22 to 20%. Consumers self-report some progress in other areas with fewer consumers reporting that they use the same cutting board for raw meat and chicken that they use for ready-to-eat foods (6% in 2011 compared with 10% in 1999) and fewer consumers reporting they do not wash utensils used to handle raw foods before they are used for cooked food (10% in 1999 vs 6% in 2011).

Another landmark study offered a 'window into the kitchen'. In 2011, University of California Davis researchers published results from a study in which 199 volunteers were videotaped preparing hamburgers and salad (Ho, 2011).

According to study author Dr. Christine Bruhn's view in a blog post, 'Volunteers knew we were going to take the temperature of their burgers when they considered the meat done. The majority successfully cooked their meat to 155 F (Food Code recommendation) or 160 F (consumer recommendation), including 25% of the households that cooked burgers to 180 F or higher–these families must like eating hockey pucks!'

Despite major efforts in the 15 years preceding their study to educate consumers about thermometer use, Bruhn and Phang found that only 4% of consumers used an instant-read thermometer when preparing burgers and only 13% knew the recommended internal temperature. Dr. Bruhn noted, 'People were not interested in using a food thermometer, even after we gave them one. They considered it unnecessary and noted that the cooks on TV never use one.'

FDA researchers in 2012 published a paper that analysed thermometer ownership and use from 1998 through 2010 (Lando, 2012). They concluded that thermometer usage varies by food type and that consumers are much more likely to use a thermometer for roasts than for burgers. They also found that higher levels of education were associated with lower levels of thermometer use and a preference for undercooked foods like less well-done hamburgers. Thermometer use for roasts increased from 76% in 1998 to 82% in 2010. Thermometer use when cooking hamburgers increased from 14% in 1998 to 23% in 2010.

Bruhn and colleagues in 2016 published their research evaluating different food safety education interventions. In their study, they explored the effectiveness of food safety education using positive deviance, personal story reading and through reading standard food safety education material to a total of 89 pregnant women and 93 people with diabetes. The study found the positive deviance approach to be the most effective. 'This suggests that food safety education is most effective when delivered in a supportive discussion format. Health education programmes for these vulnerable groups should endeavour to deliver safe food handling guidelines through a positive deviance approach,' the researchers concluded (Feng, 2016).

7 Global changes in awareness and behaviour

Australia's food safety efforts generated a greater shift in awareness since the 1997 launch of the country's food safety education campaign than has been documented in countries like the United States, which started its own major education efforts at approximately the same time. For example, Council polling in Australia in 1997 found that 20% of consumers would consume undercooked minced product. In 2006, polling showed that 92% recognised correctly that sausages should be cooked all the way through and 83% recognised that hamburgers should be cooked all the way through. However, the degree to which awareness translates into practice is unknown.

A 2005 study (Garayoa, 2005) of 562 Spanish citizens found that, 'there was considerable difference between knowledge and reported behavior.' In fact, researchers found that 98.6% of participants recognised the need for hand washing when preparing food, but only 24.4% confirmed that they actually wash hands with soap and water before, during and after food preparation. The researchers noted that those who were more knowledgeable about safe food handling practices showed better actual food safety behaviours, but even the most informed still took risks.

Research in 2007 among Slovenian consumers also found that they lack in food safety knowledge and actual food safety behaviours even more problematic (Jevšnik, 2008).

According to a 2014 Canadian review (Nesbitt, 2014), 14% of Canadians surveyed reported using a thermometer to determine doneness, 64% used visual clues, 33% used time, 10% used taste and 3% used 'another tool'. In terms of handling of specific foods, 69% used appropriate methods to determine when whole poultry was cooked sufficiently and 71% used appropriate methods to determine doneness of hamburgers. In another interesting observation, researchers noted that Canadians demonstrated awareness of the importance of hand washing practices to ensure food safety, but also admitted that sometimes they did not use soap or wash for the recommended 20 seconds. And while Canadians reported washing fruits and vegetables as recommended, half also reported washing poultry, which is not recommended. The researchers acknowledge that available research in Canada focused on self-reported behaviours rather than direct observation of actual behaviours.

8 Conclusion

Food-borne illnesses caused by bacteria commonly associated with beef products have declined markedly over time, according to CDC data. The degree to which this progress can be linked to reductions in bacteria on products and the degree to which this can be linked to incremental improvements in consumer handling is unknown. Many new technologies have been implemented that have reduced pathogens on beef products. At the same time, increased temperatures for cooking in food code have undoubtedly helped.

Food safety educators have been vigilant and creative, but larger social trends have hindered a complete transformation in behaviour. The phase-out of home economics classes and resulting decrease in knowledge about cooking, the need for speed when preparing food and the perpetuation of misinformation via social media – like the notion that grass-fed ground beef can be eaten rare, for example – will continue to challenge food safety educators in creating good and consistent food safety habits among consumers. Incorporation of safe food handling concepts into required health curriculum may help educate young people before they become food preparers.

It is also possible that progress in reducing pathogens and reductions in food-borne illnesses related to beef will reduce motivation. Bruhn's most recent finding about the benefits of positive deviance is encouraging, but much work remains to be done.

Much of the research evaluating the effectiveness of food safety education efforts relies upon self-reported behaviour changes, which may be misleading. More direct observation work is likely needed to guide effective efforts going forward and to highlight problematic practices that persist and that pose risks for the public health that must be addressed with fresh ideas and innovative approaches.

9 References

A Consumer Guide to Safe handling and Preparation of Ground Meat and Poultry. Food Marketing Institute and North American Meat Institute, 1993.

'A Report of the "Is It Done Yet?" Social Marketing Campaign To Promote the Use of Food Thermometers', FSIS.USDA.gov (accessed 27 July 2016).

Abbott, J.M., et. al. , 2009. Comparison of food safety cognitions and self-reported food-handling behaviors with observed food safety behaviors of young adults, *European Journal of Clinical Nutrition*, 63, pp. 572–9. doi:10.1038/sj.ejcn.1602961; published online 14 November 2007.

Anderson, J.B., Shuster, T.A., Hansen, K.E., Levy, A.S. and Volk, A., 2004. A camera's view of consumers food handling behaviors. *Journal of the American Dietetic Association* 104(2), pp. 186–191.

'Ask the Meat Scientist', Meat News Network, YouTube.com, www.YouTube.com/MeatNewsNetwork (accessed 1 March 2016).

'Australian Food Safety Information Council', *Wikipedia*. Wikimedia Foundation, n.d. Web. 27 July 2016.

Bernstein, C., 2016. Question on Consumer Education. Message to the author. 16 March. E-mail.

Bruhn, C. M., 1998. Consumer acceptance of irradiated food: theory and reality. *Radiation Physics and Chemistry* 52, pp. 129–33.

Burros, M., 1994. U.S. is upheld on tests of beef for deadly bacteria. *The New York Times*, 13 December, Web. 28 July 2016.

Consumer Knowledge of Home Food Safety Practices Survey, Academy of Nutrition and Dietetics, April 2011, http://www.eatright.org/~/media/homefoodsafety/multimedia/downloads/consumer%20

knowledge%20of%20home%20food%20safety%20practices%20survey.ashx (accessed 2 March 2016).

Fabricant, F., 1982. Cooking: a tend toward less well-done, *New York Times*, 6 January. http://www.nytimes.com/1982/01/06/garden/cooking-a-trend-toward-less-well-done.html?scp=1&sq=fabricant+julia+child+less+well+done&st=nyt (accessed 2 January 2016).

Fein, S. B., Lin, C.-T. J. and Levy, A. S., 1995, Foodborne illness: perceptions, experience, and preventive behaviors in the United States. *Journal of Food Protection*, 58(12), pp. 1294–411, 1405–11(7).

Feng, Y., Bruhn, C. M. and Marx, D., 2016. Evaluation of different food safety education interventions. *British Food Journal*, 118(4).

Final Report: Focus Groups on Barriers That Limit Consumers' Use of Thermometers When Cooking Meat and Poultry Products, Phase One, Macro International, Inc., January 1998.

Five Keys to Safer Food Manual, Wolrd Health Organization, 2001. http://www.who.int/foodsafety/publications/5keysmanual/en/(accessed 17 March 2016).

'Food Safe Families Toolkit', FoodSafety.adcouncil.org (accessed 27 July 2016).

Food Safety From Farm to Table: A National Food Safety Initiative, Report to the President, May 1997. http://www.cdc.gov/ncidod/foodsafe/report.htm#education (accessed 2 March 2016).

FSIS, 2016. *Changes in Consumer Knowledge, Behavior, and Confidence Since the HACCP Rule*. USDA Food Safety and Inspection Service, n.d. Web. 28 July 2016.

Garayoa, R., Cordoma, M., Garcia-Jalon, I., Sanchez-Villegas, A. and Vitas, A., 2005. Relationship between consumer food safety knowledge and reported behavior among student sfrom health sciences in one region in Spain. *Journal of Food Protection*, 12, pp. 2502–720, 2631–6.

'History of Consumer Food Safety Education: Focus on Beef', Beefresearch.org (accessed 27 July 2016).

Ho, S. P. and Bruhn, C. M., 2011. Observations of Consumer Salad Preparation. *Food Protection Trends*, pp. 274–8. Print.

Hodges, J., 'Fight BAC!', MeatScience.org (accessed 27 July 2016).

Hunt, M.C., Warren, K.E., Hague, M.A., Kropf, D. H., Waldner, C.L., Stroda, S.L. and Kastner, C.L., 1995. Cooked Ground Beef Color is Unreliable Indicator of Maximum Internal Temperature. Department of Animal Sciences, Kansas State University, Manhattan, KS 66506-0201. Presentation to American Chemical Society 6 April, 1995.

'Is It Done Yet?', *USDA.gov*. N.p., n.d. Web. 15 Mar. 2016.

'Jack In The Box Ignored Food Safety Regulations, Court Documents Say', The Spokesman Review, Spokesman.com (accessed 27 July 2016).

Jevšnik, M., Hlebec, V. and Raspor, P., 2008. Consumers' awareness of food safety from shopping to eating. *Food Control*, 19(8), pp. 737–45.

Lando, A.M. and Chen, C.C., 2012. Trends in ownership and use of food thermometers in the United States, 1998 through 2010. *Journal of Food Protection*, 73(3), pp. 556–62.

Lyon, B. G., Berry, B. W., Soderberg, D. and Clinch, N., 2000. Visual color and doneness indicators and the incidence of premature brown color in beef patties cooked to four end point temperatures. *Journal of Food Protection*, 10, pp. 1315–449.

National Food Safety Month. http://www.foodsafetymonth.com/(accessed 1 March 2016).

Nesbitt, A, et. al., 2014. *Food Control*, 38, pp. 157–73.

Newspoll 2006 Food Safety Week Study for the Food Safety Information Council, http://www.foodsafety.asn.au/the-science/food-safety-information-council-report-card-on-australian-food-safety-knowledge-1997-to-2006/(accessed 26 April 2016).

'Real Progress in Food Code Adoption', http://www.fda.gov/downloads/Food/GuidanceRegulation/RetailFoodProtection/FoodCode/UCM476819.pdf (accessed 2 January 2016)

Riley, L. W., Remis, R. S., Helgerson, S. D., McGee, H. B., Wells, J. G., Davis, B. R., Hebert, R. J., Olcott, E. S., Johnson, L. M., Hargrett, N. T., Blake, P. A. and Cohen, M. L., 1983. Hemorrhagic colitis associated with a rare Escherichia coli serotype. *The New England Journal of Medicine*, 308, pp. 681–5.

Teague, J. and Anderson, D. W., 1995. Consumer preferences for safe handling labels on meat and poultry. *Journal of Consumer Affairs*, 29 (1), pp. 108–27.
'Ten Years After the Jack-in-the-Box Outbreak – Why Are People Still Dying From Contaminated Food?', Safe Tables our Priority, http://www.safetables.org/pdf/STOP_Report.pdf February 2003 (accessed 3 January 2016).
'The Five Keys to Safer Food Programme', *World Health Organization*. N.p., n.d. Web. 27 July 2016.
'Thermy Campaign Kickoff, USDA Headquarters', Washington, DC, 25 May 2000. http://www.fsis.usda.gov/wps/portal/frame-redirect?url=http://www.fsis.usda.gov/oa/thermy/kickoff.htm (accessed 25 February 2016).
'Timeline of Events Related to E. Coli O157:H7', *USDA.gov*. N.p., n.d. Web. http://www.nytimes.com/1994/11/03/us/food-industry-groups-sue-to-stop-us-testing-of-ground-beef.html
'Unsafe Foods Cause over 200 Illnesses – PAHO/WHO Notes', *Kaieteur News*. N.p., 08 Apr. 2015. Web. 27 July 2016.
Update on Movement to Stop Irradiated Foods In Schools, Organic Consumers Association and Public Citizen. https://www.organicconsumers.org/old_articles/Irrad/irradiated20805.php (accessed 1 March).
'USDA and Michigan State University Join Forces To Promote Food Thermometer Use In Michigan', FSIS.USDA.gov, 2 August 2004 (accessed 27 July 2016).
'USDA Hotlines Post Milestones for 28 Years', Food Safety.gov (accessed 27 July 2016).
'USDA Issues Final Rule on Meat and Poultry Irradiation', FSIS.USDA.gov, December 1999 (accessed 27 July 2016).
'Where's the (irradiated) beef? Getting it in your grocery store is getting to be big business', AgriMarketing.com, February 2013 (accessed 26 July 2016).
'WHO Golden Rules for Safe Food Preparation', *Paho.org*. N.p., n.d. Web. 27 July 2016.
Williamson, D.M., Gravani, R.B. and Lawless, H.T., 1992. Correlating food safety knowledge with home food-preparation practices. *Food Technology*, 45 (6).
Yang, S., Angulo, F. and Altekruse, S., 2016. Evaluation of safe food-handling instructions on raw meat and poultry products. *Journal of Food Protection*, n.d. Web. 28 July 2016.

Chapter 11

Traceability in the beef supply chain

Daniel D. Buskirk and Tristan P. Foster, Michigan State University, USA

1 Introduction

Modern food supply chains have become larger, faster and more complex, due to the globalization and outsourcing of food products. Beef safety crises, such as bovine spongiform encephalopathy (BSE) in the United Kingdom and illness outbreaks associated with *Escherichia coli* O157:H7 in the United States, have raised consumer concerns about food safety. The economic cost of food-borne illness in the United States alone is $50–80 billion annually, which includes health care costs, lost productivity and diminished quality of life (Scharff, 2012). The food industries face growing challenges to ensure consumer health against a wide range of food safety threats (e.g. introduction of pathogens, contamination and food adulteration, degradation, mislabelling and counterfeiting) (Farooq et al., 2016). These safety concerns, combined with an increasing quest for information on the origin of food (as populations are further removed from agricultural backgrounds), are acting as major drivers for the development of traceability systems (Bloom and Hinrichs, 2010; Olynk, 2012). Recently, just 28% of the US consumers surveyed strongly agreed with the statement: 'I have access to all the information I want about where my food comes from, how it's produced and its safety' (The Center for Food Integrity, 2016; p. 18).

Motivation for designing traceability systems for cattle and beef are varied and include establishing ownership and deterring theft, use in production management (McAllister et al., 2000; Voulodimos et al., 2010; Barge et al., 2013), registration, insurance, controlling and eradicating foreign animal disease (Disney et al., 2001), controlling and eradicating domestic animal disease (NIAA-USAHA, 2013), achieving compliance for export market

http://dx.doi.org/10.19103/AS.2016.0008.15

requirements (Stanford et al., 2001; Smith et al., 2005), re-establishing ownership after an emergency or natural disaster (Blasi et al., 2003), communicating credence attributes to consumers (Caswell, 1998; Markus et al., 2011; Buskirk et al., 2013), isolating quality defects, detecting environmental contamination (Augsburg, 1990; McKean, 2001), resolving food safety issues (Augsburg, 1990; Tonsor et al., 2009; Andree et al., 2010) and minimizing product recalls (Smith et al., 2005).

Modern beef production provides multifaceted challenges for implementation of full supply chain traceability. Beef production is a highly segmented business, where animals are often managed by multiple owners within cow/calf, stocker and finishing sectors at multiple, widely dispersed sites. Beyond the farm-gate, the segmentation often continues through processor, purveyor, distributor and retail or food service. To add complication, during this process, the carcass is often separated into nearly 200 cuts of beef, with the remaining trim comingled and ground. It may be no wonder that the beef industry has been challenged with developing comprehensive traceability systems, particularly in large-scale, fast-moving supply chains. Here, we summarize the various pieces to the traceability puzzle (unique identification of cattle, beef and locations; record-keeping; and interoperable communication systems) that are being assembled.

2 Traceability defined

'Trace' originates from Latin, 'tractiare', to delineate, and from the Old French 'tracier', to look for, follow or pursue (Harper, 2016). Today, the word trace retains a similar definition as it did in the fourteenth century, yet the intended meaning of the term 'traceability' may vary depending on the specific application or industry to which it is applied. Golan et al. (2004) contended that the definition of traceability is necessarily broad, because traceability is a tool for achieving a number of different objectives, and no single approach is adequate for every objective. Supply chain management literature suggests that traceability is 'the information system necessary to provide the history of a product or a process from origin to point of final sale' (Smyth and Phillips, 2003: p. 32). Within a food industry context, Golan et al. (2004: p. 1) defined traceability as 'record keeping systems that are designed to track the flow of product or product attributes through the production or supply chain', and the International Organization for Standardization defined traceability as the 'ability to follow the movement of a food through specified stage(s) of production, processing and distribution' (ISO, 2007: p. 2). In a survey related to beef traceability, Mennecke et al. (2006: p. 32) defined traceability as 'the ability to retrieve the history, treatment, and location of the animal that a cut of meat comes from, through a recordkeeping and audit system or registered identification program'.

To be able to trace products and retrieve related information, producers must collect information and keep track of products during all stages of production (primary production, processing, distribution, retailing and consumer). Therefore, traceability can be divided into two key functions: tracking and tracing (Fig. 1). Although the terms may be used interchangeably in some literature, tracking (or forward traceability) is typically defined as the ability to follow the path of an item as it moves downstream through the supply chain from the beginning to the end, whereas tracing (or backward traceability) is the ability to identify the origin of an item or group of items, through records, upstream. Methodologies for the analyses of the food and feed materials, combined with information technology systems, are essential to delivering a working tracking and tracing system (Schwagele, 2005).

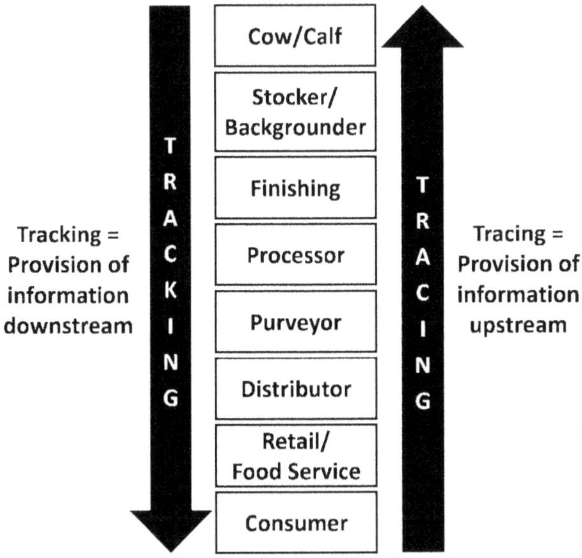

Figure 1 Tracking and tracing along the beef supply chain (adapted from Schwagele (2005)).

Traceability systems can vary with the number of data points tracked, breadth and depth of tracking throughout the supply chain and the particular technologies used to track product or characteristics (Golan et al., 2004; Souza-Monteiro and Caswell, 2004). Bulut and Lawrence (2007) summarized the definition of various types of traceability, such as 'external traceability', which refers to tracing product or product attributes through the successive stages of production (i.e. cow–calf producer, auction barn, feedlot, harvest and processing), while 'internal traceability' refers to tracing within a plant or production unit, such as that required for hazard analysis critical control point plans. Traceability is most often restricted to this type of internal traceability within beef supply chains.

Shackell and Dodds (2008) drew the analogy that the principles of meat traceability adhere closely to those of 'chain of custody' in forensic science. Chain of custody is a legal term that refers to the ability to guarantee the identity and integrity of a specimen by documenting its unbroken control and chronological history, from acquisition through analysis to storage, for future reference and/or disposal (Giannelli, 1996; Tomlinson et al., 2006). 'By referring to custody, the definition of traceability implies the same level of confidence in all the identification and record keeping from when an animal is born, while it is being raised, during processing and its transfer, as meat, into the marketplace and ultimately on to the consumer's plate' (Shackell and Dodds, 2008: p. 61).

Golan et al. (2004) advocated that a traceability system can be characterized by its breadth, depth and precision. The breadth is the amount of information recorded; the depth is how far back or forward the system tracks; and the precision is the degree of assurance by which the system can accurately recall the movement of a particular product. The level of identification determines traceability precision (or resolution) (Jansen-Vullers et al., 2003). The desired traceability resolution necessarily delineates which items and locations must be uniquely identified in the system. Moving from most precision to least, identification may take place at the individual object, case, pallet or batch/lot level, for example. As identification resolution is increased, traceability accuracy increases, but with a concomitant increase in management costs (Kelepouris et al., 2007).

The underpinning of all successful traceability systems is maintaining identification integrity of products and places throughout the supply chain. McKean (2001: p. 363) simply defined traceability as 'the ability to maintain a credible custody of identification for animals or animal products through various steps within the food chain from the farm to the retailer'. Without conservation of unique identification, tracing of meat products during a food-borne pathogen recall, for example, would be all but impossible. To a large extent, identification technologies determine the potentials and limits of a traceability system; therefore, the following sections provide a synopsis of the emerging methods for identification of cattle, beef and locations in a supply chain.

3 Methods of unique cattle identification

Cattle identification has been accomplished in numerous ways for centuries. Early motivation for group identification was proof of ownership. As herds grew in size, a unique identifier became an important part of production management record-keeping. Now, as the call for traceability intensifies, so does the need for all cattle to be identified as part of the beef supply chain.

Means of identifying cattle include a) brands (hide or horn); b) ear notches; c) tattoos (ear, shoulder or lip); d) ear or tail tags (visual, barcoded or radio-frequency identification (RFID) transponders); e) transponders (neck chain, implants or rumen bolus); and f) biometrics (DNA fingerprinting, iris scanning, retinal imaging, nose print matching and facial recognition) (Smith et al., 2005; Cai and Li, 2013; Awad, 2016). There are significant limitations with the use of conventional cattle identification methods (i.e. brands, tattoos and visual ear tags) in traceability systems that include lack of uniqueness, illegibility, vulnerability to losses and subjectivity to deformation and fraud (Awad, 2016). Barcoded ear tags offer unique identification; however, they can become dirty or manure contaminated, faded or damaged, making them unreadable (Stanford et al., 2001; Lawrence et al., 2003; Shanahan et al., 2009).

Electronic identification methods, such as RFID and biometrics, offer rapid and more secure means for providing unique and dependable animal identification (Kelepouris et al., 2007; McCarthy et al., 2011; Awad, 2016) that also allow for automated information and communication technology (ICT)-enabled systems, thereby enhancing the resolution or granularity of traceability (Dabbene et al., 2016).

Biometrics show promise as unique and reliable animal identification; however, many have data-capturing challenges, making their fit in automated traceability questionable. Iris scanning and retinal imaging require up to several minutes of animal restraint, the camera must be in line of sight and close proximity to the eye, and strong ambient light must be dimmed (Allen et al., 2008; Gonzales Barron et al., 2008; Lu et al., 2014). Challenges with face recognition are similar in that restraint, standardizing lighting conditions and camera positioning are key elements (Corkery et al., 2007). Disadvantages of muzzle prints include deformations during animal growth and the length of time required to obtain a good-quality print (Barry et al., 2007), although work continues on converting muzzle images into binary data for the purposes of identification (Barry et al., 2007; Tharwat et al., 2014). Eye-based identity, facial recognition and muzzle prints are all terminated at harvest, similar to conventional or electronic forms of identification.

The animal identity methods that are gaining the most acceptance for widespread use in automated meat traceability chains are the RFID (various forms) and DNA fingerprinting. Following is a summary of research investigating the use of these identification methods.

3.1 Radio-frequency identification

RFID is a wireless sensor technology based on electromagnetic signal detection (Landt, 2005; Domdouzis et al., 2007). The use of RFID complements traceability, logistics, supply chain management and quick response (QR) systems, because RFID is a means of automatic identification (Auto-ID) where the product can be identified without human intervention or line of sight, while in motion, and simultaneously with other items (Dabbene et al., 2016). A typical RFID system includes three components: an antenna or coil, a transceiver with decoder (reader) and a transponder (RFID tag) electronically programmed with unique information (Blasi et al., 2003; Domdouzis et al., 2007; Singh et al., 2014). Unlike barcodes, RFID can be read through a variety of materials (Roberts, 2006), or in the case of injectable or bolus tags, through the animal itself (Caja et al., 1999; Eardprab et al., 2008; Cooper-Prado et al., 2011). RFID transponders are manufactured in a variety of form factors, including injectable transponders, neck bands, boluses and ear tags.

Injectable transponders, also referred to as implants or microchips, may be covered by a capsule of biomedical glass and injected just under the skin. Injectable transponders have been implanted in a number of different body sites in cattle, including base of the ear, scutulum of the ear (Fallon and Rogers, 1991; Oakley, 1993; Hasker and Bassingthwaighte, 1995), anal region (Hasker et al., 1992), upper lip, armpit (Lambooij et al., 1999), penile sheath and dew claw (Lambooij, 1991). Studies have demonstrated that an injectable transponder can be a reliable method for identification; however, the transponder can migrate, resulting in difficult retrieval (Conill et al., 2000; Blasi et al., 2003) and the potential for the transponder to enter the food chain (Fallon and Rogers, 1999).

Transponders may be incorporated into neck bands or collars, but major disadvantages of this form factor for beef cattle are that collars must be frequently adjusted on growing animals and present a choking hazard, especially in extensive terrain.

Transponders, contained in a bolus, can be introduced orally into the forestomach of ruminants by use of a balling gun (Hasker and Bassingthwaite, 1996; Caja et al., 1999; Fallon, 2001). The RFID boluses had the lowest mean percentage losses in the European IDEA Project (Ribo et al., 2001) compared to ear tags and injectable transponders. Bolus weight, volume and specific gravity have been demonstrated to be important factors in retention (Ghirardi et al., 2006). Boluses may also be equipped with other ruminal sensors, such as temperature devices (Small et al., 2008; Cooper-Prado et al., 2011), to aid in cattle management. When a magnetic bolus (used to render accidentally ingested metallic objects harmless) is present in the reticulo-rumen in close proximity to a transponder bolus, the RFID resonant frequency may be increased, thus disrupting communication with a reader (Eardprab et al., 2008). Retrieval of rumen boluses post-harvest may be problematic (Fallon, 2001), although it is possible to retrieve and even recycle RFID rumen boluses (Mooketsi, 2013). The rumen bolus may require a coinciding external method of identification for routine management, yet it offers a relatively secure method of electronic animal identification (Fallon, 2001).

Ear tagging is one of the most widely accepted, convenient and low-cost cattle identification methods (Awad, 2016). Transponders can be placed in ear tags of numerous form factors (i.e. button tag and dangle tag) (USDA-APHIS-VS, 2016b). RFID ear tags are a temporary means of identification and are susceptible to damage, loss, unreadability and fraud. As outlined by Stanford et al. (2001), the key factors affecting retention are tag, size, design, application, position, infection rate, length of time in the ear, environmental conditions and on-farm management (i.e. feeder design, fencing type). The design considerations of RFID ear tags should also include tamper-resistance (Stanford et al., 2001). A small tag size also improves retention, while minimizing invasiveness to the animal (Blasi et al., 2003). Designing a small tag is antagonistic with the general rule that a transponder with a larger antenna requires less activation power and improves reading distance (Blasi et al., 2003), although numerous other physical and mechanical conditions can also affect the reading efficiency of transponders (Geers et al., 1997; Ryan, 2008; Ryan et al., 2010).

Although there are RFID tag types that allow information to be written to the tag by the end user (Roberts, 2006), the majority of RFID tags currently used for livestock identification are encoded by the manufacturer with a unique code that doubles as the animal identification number (AIN) and are only readable by the end user (USDA-APHIS-VS, 2016b). The RFID tags can also be active (activated by an internal power source) or passive (activated by an external power source). Although passive tags generally have a shorter communication (read) distance, they are more commonly deployed in livestock identification because of the lack of battery maintenance, smaller form factor and reduced expense compared to active tags (Hoang and Caudill, 2012). The procedure used for data transfer from RFID transponders to the reader may also differ. In the half-duplex procedure (HDX) the data transfer from the transponder to the reader (uplink) alternates with data transfer from the reader to the transponder (downlink), whereas in the full-duplex procedure (FDX) the uplink and downlink occur simultaneously (Artmann, 1999; Kampers et al., 1999; Finkenzeller, 2010). It is doubtful that transmission speed (measured in milliseconds) between HDX and FDX has a meaningful impact on the ideal transponder chosen for livestock identification in a traceability system (Wallace et al., 2008). However, Ryan et al. (2010) clearly demonstrated that transponder and transceiver manufacturer may interact to affect both read distance and variation in read distance within RFID systems. Ryan (2008) found that electromagnetic interference (EMI) can also affect the read distance of RFID ear tags. Of the various areas where RFID are used in the cattle supply chain, the highest recordings of EMI were in feedlot handling and processing areas, and cattle abattoirs (Ryan, 2008). This means that the environment must be considered when there is concern about RFID system performance.

The various radio-frequency bands on which RFID transponders operate differ in their electromagnetic characteristics and impart relative advantages or disadvantages, depending on the particular identification application. The low-frequency (LF) and ultra-high-frequency (UHF) bands are the most commonly used in animal identification (Table 1).

LF RFID devices that operate at 134.2 kHz are commonly used in the beef industry (USDA-APHIS-VS, 2016c). There are two key international standards for animal RFID devices that operate at 134.2 kHz (low frequency), which were established in 1996 by the ISO technical committee, ISO/TC 23 SC 19; one standard contains the technical concept of RFID of animals (ISO, 11785) and the other describes the standardization of the code structure for information contained in transponders (ISO, 11784).

In 2010, the ISO technical committee, ISO/IEC JTC 1/SC 31, first defined a standard for the air interface for RFID devices operating in the 860 to 960 MHz industrial, scientific and medical band used in item management applications (ISO/IEC, 18000–6). USDA-APHIS-VS has approved several ear tags that incorporate UHF RFID EPC Gen 2 (v1.2.0), which operate

Table 1 A summary comparison of low- and ultra-high-frequency transponder attributes for cattle identification[a]

Item	Radio-frequency band Low (LF)	Ultra-High (UHF)
Operating frequency	134.2 kHz	860–960 MHz
International technical standards	ISO11785	ISO/IEC 18000–6C EPC Global Gen2 (v 1.2.0)
International code structure standards	ISO11784	None[b]
Data capacity	112 bits (full-duplex)[c] 128 bits (half-duplex)[c]	64 bits, tag identification (TID) 96 bits, EPC memory (extendable into user memory) 512 bits, user memory[d]
Data writing	Generally only by issuer	By issuer to TID and EPC memory By user to user memory
Relative data transfer speed[e]	Slow	Fast
Simultaneous reading of tags[f]	No (via ISO11785)	Yes
Reading distance	Less than 1 m	Up to several metres

[a] Adapted from (Roberts, 2006, ScotEID, 2014). [b] On the basis of the ISO11784 LF standard, UHF interim coding structure standards have been adopted in the United States (USDA-APHIS-VS, 2016a), Scotland (ScotEID, 2014). [c] Restricted to 64 bits by ISO11784 (ISO, 1996). [d] According to EPC tag data standard (GS1, 2014). [e] Data transfer speed is proportional to the radio-frequency and affects the amount of information that can be read or written to a transponder in a given time. [f] The ISO standards for LF specify that only one tag at a time can register in an antenna field. The EPC Gen-2 UHF RFID standard includes protocols for anti-collision, which allows simultaneous reading of multiple tags (GS1, 2015).

in the 902–928 MHz range (USDA-APHIS-VS, 2016b). However, there is no international standard for the coding structure (tag data standard) for UHF ID devices. Research using UHF transponders, readers and ID code structure in Scotland has emulated the existing LF RFID ISO standards (ScotEID, 2014), whereas the USDA has defined an interim standard to achieve uniformity among manufacturers that are authorized to encode USDA animal numbers into UHF identification devices (USDA-APHIS-VS, 2016a).

In the absence of a government mandate, RFID technology has been only minimally adopted in the US breeding herd. In 1997, RFID use in commercial beef calves was negligible (USDA-APHIS-VS, 1997), and a decade later, use was still less than 3% (USDA-APHIS-VS, 2008). However, uptake of RFID in the finishing sector was found to be approximately 20% in 2011 (McKeith et al., 2011). The greater use of RFID in the US feedlot industry is presumably connected to enrolment in source and age verification programmes needed for export access and carcass data feedback.

3.2 DNA profiling

Biomarkers, which most prominently includes DNA profiling (or fingerprinting), play a role in many food traceability systems by providing secure unique identifiers (Pettitt, 2001; Raspor, 2005; Galimberti et al., 2013). A sample of DNA taken from the live animal can be matched with meat that is found later to contain a safety or palatability issue (Smith, 2004). DNA-based identification has the advantage over other identification methods (i.e. branding, tagging and biometrics), because DNA remains an integral and identifiable component of animal products after they are separated from the carcass (Heaton et al., 2002; IdentiGEN, 2004). The DNA may be readily extracted and analysed from samples of fresh, frozen or cooked animal products (Meyer et al., 1993; Tartaglia et al., 1998; Wang et al., 2000). Traceback through DNA sampling can be useful for 1) auditing traceback and compliance programmes (Smith et al., 2005), 2) identifying specific process control problems, 3) rapid response to a food safety event and 4) marketing claim validation (Smith et al., 2008).

Robb et al. (2006) described a typical traceability process for DNA fingerprinting: 1) a sample of muscle is archived (identified by calendar date, time of day and carcass identification number) from each individual carcass as it enters the fabrication room; 2) as each box of fabricated product moves past the box scale, it is time-stamped and recorded in the packer's computer system; 3) at the retailer's site, the box serial number can be used to identify each retail package generated from primals/subprimals in that box; 4) if there is a need for traceback, a sample of muscle from the retail cut in question, plus the serial number of the box from which it came, is sent back to the packer, which locates the box serial number in the computer database; 5) knowing the average length of time required for that cut of beef to be produced, boxed and scaled on the fabrication floor from the time the carcass entered the fabrication room, the packer can identify a range of potential carcasses from which the primal cut originated; 6) the packer then sends the sample of the retail cut, along with the samples from that range of potential carcasses to a DNA testing laboratory, to be analysed until a DNA fingerprint match is achieved.

DNA profiling is currently limited because identity verification results are not generated in real-time (i.e. beside animal or carcass) and, therefore, profiling is unlikely to become a primary identifier for live cattle or carcasses (Tate, 2001; Allen et al., 2008). In addition, a traceability system reliant on DNA identification is of little value unless a complete birth-to-harvest history for the sampled animal is available and correlated in a data set (Smith et al., 2008). Nevertheless, DNA profiling is a powerful tool and can be used effectively in traceability systems, especially in retrospective audits, or as a counter-fraud measure (Allen et al., 2008).

4 Beef identification by barcoding

The most established technology for tracking food products is the linear or one-dimensional (1D) barcode label. The 1D barcode label is a simple form of systematic representation of data. This is done by varying the width and spacing of parallel lines to form an image for an interrogator to read (GS1, 2016c). By enabling scanning objects for their identification, supply chain participants do not rely on human visual assessment to manually capture data for database entry. For such a system of identification to be universal, standards are imperative. GS1 was established in 2005, by the merger of EAN International (European Article Numbering) and Uniform Code Council. The GS1 system provides for the use of unambiguous numbers (e.g. for use in barcodes and RFID) to identify goods, services, assets and locations worldwide in any industry or trade sector. The system is designed to aid in traceability by overcoming the limitations of using company-, organization- or sector-specific coding systems and to make trading more efficient and responsive to customers (GS1, 2016b).

Although the 1D barcode, such as a UPC barcode, can adequately represent a category of products, it has serious limitations in data capacity and mobility (Kato and Tan, 2005) for use in whole-chain traceability systems (Table 2). Starting in 2010, adoption of the GS1 DataBar was first implemented with perishables, pharmaceuticals and coupons. The GS1 DataBar will enable the meat and poultry industry to move away from the UPC-type barcodes, which do not support the Global Trade Item Number (GTIN) or other data elements that are crucial for traceability (mpXML, 2010).

As part of the Global Traceability Standard (GTS), adoption of 1D barcode formats (GS1 DataBar and GS1–128) encoding the GTIN and batch or lot number is designed to enhance the effectiveness of meat and poultry product recalls (GS1, 2012). In addition to providing unique identification numbers, the GS1 system provides for supplementary information, such as best before dates, serial numbers and batch numbers, which can appear in barcode form. Currently, barcodes are used as the data carriers to enable their electronic reading wherever required in business processes (GS1, 2016b).

More recently, two-dimensional (2D) barcodes have been used in a variety of applications. A 2D barcode symbol consists of guide and data areas (Fig. 2). The guide area has a finder

Table 2 GS1 barcode examples[a]

		Data capacity	
Code type	Symbology example	Numeric, digits	Alphanumeric, characters or digits
GS1 UPC-A	7 88581 01497 4	12	No alpha
GS1–128	(01) 0 9501101 53000 3 (17) 140704 (10) AB-123	48	48
GS1 DataBar Expanded	(01) 0 9501101 53000 3 (17) 140704 (10) AB-123	74	41
GS1 QR Code		7,089	4,296
GS1 DataMatrix		3,116	2,335

[a] GS1, 2016c

pattern, which is a specific pattern that a scanner uses to locate the barcode symbol (Kato et al., 2008). The data area contains geometric patterns, such as rectangles, dots and hexagons (Burian et al., 2008), which encode for alphanumeric data, including letters, numbers and punctuation marks. Compared with linear barcodes, which hold very limited information data, 2D barcodes have a much larger data-carrying capacity (Al-Khalifa, 2008). For example, a QR code can encode up to 7089 digits (4296 letters and 2953 binary data) (Gao et al., 2007). To date, many consumer applications of 2D codes include simple encoding of a uniform resource locator (URL) that reduces keystroke entry. However, a powerful characteristic of 2D codes is their ability to serve as mobile databases that are not internet dependent.

When the data cell sizes are larger and have a more distinct pattern, barcode readers tend to produce reduced error rates when scanning 2D barcodes (Kato and Tan, 2007). Likewise, the size of the finder pattern and its cells has been found to be directly related to the 2D barcode user experience (Tan and Chai, 2010). Kato et al. (2009) suggested that the user experience was correlated to the first read rate (the number of successful first reads, divided by the number of attempted first reads).

In a recent survey, US consumers were asked what information they would like to know about their beef and how they would like to access the information. The six highest-ranking topics that consumers wanted to know were: 1) impact of food on health, 2) food safety, 3) environmental impact, 4) labour and human rights, 5) animal well-being and 6) business ethics. The methods through which the consumers wanted to access this information was directly on the package, third-party website, company website and QR code on the package (The Center for Food Integrity, 2015).

Although many consumers are motivated to scan a barcode for additional product information (Comscore, 2011), because of the smart device requirement, QR code use eliminates many mobile device users (Shin et al., 2012). Nevertheless, scanning of QR codes continues to grow in popularity as a method of accessing product information. One QR code scanning application is reporting more than 20 million scans each business quarter (ScanLife, 2014, 2015). Two-dimensional barcodes offer some interesting benefits

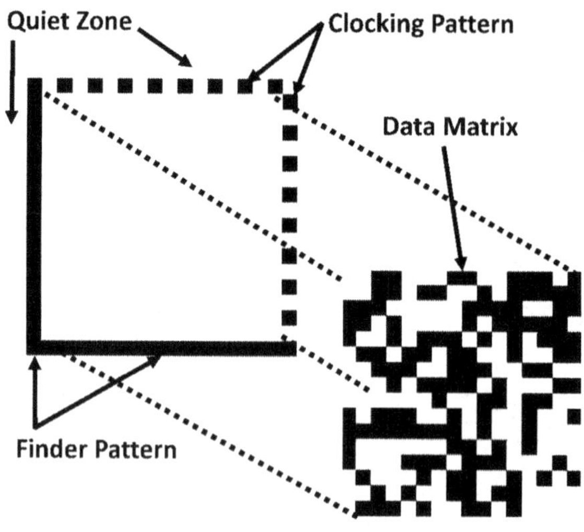

Figure 2 Major components of a 2D DataMatrix barcode.

in providing interaction with consumers (Tan et al., 2010), particularly as it relates to opportunities to communicate safety recall information.

In addition to barcodes, RFID can also be used as beef makes its way downstream in the supply chain. However, for many uses, a printed barcode label is sufficient and less expensive than carrying the same information in a RFID tag. Which technology is preferred will depend upon the identification characteristics required within the given environment (e.g. line-of-site, potential contamination).

RFID, 1D and 2D barcodes provide the basic means for identification in traceability information sharing (Food Standards Agency, 2002). In food traceability systems, the Electronic Product Code Information Services (EPCIS) standard is increasingly used in capturing events of food items passing through a supply chain network, storing these on one or more EPCIS-repositories and enabling query of these events using appropriate security mechanisms (GS1, 2016a). Events contain data about the identity of the product, the date and time of event occurrence, the location where it occurred and the reason why the event occurred. These events commonly use Auto-ID technologies, such as barcode scanning and RFID, to identify objects and locations (Verdouw et al., 2016).

5 Traceability information systems

Verdouw et al. (2016) described the information systems architecture for what they referred to as a virtualization of supply chains. Virtual supply chains build on food traceability systems and provide the information to track the location of objects and trace their history. Chain-wide traceability systems can be based on a centralized, linear or distributed approach (Kassahun et al., 2014). In a centralized approach, traceability data is collected in shared databases, for example, the national bovine animal registration systems in Europe (EC Regulation No. 911/2004 of 29 April 2004) or Animal Disease Traceability system in the United States (USDA-APHIS-VS, 2013). In a linear approach, traceability data are passed from one partner to the next one, while each partner records the supplier and customer of specific products. This approach is also referred to as the one step forward and one step back principle, as required by European General Food Law (Beulens et al., 2005). In a distributed approach, the participants of food supply chains interconnect their own traceability systems to exchange traceability data (Verdouw et al., 2016).

The linear approach to traceability lacks robustness because, in practice, not all supply chain participants are able to implement state-of-the-art systems and the benefits of gathering detailed data by one operator can be quickly lost when subsequent operators are not able to pass on the data. For example, the business processes suggested by the Meat and Poultry Data Standards group (mpXML, 2010) in implementing a meat traceability system rely upon each supply chain participant to take responsibility for supplying one partner upstream and one partner downstream the appropriate data. The practices delineated are dependent upon the GS1 GTS (GS1, 2012: p 7), which defines the data that must be collected, recorded and shared to ensure 'one step up, one step down' traceability. Supply chain traceability in this standard relies on each partner maintaining and sharing the appropriate records. It is not difficult to imagine ways in which whole-chain transparency in traceability could breakdown in a linear scenario.

The centralized approach to transparency is relatively simple to implement, but requires trust among supply chain actors or regulatory mandate, and a trusted third party that manages the centralized system. The centralized approach is often mandated in government-controlled

Table 3 Cattle and beef traceability systems implemented by countries and regions of the world

Region	Number of cattle[a] / Beef production, metric tons/yr[b]	Traceability details	References
The United States	88,526,000 11,019,600 MT	Effective 11 March 2013, the US Department of Agriculture (USDA) updated regulations for traceability of US livestock moving interstate. Adult cattle and bison (>18 months of age) must be officially identified and accompanied by an interstate certificate of veterinary inspection or other approved documentation. Official identification includes visual and RFID ear tags (containing an Animal Identification Number) or metal tags (identified by the National Uniform Eartagging System); however, brands, tattoos and brand registration may also be accepted when approved by the shipping and receiving States or Tribe. Cattle less than 18 months of age, that are moved interstate for shows, exhibitions, rodeos or recreational events, must also meet the requirements of the rule. Cattle moved interstate to a custom harvest facility are exempt from the regulations. Livestock premises in the states of Wisconsin (effective 1 November 2005) and Indiana (effective 1 September 2006) must be registered with their respective states. Effective 1 March 2007, all cattle in Michigan must be identified with an official RFID ear tag prior to movement from a premises. In Texas, effective 7 October 2013, all sexually intact adult (>18 months) beef cattle, and all dairy cattle (except cattle going directly to harvest), must be permanently identified if ownership is changed.	(State of Indiana, 2005; Kirk and Buskirk, 2006; USDA-APHIS-VS, 2013; State of Wisconsin, 2016; Texas Animal Health Commission, 2016)
Brazil	212,343,932 9,832,874 MT	Effective 9 January 2002, the Brazilian System of Identification and Certification of Bovine and Buffaloes (SISBOV) is a mandatory traceability and identification system, but only for live animal exports and select beef export markets. Roughly 0.5% of Brazilian cattle farms are certified.	(MAPA Brazil Ministry of Agriculture Livestock and Supply, 2016; Vinholis et al., 2016)

European Union	89,440,741 7,706,368 MT	The basis for cattle identification in the European Union (EU) comes from Council Directive 92/102/EEC (EEC 1992) which requires all cattle to be uniquely earmarked and registered. An updated regulation was adopted in 1997, which stated that cattle must be identified by an ear tag applied to each ear and accompanied by a passport during movement. These requirements were later expanded to include a beef labelling system (Regulation (EC) No. 1760/2000). Beef labelling requires provision of information on the animal identification, harvest facility and cutting facility. EU-member countries have latitude in system design to meet these requirements.	(Passantino, 2013)
China	117,409,587 6,386,137 MT	In 2010, the Chinese Ministry of Commerce introduced a voluntary programme called the Meat and Vegetable Distribution Traceability System and Safe Meat Service to track the movement of livestock from processors and wholesale markets to retailers. The pilot system uses barcodes and RFID tags to transmit lot/batch information to a central database. The system does not track from the farm of origin.	(Charlebois et al., 2014)
Australia	29,103,000 2,480,458 MT	Introduced in 1999, the National Livestock Identification System (NLIS) is Australia's system for the identification and traceability of cattle. Three elements to enable lifetime cattle traceability include individual identification (by visual or electronic ear tag), identification of a physical location (by Property Identification Code) and entry of data into a web-accessible database to store and correlate movement data and associated details. The standards associated with NLIS dictate that given a specified animal's identification number, within 48 hours, it should be possible to establish 1) the location(s) where the specified animal has resided for its lifetime, 2) a listing of all cattle that have lived on the same property(s) and 3) the current location of that listing of cattle.	(NLIS, 2016)
Mexico	32,939,529 2,015,300 MT	Mexico's National Livestock Individual Identification System (Sistema Nacional de Identificación Individual de Ganado; SINIIGA) emerged from the Livestock Productivity Enhancement Program (PROGAN), with rules developed in 2003. SINIIGA is voluntary, with participation required to receive payments under the PROGAN. Although many animals have been tagged, much of the producer-supplied data is incomplete.	(Ortega and Peel, 2010)

(Continued)

Table 3 *(Continued)*

Region	Number of cattle[a] Beef production, metric tons/yr[b]	Traceability details	References
India	187,000,000 966,576 MT	To date, there is no mandatory cattle traceability system in India; however, the Department of Animal Husbandry, Dairying and Fisheries is developing plans for an Animal Identification and Traceability System that would provide identification of cattle and buffalo.	(Indianbureaucracy.com, 2016)
Colombia	24,205,969 897,835 MT	The Colombian government has appointed the Colombian Ranchers Federation (FEDEGAN) for the administration, implementation and commissioning of SINIGAN (Sistema Nacional de Identificación e Información del Ganado Bovino), National Cattle Identification and Information System, to provide a voluntary traceability system for cattle and beef on the basis of the one mandated in Uruguay.	(Sonda, 2016)
Canada	12,220,000 876,380 MT	The Canadian Cattle Identification Agency (CCIA) was first established by producers in 1998. Beginning in 2006, all cattle leaving their herd of origin were required to have a CCIA-approved RFID tag, and by July 2010, this requirement extended to cattle leaving their current location. The Canadian Livestock Tracking System (CLTS) database was developed and is wholly owned by CCIA. The CLTS database allows producers to record animal identification, premises identification and movement, as well as birth dates for age verification. Provincial governments are responsible for the management and maintenance of premises identifiers within their jurisdictions.	(CIAA, 2015)
New Zealand	10,368,188 572,629 MT	New Zealand's National Animal Identification and Tracing Scheme (NAIT) became operational in July 2011, and as legislated under the National Animal Identification and Tracing Act 2012, became mandatory to register cattle and record movements. The cattle owner is assigned a NAIT number, which is associated with a FarmsOnLine (FOL) identifier for their farm(s). RFID ear tags are applied to cattle and their movements must be recorded via a web interface by both the source and destination person in charge of animals. The double-entry system provides a data validation mechanism.	(Jewell et al., 2016)

Uruguay	11,600,000 540,811 MT	Bill 17.997 was passed on 2 August 2006, which dictated use of the National Livestock Information System (Sistema Nacional de Información Ganadera; SNIG), although it did not become fully mandatory until 2013. The system requires any cattle born or transported in the national territory to wear two ear tags (one of them an RFID; both showing a unique identification number) for linkage to a system of databases managed by the government. The SNIG provides consumers information about all the places where the animal has been since birth (farms, fairs and processors); the date when the animal entered and left each place; the identification of the animals that shared the same locations and the current location of the animals that were in contact with a specific animal.	(IICA, 2009; Andres, 2015)
Japan	3,962,000 501,847 MT	The Act on Special Measures Concerning Traceability of Beef Products, commonly known as the Beef Traceability Act, was established in June 2003, following detection of bovine spongiform encephalopathy (BSE) in September 2001. All cattle are given an ear tag with a 10-digit identification number. Cattle identification numbers and packaged meat products are coded using the GS1–128 barcode system. The number is recorded with the National Livestock Breeding Center (NLBC), which is in charge of maintaining the database. Cattle data associated with the identification number include birth date, sex, breed, dam's identification, movement dates, date of harvest and processor location. Animal health and feeding information may also be added to the database. Producers can provide data to NLBC's database by fax, push-button telephone, website entry, 'handy terminal' which also reads barcoded ear tags, email batch files, electronic file transfer or mobile phone application. The 10-digit identification number, which is maintained through every level of the processing and distribution chain, must be present on the invoice or final packaged beef product. If the retailer chooses to include the number on the retail package label, consumers will be able to enter the 10-digit number in a web-accessible database and view specific information regarding that animal or product. Retailers have also included a QR code on beef packages, which contains a hyperlink to the national database website.	(Souza-Monteiro and Caswell, 2004; GS1 Japan, 2011; Jin and Zhou, 2014; Godo, 2015)
World	1,482,144,415 63,361,528 MT		

[a]2014 data (FAOSTAT, 2016). [b]2013 data (FAOSTAT, 2016).

systems. Table 3 outlines cattle and beef traceability systems that have been implemented by the largest beef-producing regions in the world. There is an observable and universal trend to utilize RFID ear tags for unique cattle identification. However, the database systems (particularly those mandated by governments) are often centrally controlled and vary widely in their standards, architecture, interfaces and capabilities.

Distributed systems require that each food operator maintains its own system or alternatively uses a cloud-based computing and standards-based approach for realizing chain-wide traceability (Kassahun et al., 2014). EPCglobal Inc., a subsidiary of GS1, was launched to create updated global standards designed to connect existing communications network infrastructure, with RFID technology, and electronic product code (EPC), that enables immediate Auto ID and tracking of an item through the whole supply chain (Schwagele, 2005; Shanahan et al., 2009).

With regard to beef traceability, it is logical to implement a traceability system that utilizes the EPCglobal Network, as it is a universally accepted identification system that facilitates efficient global commerce and improves the effectiveness of recording and exchanging traceability information between stakeholders in the supply chain (Shanahan et al., 2009; Hartley, 2013; Kassahun et al., 2016).

The EPCglobal Network consists of the EPC, RFID tags, tag readers and software that communicate in a common standard (Fig. 3). The tag reader passes the number to a computer containing the middleware that accesses the object name service (ONS). The ONS tells the computer systems where to locate information on the secure network about the EPC identified object, by converting the EPC to a URL. The communications protocol infrastructure, called the EPC Information Services (EPCIS), provides a common set of data elements, messaging structure for communication and defined messages for stakeholders to use for storing, accessing and communicating data on objects moving in the supply chain (Shanahan et al., 2009). Although there are several RFID deployment architectures, the EPCglobal Network

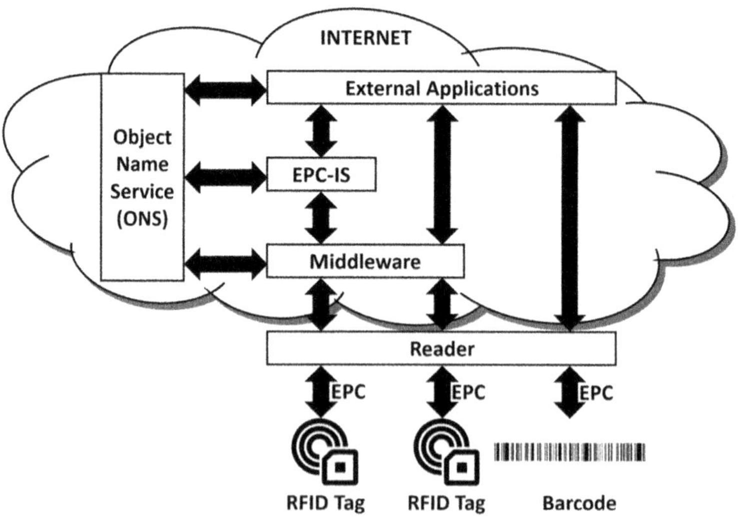

Figure 3 EPCglobal Network architecture (adapted from Zarokostas et al. (2007), Shanahan et al. (2009)).

uses open, worldwide standards that are RFID-centric, thereby significantly simplifying the device layer for supply chain participants (Musa et al., 2014; Liang et al., 2015).

6 Case study: using the EPCglobal Network for cattle and beef traceability

A proof-of-concept trial was conducted jointly by GS1 New Zealand and ANZCO Foods, which demonstrated that the EPCglobal Network could be used effectively as a traceability system for the movement of cattle from a farm, through the beef production process, to a retail outlet (Hartley and Sundermann, 2010).

Cattle used in the project were identified with UHF ear tags on a farm in the Canterbury region of New Zealand. RFID ear tags were scanned at four read 'event' locations, including at tagging, load out, processing plant arrival and at the stun box. RFID tags were placed on carcasses and read as they entered the cooler and boning room of the processing facility. Finally, RFID tags were placed on boxes of beef and read as cuts were packed in the boxes at the load-out dock door and upon arrival at a retailer's receiving dock door. All item and location identifiers, RFID tags, readers and network and process standards were EPC compliant. A total of nine read events and associated location data were transmitted to the EPCglobal Network (Table 4).

The EPCglobal Network is an internet-based global system that has an information receiving, storage, query and response component called the EPCIS. The EPCIS is the database component of the EPCglobal Network, which stores individual item data and event reads. Because each read event captures the unique item identifier, its location and the read time, the EPCglobal Network contains the components required for real-time traceability. The EPCIS enables network users to exchange EPC-associated data and thereby manage the movement, storage and presentation of this dynamic information. The information can be accessed by any appropriate supply chain participant.

A barcode on an item in the supermarket contains only the identity of the product and its manufacturer. In contrast, the EPC enables inclusion of serial numbers, which identify the item down to the 'instance' level (i.e. a box of beef tenderloins can be identified from all other boxes of the same product). Because they are serialized, the EPC scheme can be used to identify specific items (e.g. cow, carcass, box of beef), physical locations (e.g. farm, processing plant, retail store), or even a specific area within a location (carcass cooler, loading dock door) (GS1, 2014).

Ultimately, the proof-of-concept trial confirmed that the EPCglobal Network, accompanied by the use of EPC-compliant RFID tags, identifiers, readers and process standards, can be effective in tracing the movement of cattle from farms, carcasses through processing plants and boxed beef to retailers.

7 Future trends

The most widely chosen method of unique cattle identification is RFID, due to its relative ability to perform well as an Auto-ID in livestock environments. However, the ideal animal identification method would be inexpensive to use, completely unique, able to securely maintain identification, fraud-proof and preserved with the animal and its tissues constantly from the time from birth to consumption. Therefore, as technology advances,

Table 4 Radio-frequency identification (RFID) read events and corresponding data collected from farm through processing of cattle, carcasses and boxes

Read event	Process	Item	Data[1]	Business step
1	Cow tagged with UHF RFID at the farm	Cow	EPC identifier for each cow (SGTIN for cow) EPC identifier for the farm (SGLN for farm)	Commissioning
2	Cow loaded onto truck to leave farm	Cow	SGTIN for cow EPC identifier for truck loading ramp (SGLN for ramp)	Loading
3	Cow arrives at processing plant	Cow	SGTIN for cow EPC identifier for holding yards at processing plant (SGLN for yards)	Receiving
4	Cow arrives at stun box	Cow	SGTIN for cow EPC identifier for plant's stun box (SGLN for stun box)	Transforming
5	Carcass stored in cooler	Carcass	EPC identifier for carcass (SSCC for carcass) EPC identifier for plant's cooler (SGLN for cooler)	Commissioning
6	Carcass arrives in boning room	Carcass	SSCC for carcass EPC identifier for plant's boning room (SGLN for boning room)	Transforming
7	Beef cuts packed in boxes	Box of beef	EPC identifier for box of beef (SGTIN for box) EPC identifier for plant's packing line (SGLN for packing line)	Commissioning
8	Boxes loaded into containers	Box of beef	SGTIN for box EPC identifier for plant's loading-out dock door (SGLN for plant's door)	Shipping
9	Boxes received at retail outlet	Box of beef	SGTIN for box EPC identifier for retailer's receiving dock door (SGLN for retailer's door)	Receiving

[1]EPC = Electronic Product Code; SGTIN = Serialized Global Trade Item Number; SGLN = Serialized Global Location Number (With or Without Extension); SSCC = Serialized shipping container code.

research that explores application of adaptable biometric identification methodologies is warranted.

Despite numerous comprehensive beef traceability systems, it is likely that the majority of beef produced worldwide has either meagre or no traceability. This is because traceability is incrementally more difficult as the flow of the supply chain increases, especially as processing speeds increase during fabrication. Emerging systems must be scalable to large processors that harvest and fabricate ever-increasing volumes of cattle and beef.

With increasing world trade, it would be beneficial for global markets if countries would work towards the development of an interoperable and uniform global traceability system (e.g. following the model of the EU) by developing uniform requirements for identification, database systems and interoperability. These hardware standards must be coupled with and supported by globally recognized software architecture standards (Charlebois et al., 2014; Musa et al., 2014).

Because there are hundreds of inputs and processes in the production of beef, a system for tracking every input and process with a degree of precision adequate for every objective is virtually impossible (Golan et al., 2004). With the prospect of improved supply chain visibility, decision models that evaluate risk and reward will be necessary to prioritize and rank the required data entry supplying the virtual supply chains. Traceability systems can have multiple objectives such as food safety, animal health, marketing, extended methods of recall or continuous quality improvement. Research on ways to combine these divergent needs into refined visibility systems has the potential to significantly improve efficiency of resource use.

As demands increase for greater levels of breadth, depth and precision of traceability, so will the demands for autonomous and self-adaptive systems that are able to capture data without human intervention. This will involve continued work in RFID and sensor (i.e. temperature, geospatial, bacterial, toxin) networks that can feed the supply chain's demand for data. Ultimately, these complex visibility systems need to report more than the data, particularly regarding food safety, the systems need to deliver actionable information.

8 Conclusion

The beef industry needs ever-increasing supply chain visibility to support immediate supply adulteration containment and consumer demands for transparency. Electronic data collection and networking technologies allow for supply chain virtualization, such that data can be tracked forward and traced backward in near real time. Much of the world has developed tracking technology for cattle and beef, yet many systems remain voluntary, and even in the most sophisticated mandated systems there are large gaps in data entry. Legislated tracking systems tend to capture only a handful of data points, but are often much less adept at providing actionable information to supply chain participants.

RFID ear tags have emerged as the most widely accepted method of uniquely identifying live cattle. A series of one- and(or) two-dimensional barcodes are the most widely used method of identifying carcasses, beef cuts and packages. In the future, it is conceivable that biometrics could complement or supplant RFID in live cattle, and RFID may gain a more expansive role or supplant barcodes in product areas.

Relatively few systems maintain the identity chain of custody from calf birth through beef consumption. The beef industry differs from the majority of manufacturing supply chains whose primary objective is product assembly, whereas beef processing is a disassembly process. Large and fast beef supply chains encounter real challenges in transferring identity, especially from carcass to fabricated beef cuts.

Currently, the majority of government legislated beef traceability systems feed into centrally administrated databases. However, much of the current research interest is in utilization of cloud-based, distributed information systems for data entry, storage and retrieval. Distributed systems appear to offer the most promise for providing standard protocols that can be utilized by supply chain participants regardless of their size, industry segment or nation.

9 Where to look for further information

Animal identification methods

Blasi, D. A., Dhuyvetter, K. C., Spire, M. F., Epp, M. P. and Barnhardt, B. B. 2003. A guide for electronic identification of cattle, Kansas State Univ. Ag. Exp. Sta. and Coop. Ext. Serv., Manhattan.

Awad, A. I. 2016. From classical methods to animal biometrics: A review on cattle identification and tracking. *Computers and Electronics in Agriculture*, 123: 423–35.

Object and location identification, and network communication standards

GS1 http://www.gs1.org/

Beef traceability systems throughout the world

Bowling, M. B., Pendell, D. L., Morris, D. L., Yoon, Y., Katoh, K., Belk, K. E. and Smith, G. C. 2008. Review: Identification and Traceability of Cattle in Selected Countries Outside of North America. *The Professional Animal Scientist*, 24: 287–94.

Charlebois, S., Sterling, B., Haratifar, S. and Naing, S. K. 2014. Comparison of Global Food Traceability Regulations and Requirements. *Comprehensive Reviews in Food Science and Food Safety*, 13: 1104–23.

Schroeder, T. C. and Tonsor, G. T. 2012. International cattle ID and traceability: competitive implications for the US. *Food Policy*, 37: 31–40.

Souza-Monteiro, D. M. and Caswell, J. A. 2004. The Economics of Implementing Traceability in Beef Supply Chains: Trends in Major Producing and Trading Countries. Available: http://dx.doi.org/10.2139/ssrn.560067

10 References

Al-Khalifa, H. 2008. Utilizing QR code and mobile phones for blinds and visually impaired people. *In:* Miesenberger, K., Klaus, J., Zagler, W. and Karshmer, A. (eds), *Computers Helping People with Special Needs*. Springer Berlin Heidelberg.

Allen, A., Golden, B., Taylor, M., Patterson, D., Henriksen, D. and Skuce, R. 2008. Evaluation of retinal imaging technology for the biometric identification of bovine animals in Northern Ireland. *Livestock Science*, 116: 42–52.

Andree, S., Jira, W., Schwind, K. H., Wagner, H. and Schwagele, F. 2010. Chemical safety of meat and meat products. *Meat Science*, 86: 38–48.

Andres, R. 2015. Traceability in the livestock and beef industry, Uruguay. *Knowledge-intensive business services in resources rich economies* [Online]. Available: http://cienciassociales.edu.uy/wp-content/uploads/sites/2/2015/04/Redlas_Trazabilidad-Rius.pdf [Accessed 10/22/16].

Artmann, R. 1999. Electronic identification systems: state of the art and their further development. *Computers and Electronics in Agriculture*, 24: 5–26.

Augsburg, J. K. 1990. The benefits of animal identification for food safety. *Journal of Animal Science*, 68: 880–3.

Awad, A. I. 2016. From classical methods to animal biometrics: A review on cattle identification and tracking. *Computers and Electronics in Agriculture*, 123: 423–35.

Barge, P., Gay, P., Merlino, V. and Tortia, C. 2013. Radio frequency identification technologies for livestock management and meat supply chain traceability. *Canadian Journal of Animal Science*, 93: 23–33.

Barry, B., Gonzales-Barron, U. A., Mcdonnell, K., Butler, F. and Ward, S. 2007. Using muzzle pattern recognition as a biometric approach for cattle identification. *Transactions of the ASABE*, 50: 1073–80.

Beulens, A. J. M., Broens, D.-F., Folstar, P. and Hofstede, G. J. 2005. Food safety and transparency in food chains and networks: Relationships and challenges. *Food Control*, 16: 481–6.

Blasi, D. A., Dhuyvetter, K. C., Spire, M. F., Epp, M. P. and Barnhardt, B. B. 2003. *A guide for electronic identification of cattle*, Kansas State Univ. Ag. Exp. Sta. and Coop. Ext. Serv., Manhattan.

Bloom, J. D. and Hinrichs, C. C. 2010. Moving local food through conventional food system infrastructure: Value chain framework comparisons and insights. *Renewable Agriculture and Food Systems*, 26: 13–23.

Burian, A., Vehvilainen, M. and Kangas, J. 2008. Camera barcode reader with automatic localization, detection of orientation and type classification. *Computers and Simulation in Modern Science*, 1: 214–19.

Buskirk, D., Schweihofer, J., Clarke, R., Grooms, D. and Foster, T. 2013. A traceability model for beef product origin within a local institutional value chain. *Journal of Agriculture, Food Systems, and Community Development*, 3: 33–43.

Cai, C. and Li, J. Cattle face recognition using local binary pattern descriptor. Signal and Information Processing Association Annual Summit and Conference (APSIPA), 2013 Asia-Pacific, Oct. 29-Nov. 1, 2013 2013. 1–4.

Caja, G., Conill, C., Nehring, R. and Ribó, O. 1999. Development of a ceramic bolus for the permanent electronic identification of sheep, goat and cattle. *Computers and Electronics in Agriculture*, 24: 45–63.

Caswell, J. 1998. How labeling of safety and process attributes affects markets for food. *Agricultural and Resource Economics Review*, 27: 151–8.

Charlebois, S., Sterling, B., Haratifar, S. and Naing, S. K. 2014. Comparison of Global Food Traceability Regulations and Requirements. *Comprehensive Reviews in Food Science and Food Safety*, 13: 1104–23.

CIAA. 2015. Canadian Cattle Identification Agency Annual Report. Available: http://www.canadaid.com/documents/Annual_Report_2015_online.pdf [Accessed 10/22/16].

Comscore. 2011. *14 million Americans scanned QR codes on their mobile phones in June 2011* [Online]. comScore.com. Available: http://www.comscore.com/Insights/Press-Releases/2011/8/14-Million-Americans-Scanned-QR-or-Bar-Codes-on-their-Mobile-Phones-in-June-2011 [Accessed 29 October 2014].

Conill, C., Caja, G., Nehring, R. and Ribó, O. 2000. Effects of injection position and transponder size on the performances of passive injectable transponders used for the electronic identification of cattle. *Journal of Animal Science*, 78: 3001–9.

Cooper-Prado, M. J., Long, N. M., Wright, E. C., Goad, C. L. and Wettemann, R. P. 2011. Relationship of ruminal temperature with parturition and estrus of beef cows. *Journal of Animal Science*, 89: 1020–7.

Corkery, G. P., Gonzales-Barron, U. A., Butler, F., Donnell, K. M. and Ward, S. 2007. A Preliminary Investigation on Face Recognition as a Biometric Identifier of Sheep. *Transactions of the ASABE*, 50: 313–20.

Dabbene, F., Gay, P. and Tortia, C. 2016. 5- Radio-Frequency Identification Usage in Food Traceability A2 - Espiñeira, Montserrat. *In:* Santaclara, F. J. (ed.), *Advances in Food Traceability Techniques and Technologies*. Duxford, UK: Woodhead Publishing.

Disney, W. T., Green, J. W., Forsythe, K. W., Wiemers, J. F. and Weber, S. 2001. Benefit-cost analysis of animal identification for disease prevention and control. *Revue Scientifique et Technique de l'OIE*, 20: 385–405. doi.org/10.20506/rst.20.2.1277

Domdouzis, K., Kumar, B. and Anumba, C. 2007. Radio-Frequency Identification (RFID) applications: A brief introduction. *Advanced Engineering Informatics*, 21: 350–5.

Eardprab, S., Wounchoum, P., Phongcharoenpanich, C. and Torrungrueng, D. Effects of ruminal magnets in the vicinity of a bolus tag of an LF-RFID system. Electrical Engineering/Electronics, Computer, Telecommunications and Information Technology, 2008. ECTI-CON 2008. 5th International Conference on, May 14–17, 2008 2008. 749–52.

EC Regulation No. 911/2004 Of April 29, I. R. 2004. Regulation No. 911/2004 of 29 April 2004 implementing regulation (EC) No 1760/2000 of the European Parliament and of the Council as regards ear tags, passports and holding registers. *Official Journal of the European Union*, 163: 65–70.

Fallon, R. J. 2001. The development and use of electronic ruminal boluses as a vehicle for bovine identification. *Revue Scientifique et Technique de l'OIE*, 20: 480–90. doi.org/10.20506/rst.20.2.1285

Fallon, R. J. and Rogers, P. A. M. 1991. Use and recovery of implantable electronic transponders in beef cattle. *Automatic electronic identification systems for farm animals* [Online].

Faostat, 2016. Food and Agriculture Organization of the United Nations - Statistics Division.

Farooq, U., Tao, W., Alfian, G., Kang, Y.-S. and Rhee, J. 2016. ePedigree Traceability System for the Agricultural Food Supply Chain to Ensure Consumer Health. *Sustainability*, 8: 839–55.

Finkenzeller, K. 2010. Fundamental Operating Principles. *RFID Handbook*. John Wiley and Sons, Ltd.

Food Standards Agency, 2002. Traceability in the food chain. Available: http://multimedia.food.gov.uk/multimedia/pdfs/traceabilityinthefoodchain.pdf [Accessed 6 December 2014].

Galimberti, A., De Mattia, F., Losa, A., Bruni, I., Federici, S., Casiraghi, M., Martellos, S. and Labra, M. 2013. DNA barcoding as a new tool for food traceability. *Food Research International*, 50: 55–63.

Gao, J. Z., Prakash, L. and Jagatesan, R. Understanding 2D-barcode technology and applications in M-commerce - design and implementation of a 2D barcode processing solution. Computer Software and Applications Conference, 2007. COMPSAC 2007. 31st Annual International, 2007, pp. 49–56.

Geers, R., Puers, B., Goedseels, V. and Wouters, P. 1997. *Electronic Identification, Monitoring and Tracking of Animals*, Oxford, UK, Oxford University Press.

Ghirardi, J. J., Caja, G., Garin, D., Casellas, J. and Hernandez-Jover, M. 2006. Evaluation of the retention of electronic identification boluses in the forestomachs of cattle. *Journal of Animal Science*, 84: 2260–8.

Giannelli, P. 1996. Forensic Science: Chain of Custody. *Criminal Law Bulletin*, 32: 447–65.

Godo, Y. 2015. The beef traceability system in Japan. *FFTC Agricultural Policy Platform* [Online]. Available: http://ap.fftc.agnet.org/ap_db.php?id=530 [Accessed 10/20/16].

Golan, E., Krissoff, B., Kuchler, F., Calvin, L., Nelson, K. and Price, G. 2004. Traceability in the U.S. food supply: Economic theory and industry studies. *Agricultural Economic Report Number 830*. United States Department of Agriculture.

Gonzales Barron, U., Corkery, G., Barry, B., Butler, F., Mcdonnell, K. and Ward, S. 2008. Assessment of retinal recognition technology as a biometric method for sheep identification. *Computers and Electronics in Agriculture*, 60: 156–66.

GS1 2012. GS1 standards document GS1 global traceability standard. GS1.

GS1 2014. EPC tag data standard.

GS1 2015. EPC™ radio-frequency identity protocols generation-2 UHF RFID.

GS1 2016a. EPC Information Services (EPCIS) Standard.

GS1 2016b. General specifications.

GS1. 2016c. *GS1 barcodes* [Online]. Available: http://www.gs1.org/barcodes [Accessed 9 October 2016].

GS1 Japan, 2011. *Barcodes and identification* [Online]. GS1 Japan. Available: http://www.gs1jp.org/2011/barcodes_identification/1_5.html [Accessed 10 July 2014].

Harper, D. 2016. 'trace'. *Online Etymology Dictionary.* <http://www.etymonline.com/index.php?term=trace>.

Hartley, G. 2013. The use of EPC RFID standards for livestock and meat traceability. Available: http://www.rfid-pathfinder.org.nz/wp-content/uploads/2012/08/EPCIS-Final-Report5.pdf [Accessed 1 October 2016].

Hartley, G. and Sundermann, E. 2010. The efficacy of using the EPCglobal Network for livestock traceability: A proof of concept. Available: http://www.gs1nz.org/files/2313/7947/8796/Livestock_Traceability.pdf.

Hasker, P. J. S. and Bassingthwaighte, J. 1995. Implanting electronic identification transponders under the scutifon cartilage of beef cattle is inappropriate under Australian conditions. *Australian Journal of Experimental Agriculture*, 35: 15–18.

Hasker, P. J. S. and Bassingthwaite, J. 1996. Evaluation of electronic identification transponders implanted in the rumen of cattle. *Australian Journal of Experimental Agriculture*, 36: 19–22.

Hasker, P. J. S., Round, P. J. and Slack, D. J. 1992. Implantion and recovery of identification transponders in the anal region of steers. *Australian Journal of Experimental Agriculture*, 32: 689–91.

Heaton, M. P., Harhay, G. P., Bennett, G. L., Stone, R. T., Grosse, W. M., Casas, E., Keele, J. W., Smith, T. P., Chitko-Mckown, C. G. and Laegreid, W. W. 2002. Selection and use of SNP markers for animal identification and paternity analysis in U.S. beef cattle. *Mammalian Genome*, 13: 272–81.

Hoang, B. and Caudill, A. 2012. *RFID* [Online]. IEEE Emerging Technology. Available: https://origin.www.ieee.org/about/technologies/emerging/rfid.pdf [Accessed 6 June 2014].

Identigen 2004. Farm to fork traceability through DNA traceback. Identigen, Ltd., Dublin, Ireland.

IICA. 2009. Uruguay's experience in beef cattle traceability. Available: http://www.inac.gub.uy/innovaportal/file/6323/1/libro_trazabilidad_ingles.pdf [Accessed 20 October 2016].

Indianbureaucracy.com. 2016. Animal Identification and Traceability System. Available: http://www.indianbureaucracy.com/animal-identification-and-traceability-system/ [Accessed 15 August 2016].

ISO 1996. ISO 11784:1996. *Radio frequency identification of animals – Code structure.*

ISO 2007. ISO 22005:2007 Traceability in the feed and food chain – General principles and basic requirements for system design and implementation. First edition ed.: The International Organization for Standardization.

Jansen-Vullers, M. H., Van Dorp, C. A. and Beulens, A. J. M. 2003. Managing traceability information in manufacture. *International Journal of Information Management*, 23: 395–413.

Jewell, C. P., Van Andel, M., Vink, W. D. and Mcfadden, A. M. 2016. Compatibility between livestock databases used for quantitative biosecurity response in New Zealand. *New Zealand Veterinary Journal*, 64: 158–64.

Jin, S. and Zhou, L. 2014. Consumer interest in information provided by food traceability systems in Japan. *Food Quality and Preference*, 36: 144–52.

Kampers, F. W. H., Rossing, W. and Eradus, W. J. 1999. The ISO standard for radiofrequency identification of animals. *Computers and Electronics in Agriculture*, 24: 27–43.

Kassahun, A., Hartog, R. J. M., Sadowski, T., Scholten, H., Bartram, T., Wolfert, S. and Beulens, A. J. M. 2014. Enabling chain-wide transparency in meat supply chains based on the Epcis global standard and cloud-based services. *Computers and Electronics in Agriculture*, 109: 179–90.

Kassahun, A., Hartog, R. J. M. and Tekinerdogan, B. 2016. Realizing chain-wide transparency in meat supply chains based on global standards and a reference architecture. *Computers and Electronics in Agriculture*, 123: 275–91.

Kato, H. and Tan, K. 2D barcodes for mobile phones. 2nd International Conference on Mobile Technology, Applications and Systems, Nov. 15–17, 2005 2005. 8.

Kato, H. and Tan, K. 2007. Pervasive 2D barcodes for camera phone applications. *IEEE Pervasive Computing*, 6: 76–85.

Kato, H., Tan, K. and Chai, D. Development of a novel finder pattern for effective color 2D-barcode detection. International Symposium on Parallel and Distributed Processing with Applications, Dec. 10–12, 2008, pp. 1006–13.

Kato, H., Tan, K. and Chai, D. Novel colour selection scheme for 2D barcode. International Symposium on Intelligent Signal Processing and Communication Systems, Jan. 7–9, 2009, pp. 529–32.

Kelepouris, T., Pramatari, K. and Doukidis, G. 2007. RFID-enabled traceability in the food supply chain. *Industrial Management and Data Systems*, 107: 183–200.

Kirk, K. and Buskirk, D. 2006. Mandatory radio frequency identification (RFID) of cattle in Michigan. E. Lansing, MI: Michigan State Univ. Extension.

Lambooij, E. 1991. Automatic electronic identification systems for farm animals. Available: http://agris.fao.org/agris-search/search.do?recordid=xf2015011316 [Accessed 22 October 2016].

Lambooij, E., van't Klooster, C. E., Rossing, W., Smits, A. C. and Pieterse, C. 1999. Electronic identification with passive transponders in veal calves. *Computers and Electronics in Agriculture*, 24: 81–90.

Landt, J. 2005. The history of RFID. *IEEE Potentials*, 24: 8–11.

Lawrence, J. D., Strohbehn, D. R., Loy, D. D. and Clause, R. J. 2003. Lessons learned from the Canadian cattle industry: National animal identification and the mad cow. *Matric Research Paper 03-MRP 7* [Online]. Available: http://lib.dr.iastate.edu/cgi/viewcontent.cgi?article=1006&context=matric_researchpapers [Accessed 15 October 2016].

Liang, W., Cao, J., Fan, Y., Zhu, K. and Dai, Q. 2015. Modeling and Implementation of Cattle/Beef Supply Chain Traceability Using a Distributed RFID-Based Framework in China. *PLoS One*, 10, e0139558.

Lu, Y., He, X., Wen, Y. and Wang, P. S. P. 2014. A new cow identification system based on iris analysis and recognition. *International Journal of Biometrics*, 6: 18–32.

Mapa Brazil Ministry of Agriculture Livestock And Supply. 2016. *Sisbov* [Online]. Brazil Ministry of Agriculture, Livestock and Supply (MAPA). Available: http://www.agricultura.gov.br/animal/rastreabilidade/sisbov [Accessed 14 October 2016].

Markus, S. B., Aalhus, J. L., Janz, J. A. M. and Larsen, I. L. 2011. A survey comparing meat quality attributes of beef from credence attribute-based production systems. *Canadian Journal of Animal Science*, 91: 283–94.

Mcallister, T. A., Gibb, D. J., Kemp, R. A., Huisma, C., Olson, M. E., Milligan, D. and Schwartzkopf-Genswein, K. S. 2000. Electronic identification: Applications in beef production and research. *Canadian Journal of Animal Science*, 80: 381–92.

Mccarthy, U., Ayalew, G., Butler, F., Mcdonnell, K. and Ward, S. 2011. The case for UHF RFID application in the meat supply chain in the Irish context: A review perspective. *Agric Eng Int: CIGR Journal*, 13.

Mckean, J. D. 2001. The importance of traceability for public health and consumer protection. *Revue Scientifique et Technique de l'OIE*, 20: 363–71. doi.org/10.20506/rst.20.2.1280

Mckeith, R., Moore, M., Gray, G., Hale, D., Griffin, D., Kerth, C., Savell, J., Raines, C., Belk, K., Woerner, D., Tatum, J., Igo, J., Vanoverbeke, D., Mafi, G., Lawrence, T., Delmore, J., R., Christensen, L., Shackelford, S., King, D., Wheeler, T., Meadows, L. and O'connor, M. 2011. National Beef Quality Audit - 2011: Harvest-floor assessments, carcass characteristics and instrument-grading information. Available: http://www.bqa.org/Media/Bqa/Docs/harvest-floor_assessments_carcass_characteristics.pdf [Accessed 4 August 2015].

Mennecke, B., Townsend, A., Hayes, D. J. and Lonergan, S. 2006. A study of the factors that influence consumer attitudes toward beef products using the conjoint market analysis tool. *Card Working Paper 06-WP 425* [Online]. [Accessed 15 October 2016].

Meyer, R., Candrian, U. and Lüthy, J. 1993. Detection of pork in heated meat products by the polymerase chain reaction. *Journal of AOAC International*, 77: 617–22.

Mooketsi, B. E. 2013. Optimisation of livestock identification and trace-back system Lits database to meet local needs: Case study of Botswana. *The Journal of Community Informatics*, 9.

Mpxml, I. 2010. Traceability for meat and poultry: US implementation guide. Available: http://www.eurocarne.com/daal?a1=informes&a2=trazabilidad_EE.UU.pdf [Accessed 22 October 2016].

Musa, A., Gunasekaran, A. and Yusuf, Y. 2014. Supply chain product visibility: Methods, systems and impacts. *Expert Systems with Applications*, 41: 176–94.

Niaa-Usaha. 2013. Joint strategy forum on animal disease traceability: Bringing industry and regulatory leaders together to create sensible solutions. Available: http://www.animalagriculture.org/Resources/Documents/Conf%20-%20Symp/Symposiums/2013%20adt/adt%20White%20Paper_Final_paper%20only.pdf [Accessed 22 October 2016].

Nlis. 2016. *National Livestock Identification System* [Online]. Available: https://www.nlis.com.au/ [Accessed 10/18/16].

Oakley, D. 1993. Evaluation of the use of electronic transponders to identify cattle. *Final trial report* [Online].

Olynk, N. J. 2012. Assessing changing consumer preferences for livestock production processes. *Animal Frontiers*, 2: 32–8.

Ortega, C. and Peel, D. S. 2010. The Mexican Animal Identification System: Current situation, problems, and potential. *Journal of Agricultural and Applied Economics*, 42: 551–5.

Passantino, A. 2013. The current EU rules on bovine electronic identification systems: State of the art and its further development. *Archiv Tierzucht*, 56: 344–53.

Pettitt, R. G. 2001. Traceability in the food animal industry and supermarket chains. *Revue Scientifique et Technique de l'OIE*, 20: 584–97. doi.org/10.20506/rst.20.2.1299

Raspor, P. 2005. Bio-markers: traceability in food safety issues. *Acta Biochimica Polonica*, 52: 659–64.

Ribo, O., Korn, C., Meloni, U., Cropper, M., de Winne, P. and Cuypers, M. 2001. Idea: a large-scale project on electronic identification of livestock. *Revue Scientifique et Technique de l'OIE*, 20: 426–36. doi.org/10.20506/rst.20.2.1281

Robb, J., Lawrence, A. E. and Rosa, E. 2006. Issues related to beef traceability: A discussion of transforming cattle into products. Available: http://lmic.info/sites/default/files/publications/fs07.pdf [Accessed 26 February 2015].

Roberts, C. M. 2006. Radio frequency identification (RFID). *Computers and Security*, 25: 18–26.

Ryan, S. E. 2008. *Evaluation of ISO 11785 low-frequency radio identification devices and the characterization of electromagnetic interference in practical cattle management scenarios.* Master of Science M.S. Thesis, Kansas State Univ., Manhattan.

Ryan, S. E., Blasi, D. A., Anglin, C. O., Bryant, A. M., Rickard, B. A., Anderson, M. P. and Fike, K. E. 2010. Read distance performance and variation of 5 low-frequency radio frequency identification panel transceiver manufacturers. *Journal of Animal Science*, 88: 2514–22.

Scanlife. 2014. Mobile barcode trend report: Q1 2014. Available: http://www.scanlife.com/assets/images/pdf/ScanLife_Trend_Report_Q1.2014.pdf.

Scanlife. 2015. ScanLife Mobile Trend Report: Q3 2015.

Scharff, R. L. 2012. Economic burden from health losses due to foodborne illness in the United States. *Journal of Food Protection*, 75: 123–31.

Schwagele, F. 2005. Traceability from a European perspective. *Meat Science*, 71: 164–73.

Scoteid. 2014. Cattle EID in Scotland: A research update. Available: https://www.scoteid.com/Public/Documents/Cattle%20EID%20in%20Scotland%20research%20update%20190514.pdf [Accessed 22 October 2016].

Shackell, G. H. and Dodds, K. G. 2008. DNA-Based Traceability of Meat. *In:* Toldrá, F. (ed.), *Meat Biotechnology*. New York, NY: Springer New York.

Shanahan, C., Kernan, B., Ayalew, G., Mcdonnell, K., Butler, F. and Ward, S. 2009. A framework for beef traceability from farm to slaughter using global standards: An Irish perspective. *Computers and Electronics in Agriculture*, 66: 62–9.

Shin, D.-H., Jung, J. & Chang, B.-H. 2012. The psychology behind QR codes: User experience perspective. *Computers in Human Behavior*, 28: 1417–26.

Singh, A. K., Ghosh, S., Roy, B., Tiwari, D. K. & Baghel, R. P. S. 2014. Application of radio frequency identification (RFID) technology in dairy herd management. *International Journal of Livestock Research*, 4: 10–19. doi.org/10.5455/ijlr.20140208085018

Small, J. A., Kennedy, A. D. and Kahane, S. H. 2008. Core body temperature monitoring with passive transponder boluses in beef heifers. *Canadian Journal of Animal Science*, 88: 225–35.

Smith, G. 2004. Tracing US progress on individual ID. *In Feedback, Meat & Livestock Journal*, Nov./Dec. Issue, Meat & Livestock Australia, Sydney, p. 11.

Smith, G. C., Pendell, D. L., Tatum, J. D., Belk, K. E. and Sofos, J. N. 2008. Post-slaughter traceability. *Meat Science*, 80: 66–74.

Smith, G. C., Tatum, J. D., Belk, K. E., Scanga, J. A., Grandin, T. and Sofos, J. N. 2005. Traceability from a US perspective. *Meat Science*, 71: 174–93.

Smyth, S. and Phillips, P. W. B. 2003. Product differentiation alternatives: Identity preservation, segregation, and traceability. *AgBioForum*, 5: 30–42.

Sonda. 2016. *Livestock Traceability* [Online]. Available: http://www.sonda.com/en/livestock-traceability/ [Accessed 10/22/16].

Souza-Monteiro, D. M. and Caswell, J. A. 2004. The Economics of Implementing Traceability in Beef Supply Chains: Trends in Major Producing and Trading Countries. Available: http://dx.doi.org/10.2139/ssrn.560067

Stanford, K., Stitt, J., Kellar, J. A. and Mcallister, T. A. 2001. Traceability in cattle and small ruminants in Canada. *Revue Scientifique et Technique de l'OIE*, 20: 510–22. doi.org/10.20506/rst.20.2.1291

State of Indiana 2005. Title 345 Indiana State Board of Animal Health. Indiana Register.

State of Wisconsin 2016. Animal Health - Updated 2013–14 Wis. Stats. - s. 35.18.

Tan, K. and Chai, D. 2010. A new perspective on first read rate of 2D barcodes in mobile applications, in Proceedings of the 2010 6th IEEE International Conference on Wireless, Mobile, and Ubiquitous Technologies in Education. *6th IEEE International Conference on Wireless, Mobile and Ubiquitous Technologies in Education*. Kaohsiung, Taiwan: IEEE Computer Society.

Tan, S.-Y., Foo, L. Y. and Idrus, R. Application of quick response (QR) codes in mobile tagging system for retrieving information about genetically modified food. Proceedings of the 9th WSEAS international conference on data networks, communications, computers, 2010 Stevens Point, Wisconsin, pp. 114–18.

Tartaglia, M., Saulle, E., Pestalozza, S., Morelli, L., Antonucci, G. and Battaglia, P. A. 1998. Detection of bovine mitochondrial DNA in ruminant feeds: a molecular approach to test for the presence of bovine-derived materials. *Journal of Food Protection*, 61: 513–18.

Tate, M. Traceability of meat products- application of DNA technology. Proceedings of the New Zealand Grassland Association, 2001, pp. 255–8.

Texas Animal Health Commission. 2016. *Animal Disease Traceability (ADT)* [Online]. Available: http://www.tahc.state.tx.us/adt/ [Accessed 10/15/16].

Tharwat, A., Gaber, T., Hassanien, A. E., Hassanien, H. A. and Tolba, M. F. Cattle identification using muzzle print images based on texture features approach. *In:* Kömer, P., Abraham, A. and Snášel, V., (eds), *Proceedings of the Fifth International Conference on Innovations in Bio-Inspired Computing* and *Applications*, 2014 Ostrava, Czech Republic. Springer International Publishing, pp.217–27.

The Center for Food Integrity. 2015. A clear view of transparency and how it builds consumer trust. 2015 *Consumer trust research* [Online]. Available: http://www.foodintegrity.org.s3.amazonaws.com/wp-content/uploads/2014/12/CFI-2015-Consumer-Trust-Research-Booklet.pdf [Accessed 30 November 2015].

The Center for Food Integrity. 2016. Inside the minds of influencers: The truth about trust. Available from: http://www.foodintegrity.org/wp-content/uploads/2015/11/CFI_PublicFacing_together_final.pdf [Accessed 15 February 2017].

Tomlinson, J. J., Elliott-Smith, W. and Radosta, T. 2006. Laboratory information management system chain of custody: Reliability and security. *Journal of Automated Methods and Management in Chemistry*, 2006, Article ID 74907, pp. 1–4.

Tonsor, G. T., Schroeder, T. C. and Pennings, J. M. E. 2009. Factors Impacting Food Safety Risk Perceptions. *Journal of Agricultural Economics*, 60: 625–44.

Usda-Aphis-V. S. 1997. Part 1: Reference of 1997 beef cow-calf management practices. Fort Collins,CO: National Animal Health Monitoring System.

Usda-Aphis-V. S. 2008. Beef 2007–08 Part I: Reference of beef cow-calf management practices in the United States, 2007–08. Fort Collins,CO: National Animal Health Monitoring System.

Usda-Aphis-V. S. 2013. Final Rule: Traceability for Livestock Moving Interstate. Available: https://www.aphis.usda.gov/traceability/downloads/Adt_summary_revisions.pdf [Accessed 22 October 2016].

Usda-Aphis-V. S. 2016a. Interim tag data standard for UHF animal identification. Available: https://www.aphis.usda.gov/traceability/downloads/uhf-interim-tag-data-standard.pdf [Accessed 15 February 2017].

Usda-Aphis-V. S. 2016b. Official animal identification number (AIN) devices with '840' prefix. Available: https://www.aphis.usda.gov/traceability/downloads/Adt_device_ain.pdf [Accessed 22 October 2016].

Usda-Aphis-V. S. 2016c. Ultra-High radio frequency identification demonstration project summary and state reports. Available: https://www.aphis.usda.gov/traceability/downloads/uhrf-id-demo-pro-summary-st-rpt.pdf [Accessed 22 October 2016].

Verdouw, C. N., Wolfert, J., Beulens, A. J. M. and Rialland, A. 2016. Virtualization of food supply chains with the internet of things. *Journal of Food Engineering*, 176: 128–36.

Vinholis, M. D. M. B., Filho, H. M. D. S., Carrer, M. J., Barioni, W. J. and Chaddad, F. R. 2016. Complementarity in the adoption of traceability of beef cattle in Brazil. *Production*, 26: 540–50.

Voulodimos, A. S., Patrikakis, C. Z., Sideridis, A. B., Ntafis, V. A. and Xylouri, E. M. 2010. A complete farm management system based on animal identification using RFID technology. *Computers and Electronics in Agriculture*, 70: 380–8.

Wallace, L. E., Paterson, J. A., Clark, R., Harbac, M. and Kellom, A. 2008. Readability of thirteen different radio frequency identification ear tags by three different multi-panel reader systems for use in beef cattle. *The Professional Animal Scientist*, 24: 384–91.

Wang, R. F., Myers, M. J., Campbell, W., Cao, W. W., Paine, D. and Cerniglia, C. E. 2000. A rapid method for PCR detection of bovine materials in animal feedstuffs. *Molecular and Cellular Probes*, 14: 1–5.

Zarokostas, N., Dimitropoulos, P. D. and Soldatos, J. 2007. RFID middleware design for enhancing traceability in the supply chain management. *The 18th Annual IEEE International Symposium on Personal, Indoor and Mobile Radio Communications*. Athens, GR: IEEE.

Index

Lightning Source UK Ltd.
Milton Keynes UK
UKOW06n0724070617
302802UK00001B/1/P

9 781786 760562